━ *The History of Pain* ━

# The History of Pain

## Roselyne Rey

Translated by Louise Elliott Wallace,
J. A. Cadden, and S. W. Cadden

*Harvard University Press*
Cambridge, Massachusetts
London, England
1995

First published as *Histoire de la Douleur,* Paris:
Editions La Découverte, 1993

*Library of Congress Cataloging-in-Publication Data*
Rey, Roselyne.
[Histoire de la douleur. English]
The history of pain / Roselyne Rey: translated by Louise Elliott
Wallace, J. A. Cadden, and S. W. Cadden.
p. cm.
Originally published: Paris: Decouverte, 1993, in series:
Histoire des sciences.
Includes bibliographical references and index.
ISBN 0-674-39967-6 (alk. paper)
1. Pain—History. I. Title.
[DNLM: 1. Pain—history. 2. History of Medicine.
WL 11.1 R456h 1995a]
RB127.R4913 1995
616'.0472'09—dc20
DNLM/DLC
for Library of Congress
94-31948

# ⇒ *Contents* ⇒

# ⊸ *Acknowledgments* ⊱

This work was conceived as a result of a convergence between my preoccupations as an historian and the interest that Daniel Le Bars (Inserm) and Jean-Claude Willer (Pitié-Salpêtrière) showed—as both are involved in solving current problems regarding pain—in gaining a better knowledge of the history of their particular discipline. Thanks to them, and to the small group of colleagues from the Société française de la douleur who were concerned with this undertaking, a fruitful dialogue developed and has continued for several years now. I also wish to extend my heartfelt gratitude to D. Albe-Fessard who granted me a number of interviews which were always highly stimulating and edifying. Finally, this book would not have been possible without the competence and unstinting assistance of B. Molitor and R. Rivet, Head Librarians at the Bibliothèque interuniversitaire de médecine, where I always found exceptional working conditions.

# ⊷ *Introduction* ⊷

I wandered far and wide through Césarée . . .," this verse from Racine, which haunted Aurélien in Aragon's novel, reminds me of another phrase which is undoubtedly the remote basis of this work: *"Omne animal, simul atque natum sit, voluptatem appetere eaque gaudere ut summo bono, dolorem aspernari ut summum malum et, quantum possit, a se repellere, idque facere nondum depravatum, ipsa natura incorrupte atque integre judicante,"* in other words, "Every living being from its very moment of birth seeks pleasure, enjoying it as the ultimate good while rejecting pain as the ultimate adversity and, insofar as is possible, doing his best to avoid it; he behaves in this fashion to the extent that he has not yet been conditioned and insofar as his basic nature has been left intact to judge naturally and with integrity." If instinct prompts people and animals alike to repel pain with all their energy, then from an historian's point of view, the most pressing question might be to attempt to understand and trace man's long struggle with pain. However, despite his apparently straightforward flight and fear response when confronted with pain, or perhaps precisely because of it, pain's composite nature makes undertaking an historical analysis of it quite problematical. The task of drawing up a complete history of pain does not simply entail carefully probing unexplored historical "territories," in the same sense one might delve into the history of the body, of food preparation, or of clothing, with each endeavour providing precious new topics of interest and reshuffling our analytical categories and representational systems; nor does it just entail turning deliberately towards a history of cultural sensibilities or mores, but it does require instead that one pursue an evasive subject with a dual nature, at the crossroads between biology and cultural or social conventions.

Pain is indeed certainly a combination of cultural and social factors: it has not had the same significance throughout the ages nor in the various different civilisations; even within the framework of Western civilisation itself, the collective memory recalls various episodes or circumstances where the limits of endurance were strangely removed, or virtually obliterated. Such examples include the processions of penitent flagellants during the Middle Ages, Napoleon's soldiers during the Russian campaign going back into battle on horseback after having had limbs amputated, Saint-Médard's "convulsionaries" during the 18th century who chastised themselves with varied torments (live coals, red-hot branding irons, blows and bruising), processions of martyrs, and the accounts of the lives of mystics—"to suffer or to die," believed St. Teresa of Avila. The examples and testimonials all reveal how man's relationship to pain is affected by his beliefs as well as by the context of differing philosophical or religious backdrops. There is hardly a need to search for further examples of exotic rites of initiation or special ceremonies designed to ensure fertility or abundant harvests, such as the "swinging hooks" by which celebrants are sometimes suspended in certain regions of India. It is not so much the altered meaning of pain conferred by a society that is of interest here, but rather the consequences of such a reinterpretation on the individual's own experience of pain. The different meanings attributed to pain—necessary trial, unpleasantness preceding some greater good, punishment, or fate—undoubtedly have an effect on the way the subject views it and thereby raises or lowers his resistance threshold; could this endurance capacity also be affected by moral courage and the spirit's control over the body, or a combination thereof, by establishing a state in which willpower and heroism play an important role in an individual's physiological capacity to resist?

Isn't an in-depth analysis of pain also a means of probing the relationship between mind and body, and of examining the dualism that somehow underlies our various ways of thinking? It is spontaneously evident in the opposition between pain, which is physical, and suffering, which can be considered moral. If we may temporarily accept the pertinence of this distinction, then the brief historical insights presented in this book are clearly circumscribed within the field of physiological pain. But if we take a closer look at the linguistic meaning of the terms pain and suffering, a second distinction can be superimposed on the first: the word suffering seems more to refer to the subject while pain seems more the objectification of this suffering, which legal parlance translates perfectly when it evaluates the *"pretium doloris."* When a doctor questions a patient, he is more likely to ask, "Where does it hurt?" or "Are you suffering?" or even "What seems to be the trouble?" rather than to ask him directly what type of pains he feels; however, he transcribes in his patient's file "abdominal pain" or "lower back pain." The etymology of the verbs from which the nouns pain and suffering are derived provides another perspective on their specialised meanings: to suffer, for instance, from the Latin *sufferre*, means to bear, to endure, to allow, or so many verbs which necessitate an active subject or even more, a person; the older French verb *se douloir* (to feel physical pain, to complain) could be constructed with an inanimate subject because its Latin root, *doleo*, could use an impersonal form or, more often, the painful member itself became the subject: *"caput dolet,"* i.e. my head feels pain or, in other words, I have a headache. Though in our current use of language we are seldom concerned with its Latin heritage, our linguistic roots can help us in these circumstances determine in what exact sense we are referring to pain, rather than to suffering. Thus, in terms of our latter "objectification" criterion, this historical analysis definitely concerns pain rather than suffering.

Nonetheless, when pain is intense and persistent or simply chronic, it always involves the entire being. It does not only limit itself to the painful region, but it is the whole person as an individual entity who then becomes affected as a result; his entire personality becomes doleful and his intellect becomes dulled. As Montaigne wrote, "The types of suffering which simply afflict us by tormenting our soul are much less distressful to me than they are to most other men: . . . but truly basic, corporeal suffering, I experience fully." The pain, or types of pains, being discussed in this book include the pain of sciatica and of migraine, the pain of gout, of war injuries, of childbirth and of stones—all distinctly physical pains if any there are—but also pain which clearly afflicts the entire being. And it is the suffering individual whom the physician must face rather than just his pain. Thus, this book is dedicated to discovering the ways in which physicians, physiologists, and neurologists have throughout the ages attempted to understand the practical mechanisms of pain and to find appropriate remedies for it.

Another approach of interest might have encompassed both the physician's point of view along with the patient's point of view, but that's a
different story altogether, another interesting project to be undertaken
for its own merits, although a singularly arduous one at that, as eye-
witness accounts from the most ancient periods are hard to come by.

However, because those bygone sufferers—whose views have yet to
be explored—left few accounts, it seemed necessary at times to consult
various literary works likely to shed new light on the debates and
reactions about pain in the past. The art and literature of an era cannot
be treated as if it were just another form of documentation or reference
material, though, because it is not a straightforward reflection or factual
account of "reality," even when it claims to be either "realist" or
"naturalist." It is always a "translation" in the Proustian sense, if you
will, an elaborate construction that follows its own rules and, at least
in part, escapes the standards and social conditions from which it
springs. But a work of art often provides something of greater value
than a reflection of reality, a truth whose symbols it presents to those
who are willing to decipher them.

Moreover, pain involves a codified form of social behaviour which
sets the parameters of allowable overt manifestations and regulates the
expression of such innermost personal experiences, whether endured in
the family bosom or alone in a solitary confrontation with the self. Pain
always has a specific language, whether it is a cry, a sob, or a tensing
of the features, and it is a language in itself as well. As such, it is defined
by society's standards of permissiveness or its notions of transgression,
between what can be shown or what must be kept quiet or hidden;
these norms or behavioural codes depend upon the cultural foundations
of the societies in which they arise. Is the pain of mourning greater for
having been accompanied by a chorus of wailing women lacerating their
chests and pulling out their hair, or when the grief has been contained
and released in private? Pain's expression, whatever form it takes, does
not escape the dialectic concerning what must be concealed and what
may be revealed, and one of the hypotheses presented in this work
suggests that the manner in which pain is expressed—either in a reserved, contained fashion, or disclosed in an explosion of wails and
moans—has a direct relation to the way in which pain is actually borne
and, in the fullest sense of the term, to what is actually felt. The very
act of proclaiming one's pain, beyond what is actually manifested and
beyond the meaning it projects, has a direct effect on the reality of the

experience without our being able to fully determine whether the actual expression brings relief by liberating, or perhaps amplifies the feeling through an echoing phenomenon.

The various "idioms of pain," beyond the symptoms they describe and the diagnoses they serve to establish, remain a fruitful province for further consideration by physicians and sufferers alike. The recent attempt to draw up an international questionnaire to be used during pain consultations, by taking into account every qualitative and emotional dimension of the sensation, and the way in which consultations are now carried out in pain clinics certainly seem to confirm the validity of placing more emphasis on greater communication as the basis of medical consultations. Just as two leaves are never exactly alike, two pains are never quite identical even within a single subject and at different moments of existence. Earlier experiences of similar pains, the influence of memory which either attenuates or amplifies, or the state of mind when the pain actually occurs are all factors that modify the way we perceive and tolerate pain. The painful experience is therefore always coloured by subjective considerations.

All the same, despite its many individual, social, and cultural characteristics, pain is not an historical subject in the same sense as fear, or hell, or purgatory. Pain is based on an anatomical and physiological foundation, and if there is one experience where the human condition's universality and the species' biological unity is manifest, pain is certainly it. Through his skin—the boundary between the self and the world or, as Bichat put it, "The sensitive limit of our soul"—every human being is subject to a multitude of impressions by which he perceives the objects he is in contact with, their position in space, their form, and their weight; such objects may also at times harm him when they strike him violently, burn or freeze him, or because they penetrate his flesh or abrade it. At other times, it is his own body which awakens and calls for attention. Hasn't the innermost murmur of the organs and the relationship to the world—which denotes pain as much as it does pleasure—been interpreted by certain philosophers of the phenomenological tradition as those manifestations through which each individual assumes his corporeal identity, and without which he could not define his immediate "life-world"? But how do these sensations convert into perceptions and how do they reach our awareness? What are the requisite transmission conditions from the periphery to the nerve centres and from the inner depths to the cortex necessary to experience pain?

This work is primarily devoted to clarifying the pain mechanism by providing a history of all the scientific theories and hypotheses (those developed by scientists in other words) on the physiology of sensation. In this retrospective, which takes us far into the distant past, a major finding emerges: the physiological study of pain and the actual consideration of it as a topic worthy of study in its own right has also required something of a struggle. Its specificity as an object of scientific interest went unrecognised for a long time and for a variety of frequently contradictory reasons: deemed the inevitable accompaniment of illness, pain was usually acknowledged and then relegated to a place of secondary importance, rather than studied for its own intrinsic qualities. This situation was all the more customary as medicine defined itself above all in terms of its healing intent and its aim of eradicating disorders; this necessarily assigned a hierarchical order of priorities on patients' needs and requirements: if the primary goal was to recover, then a cure had to be administered at all costs—never mind the pain inflicted to reach this objective—pain at times endured in the name of some questionable therapeutic treatment. This logic, where greater energy is expended to take care of the illness rather than to care for the patient and in which the after-effects of the disorder are disregarded (painful scars secondary to treatments, postoperative pain) became increasingly prevalent with medicine's growing successes. It stems from an optimistic view of medicine's capacities and ambitions, while the relegation of pain to a modest or insignificant rank is, in a sense, the ransom to be paid or the other side of the coin.

This situation also describes a certain type of physician/patient relationship, pointing up the virtual absence of the patient as an independent subject with free volition and confirming the generalised negation of his opinion and his will. Secondly, pain still has no clearly defined status: between being considered an emotion or a sensation, it has constantly been shunted between two equally unsatisfactory viewpoints. By analysing pain as an emotion—at the far end of the spectrum from pleasure—it tends to be precluded from its place in the physiological field and, instead, it is often assigned to the realm of psychology or of philosophy. Or conversely, in attempting to identify pain as a specific sense and thereby equating it with the five other traditional senses, hasn't physiology reduced the problem of pain to a set of questions on specificity: the specificity of pain receptors, of conduction pathways, and of the nervous centres? How can the specific physiology of pain become established without being turned into a simple annexe to the theories of sensation, or into an appendix of general neurology? In analysing pain's various processes, the problem remains to reconstruct all the multiple interactions involved, its duration, its repetitions, the influence of past experiences, as well as affectivity and sensitivity factors, automatic reflexes and deliberate reactions.

This study has attempted to disentangle the various components of the complex object we call pain. Historically speaking, knowledge about pain has usually been gathered within the framework of the strained—frequently broken, and then reestablished—relationship between experimental physiology and clinical medicine. It is along such lines that this enquiry has been made, with particular attention being paid to uncovering the obstacles in the way of establishing this topic of study's validity: investigative methods, too strictly delineated scientific disciplines, institutional problems, global views of health and disease, existing ambiguities about the "usefulness of pain," which all served as so many stumbling-blocks before finally coming to a limited number of certainties and being able to better identify any areas of doubt. A deep desire to understand the physiological or psycho-physiological nature of pain—a passion that has prompted so much research and inspired so many theories which, whenever possible, were tested through experimentation and validated or rejected on the basis of therapeutic results—has been the driving force behind this work.

Would it have been best to limit this study to a history of the physiology of pain? We felt it was impossible insofar as pathology—as well as therapeutics—have continually been called upon to validate or refute explanations about the normal mechanisms of pain. To this reciprocal elucidation of physiological factors by pathological ones—so rewarding from the methodological viewpoint—another reason should be added which stems from pain's very nature: is it normal or pathological? By which criteria should we decide that there is a good, healthy pain that, like a watchful sentinel, warns of imminent dangers, and another sort which undermines and enslaves us? The child that gets burned undoubtedly learns to remove his hand from the source of heat, but this apprenticeship also entails a smarting pain, with hypersensibility all around the zone for quite some hours, if not longer when there is infection or even mutilation. In such situations, wouldn't the simple sensation of an unpleasant contact have been sufficient? Pain is something elusive which, depending on intensity, duration and location, incessantly shifts from one category to another. Furthermore, where can pain be situated in the classical dichotomy between health and illness? More precisely put, if there is generally no illness without pain, would it be true to say that the perception of pain excludes the presence of health? It is clear that in order to answer these questions we must also undertake a critical revision of the standard definitions of health and illness, and now view those as historical categories. Finally, one further reason can be invoked to justify the close "interweaving" between physiology and pathology: the empirical treatments used to relieve pain preceded and prepared the way for the later body of knowledge about pain, while medical opinions about pain by far antedate the physiological study of pain in the modern sense. It is quite probable that medicine at its very birth was concerned with ways of alleviating pain, and modern medical art would be nonexistent without the continual quest for effective remedies, the efforts to interpret pain and identify ailing parts or predict the likely outcome of disorders. On both the semiological and therapeutic levels, pain has a rightful place in the history of medicine, and we have attempted herein to determine past methods of deciphering pathological conditions, the available therapies and ways in which they were administered by describing a few typical examples of particularly painful conditions, accounts of which have come down to us through the ages.

As a result of this approach which delves far back in time and is of necessity quite general, it is evident that this history is characterised by discontinuity and circuitous developments rather than by a more linear, cumulative evolution. To counter an illusory backward look that might tend to view early remedies with a condescending eye, we should remember the multitude of cures available in Graeco-Roman Antiquity to relieve pain, admire the pertinence of their investigations, analyse the practices of early physicians and, finally, we must ask ourselves whether the attitudes we automatically assumed were generalised—such as a resignation to pain or a reticence to relieve pain through every available means, either because of ill-founded apprehensions or indifference—are not actually fairly recent phenomena. Should we therefore demand a careful accounting of the pain-alleviating results obtained by contemporary medicine? Of course not. There are few human endeavours where the notion of progress is so compelling, where hope has such broad horizons, and where human reason is so well aware of its powers as well as of the boundaries it must still overcome.

There should be little need to stress that the chronological journey undertaken is not exhaustive and that the indicators and landmarks explored can provide a foundation for further exploration. The exceedingly long time-period this history of pain has spanned imposed certain choices and constraints: first, to present the major scientific contributions made to the body of knowledge and to the therapies used in treating pain; secondly, to analyse the scientific and institutional conditions under which pain theories and hypotheses were formulated, concentrating for the most part on France, as a study of this type must of necessity be limited; and finally, still with regard to France, and to the extent that to write a work covering "the culture of pain" throughout the world seemed hazardous, to provide a few elements for reflection on the Church's relationship to pain as well as literature's outlook on pain, to provide a valuable contrast to our primary course of investigation. If there were a fitting metaphor to describe this history which unfolds on a number of levels which have varying degrees of importance, it would be borrowed from Baudelaire's poem entitled "Le Thyrse" (The Thyrsus), this "pure staff" around which leaves and stems are wreathed: the straight line and the arabesque, one common goal and a variety of means which all add up to make an historical subject of pain.

# 1
## *Antiquity*

Although pain is experienced by everyone, it has not had the same status in every society or in every age. Despite the currently held view that all pain falls into one universal category, it seems in fact that pain or pains are not expressed in the same way in every culture and perhaps are not even felt identically. For example, we know that the threshold at which pain becomes unbearable varies not only from one person to another but also according to the culture from which the person comes.[1] Underneath a base built from anatomical and physiological data, pain also seems to have a cultural and social foundation. This is why the historian's viewpoint, when taken as we will do here from a few examples in Ancient Greece, can show up irreconcilable differences and the need to resort to the unfamiliar reasoning of Hippocratic argument in order to analyse our ways of verbalising pain and to show that they are not natural or spontaneous: the past is useful, not for its curiosity value but for its contribution to the understanding of present-day behaviour *vis à vis* pain since it is a possible means of putting our preconceptions and the fundamental basis of our understanding of pain into context.

Why choose Ancient Greece? In particular because texts such as the *Iliad* and the *Odyssey* made up part of a common Western cultural identity which has continued to the present day, and because they placed so much emphasis on pain. In the case of this particular culture, an historical perspective of the world of Homer as described in his writings (which in the *Iliad* was from around the 8th until the middle of the 4th century BC) allows one to contrast different kinds of texts from that age. For example, we can compare the epic and the tragic genres for which we have chosen some of the works of Sophocles, or we can compare secular treatises with specialised medical ones, such as that of the *Hippocratic Collection* which was partly contemporary to the tragedies of Sophocles. What is more, the complete work known as the *Hippocratic Collection* continued to have a living affinity with medical knowledge until around the middle of the 19th century. Before becoming an object of study for philologists and medical historians, the texts of the Hippocratic works were taught in medical schools for their valuable scientific doctrines, for their spirit and their method and also for their clinical observations. By giving priority to the medical point of view, our purpose is not to indulge in the dangerous game of trying to make retrospective diagnoses in order to rediscover in the Hippocratic works such a nosological entity or group of symptoms which one might be wrongly tempted to translate into modern day language.[2] If we ask ourselves about how pain is described in some of the Greek texts, we must try to understand the way in which a culture describes the experience of pain, what it conceals and what it expressly reveals about it.

It may be that the legitimate aim of every study of the structure and function of the vocabulary of a given work is to grasp the pathological truths behind the various texts and categories of statements. However, in order to encompass the limits of that study, it is important to remember the other irreplaceable means of getting at the truth, namely iconography and, indirectly, paleopathology.

## The Vocabulary of Pain in Homer's Work

The epic genre itself describes a certain experience of pain, namely that of a wound or blow received in single combat; however, there is hardly any place for chronic pains, such as those accompanying a long illness. Because this research is aimed at physical pain and not at the ordeals and suffering endured by the heroes returning home after their victory in Troy, our enquiry examined the *Iliad* before the *Odyssey*. Six groups of words can be distinguished which all have a link with pain.[3] It is possible to distinguish a vocabulary which describes mourning and its ritual with πένθος *(penthos)* and a vocabulary which describes grief and the exceptional care given to this event in life which originally indicated worry and obsession κῆδος *(kèdos)* and which can also vie with the word ἄλγος *(algos)*. Finally, there is a whole vocabulary based around ἄχος *(achos)* which additionally expresses a sudden and violent emotion and a confusion of feelings which can lead to despondency. In these three instances, there are particular connotations which allow us to understand the meaning of the terminology by looking back at social circumstances or types of feelings. However, it must be said that the distinction between physical and moral pain is not an appropriate criterion for distinguishing the terms. As for the three other words around which the vocabulary of pain is built (*odunè, pèma* and *algos*), the overlap between the physical and the moral is complete, as was emphasised by the examples which Liddel-Scott provided (1966).[4] The word *odunè*, which sometimes without due consideration has been reconciled etymologically to the vocabulary which describes the pains of childbirth ὠδῖνες *(odines)*,[5] seems to explain a sharp shooting pain which is generally well localised. The adjectives which are most often associated are ὀξύς *(oxus)* and πικρός *(pikros)*, *i.e.* sharp, pointed or cutting and biting respectively: by a sort of reversal, and not purely a rhetorical one, they therefore refer back to the instrument that caused the wound or sting, such as a javelin or an arrow. Doubtlessly for the very same reason, they are characterised by rapidity such as in book V of the *Iliad* when Aphrodite is wounded, and in which the pain of gods or goddesses is not of a different nature than that of the heroes.[6] This can be seen again in book XI.[7] More suggestive still is the use of the verbs employed to describe this experience: the flesh is pierced through and through, which is clearly denoted by the prefix *dia*, across, through and through.[8] The wounded feel a sense of exhaustion, literally of wear and tear—τείρω *(teiro)*, accentuated by the sweat that forms, whilst in other passages it is the feeling of weight or pressure linked to pain: "Heavy sufferings are eased."[9] In the *Iliad* there are numerous ways used to describe the alleviation of pain: for example, after the removal of an arrow or sharp object, the god of healing, Peon, otherwise known as Apollo, knows how to administer "remedies that relieve afflictions" although we are not told what these remedies are.

As apposed to *odunè*, the word ἄλγος *(algos)* and its derivatives represent a more general type of suffering involving the whole body: it is because this word is vague and undefined that its derivatives such as cardalgia and cephalgia give additional information as to the whereabouts of the pain. *Algos* is used not only, as its verbal contexts suggest (to endure, to put up with or to work with pain), to indicate a sub-mission to suffering and consequently a trait inherent to human destiny; it also signifies prolonged suffering by recording its duration and susceptibility to recurrence: this is indicated in verses 321–6 of book IX of the *Iliad* where Achilles declares: "It is for nought that my heart has suffered such pain and I have risked my life in combat each day." It would be possible to introduce an additional subtlety by going back to the idea that *algos* "always maintains a certain psychological distance in relation to events."[10] This distance is both temporal and psychological: when the hero of the epic talks of *algos*, he refers either to other people's pain, or he himself is not suffering at the time he speaks of it.

Although the root *algos* has been more productive for the modern, medical vocabulary of pain, in Homer's language it is the word *odunè* which is the technical term that belongs to the specialised vocabulary of medicine.

We still have a few words to say about a third grouping, namely that of πῆμα (*pèma*), which is found in contexts associated with *algos*, and often with the same verbs. We will not discuss the value of the Greek suffix *ma here: it is enough to say that the value of the result of action, rather than that of the passive state often attributed to this suffix, corresponds to a later formation of language. In this situation, the suffix *ma is the opposite to the suffix *sis, which is also used to form words of action. In Homeric texts, it would be better to compare it to the formation of the middle voice in Greek, which implies a subjective subtlety and denotes an activity in which the subject of the verb is involved. Thanks to this, the differences between the suffixes *ma and *os, and between *pèma* and *algos* quite clearly express a phenomenon which is not related to any cause and which, by this process, imposes itself on the afflicted person despite any personal volition. How then can we make sense of the double meaning of *pèma*, on the one hand adversity, scourge, a cause or subject of misfortune, on the other hand pain and suffering? In this latter use, *pèma* can sometimes mean the same as *algos* because it is tainted by its use in the expression *pèma pascho* (πῆμα πάσχω) which is the equivalent of ἄλγη πάσχω (*algè pascho*).[11] The partly social meaning of *pèma*—a scourge—does not contradict what was said above about the value of the suffix *ma, since it is possible to distinguish an objective meaning which implies a pain existing independently of the mind, or quite outside it.

It is possible to draw some conclusions from this initial study: in addition to there having been a specifically medical vocabulary in Ancient times, pain is described in ways which do not concentrate on the antagonism between the physical and the moral or upon the intensity of pain, but which follow two axes: the extent to which the subject is engrossed in the pain and how he or she perceives it with respect to time and to its origin: long-lasting and fast, sharp or cutting, *i.e.* by referring directly to the instrument which causes it and which simultaneously defines the very nature of the sensation.

### Pain As Described By the Tragedians: The Example of Sophocles

The tragic genre evidently gives pain a special place in which it finds its natural means of expression. We will now turn to the works of Sophocles, in spite of the fact that many other choices would have been suitable, such as Thucydides' account of the suffering during the "plague" of Athens. These works were contemporary to certain parts of the *Hippocratic Collection* and date from the 5th century BC. However, in order to be able to make an appropriate comparison, the vocabulary describing human pathos and misfortune is altogether too general and has had to be eliminated in order to try and concentrate on the tragedies which focus on the whole on physical pain, even if again, in this instance, physical and moral pain overlap.[12] It is for this very reason that Sophocles' *Philoctetes* has been chosen since it describes the pain caused by a wound inflicted some ten years previously, to a foot which became in some way gangrenous, a sore which may have been poisoned by the arrow which pierced the flesh and is therefore liable to paroxysms relieved by periods of intermission: we are dealing with chronic pain here and it is in this sense that it is radically different from the experiences described in the epic works. We will refer secondarily to the tale of the agony of Heracles in the *Trachiniæ,* when the hero had put on the tunic poisoned by the centaur's blood which had been sent to him by his wife Deianira, who was unaware of the disastrous effect this gift would have. Although this episode has a symbolic meaning and occurs within a mythological context, the expression of suffering in these two works reveals forms of particular sensibility, and even some ways of conceiving and managing pain. The literary way of describing things by concentrating on certain characteristics does not always give an accurate picture but is able to provide a discernible outline by using groups of metaphors which would be less significant otherwise.

Pain is perceived almost as an independent being which takes possession of the subject, invades it and takes over. Thus pain is often qualified by the word "consuming" in its true sense,[13] or "devouring."[14] It is the image of the living being who feeds on his victim the sufferer and who gradually grows stronger while the sick person grows weaker. The sufferer is worn out, exhausted: we have already encountered the use of the verb *teiro* and *Philoctetes* invokes the "painful manner of an exhausted man" by using the adjective *trusanor* in verse 209.[15] The examples given above take on a more precise significance if we relate them to an adjective which is often associated with pain or illness, ἄγριος *(agrios)*, or savage, which is also used for talking about wild animals. It is not possible to tame savage pain because of its intensity and also because one cannot predict when and how it will recur. Pain "comes and goes" (φοιτά-ω) with its attacks and paroxisms. For example, in *Philoctetes*, verse 758, "pain only returns after long days, when it is doubtlessly weary of having hurried elsewhere," is what the ancient scholar says in his analogy with a wild animal. This same image is used in verse 978 of the *Trachiniæ*: "Do not go and revive a pain such that it has dreadful recurrences." Several passages bring to mind a hunting scene where the traditional relationship between man and beast are reversed: pain stalks its prey and overcomes it at the most convenient moment, "It surges forth and leaps" in order to kill. This savage pain is so explicitly described in the tragedy because it tellingly reflects the will of the gods, and it ends up by contaminating the one it attacks. The sufferer loses his way and becomes like a diseased animal on the prowl who advances and attacks suddenly:[16] Heracles "writhes on the ground or else leaps up shouting and crying"; he roars "in the midst of his contortions."[17] When pain is at its worst, Philoctetes is in a sort of delirious state where he can neither recognise nor communicate with those closest to him. Only the cry survives, which gives vent to the pain by interjections, groans, and shouts. In order to try to explain the terrible nature of pain, Greek texts simply say that it is unapproachable, ἀποτίβατος *(apotibatos)* and intractable.[18]

The different sensations which characterise pain seem more variable than in Homer's work. In addition to the adjective πικρός *(pikros)*, pointed or sharp, or the vocabulary for describing a tear or mark made by an incision,[19] we also have the image of the gadfly's sting which sends animals mad, for example, as in verse 1253 of the *Trachiniæ*, when Heracles speaks "before a seizure or shooting pain (σπαραγμόν ἤ τίν οἶστρον) engulfs (me)." The same is true for the strength of the pain, which is expressed by a very precise word describing the weight put on a saddle (τό ἐπίσαγμα). The wealth of metaphors is not only limited to the animal kingdom, since the pain experienced by Heracles is that caused by an unbearable itching: verses 769–770 describe "the spasmodic itching which tortures him to the very bone."

Adjectives and verbs often convey a powerful meaning and seem to comprise most semantic descriptions characteristic of pain which more often than not are linked to familiar real experiences. On the other hand, the nouns, with the exception of the itching or sting described above, are more vague. The texts generally talk of ailments κακός *(kakos)*, illness νόσος *(nosos)* and pains ὄδυναι *(odunai)* in the plural. At first sight at least, the division between *algèma, odunè* and *ponos* seems less straight-forward than in the previous texts. *Ponos,* which originally indicated strain as well as work-related fatigue, can encompass pains which have nuances and different severities. We can see this in the verse where Neoptoleme says to Philoctetes: "You have experience of the whole spectrum of pain."[20] However, *algos,* which is mentioned seventeen times in Sophocles' work, does not compete with *algèma* and *algèsis* in the tragedy, as these terms evolved later.[21] It is *odunè* however that has the most technical meaning depending upon which compounds or derivatives it forms, *e.g.* νώδυνος *(nodunos)* for something that does not feel pain or which calms pain and which is used in contexts which are often therapeutic: in *Philoctetes,* mention is made of "plants which soothe pain,"[22] of what soothes or calms pain, πραύνειν *(praunein),* πάυειν *(pauein),* without our knowing which specific botanicals are being referred to. There is also reference in *Philoctetes* to the powerlessness of not being able to ease the pain.[23]

The Greek tragedy of the 5th century BC teaches us not only about pain but also about suffering in the most physical, tangible and blunt way. This is very different, for example, from the French classical tragedies of the 17th century where a rhetorical use of understatement serves to press home unseen torments. With the Greeks, however, even during narrative passages, the presence of a person in pain, badly wounded or shattered, adds weight to the truth of the story: this is the case with Hippolytus who, in Euripides' tragedy of the same name, comes back on stage suffering after his body has been broken in his frenzied attempts to escape a monstrous bull that has risen from the sea. This is also the case for the afflicted Heracles in the *Trachiniæ.* The suffering body seems only to emphasise tragic irony, in a sense, and to bring to life the antitheses whereby the curse afflicts the innocent and not the guilty person, while the gift that is meant to cement the love of husband and wife becomes a mortal gesture, and he who tries to avoid his destiny only precipitates it, thinking he has shrugged it off. Since the whole of Greek society in the 5th century BC participated in theatrical drama as a matter of course that was virtually both civil and religious, it accepted this emotional communion with the spectacle of pain; thus it can be argued that pain was accepted as part and parcel of everyday life and so was by no means something for the hero to be ashamed of. The tragic way of expressing pain was apparently heightened by the status of the text and perhaps by its role as a catharsis, revealing an explicit, brutal vision of pain, tangible and full of imagery. It may also provide an irreplaceable testimony since medical language, on the contrary, sought to reduce the infinite variety of individual experiences and their subjective details, to foster an understanding of these by cataloguing them. It is precisely in this way that medical language attempted to master pain—by separating it from its powerful emotional charge.

## The Significance of Pain in the Hippocratic Work

It is not our aim here to consider all aspects of pain by referring to the imposing body of some sixty treatises from the *Hippocratic Collection*.[24] If we did, it would be necessary to look in detail at the problems of authenticity and attribution, and to consider the substantial revival of Hippocratic studies and the current arguments.[25] It is clear that a large number of the treatises were written between 430 and 380 BC although some may have been written later and the collection itself was compiled at a much later date. However, it is also clear that they are not all by the same author, since both the aetiological concepts and the structures of the texts are different.[26] Moreover, the classification that was carried out for the treatises of the School of Cos (*The Nature of Man, Air, Water and Places, Prognostics, Epidemics I* and *III, Aphorisms* and *Cosian Prenotions,* to name a few) and the treatises from the School of Cnidus (*Diseases II, Internal Ailments, Ailments* and all the gynaecological treatises) leaves a certain number of treatises in doubt. Above all, this makes it difficult to define criteria which would allow us to distinguish one school from the other. It seems probable that we will have to return

to the excessive distinctions made between the two, *e.g.* those who in
the School of Cos see only a concern for prognosis and not for diagnosis, or who would place the theory of the four humours (phlegm,
blood, yellow and black bile) of Cos (as described in *Of the Nature of
Man*) in opposition to the four humours of the Cnidians, including
water. There is common ground between the two schools in a whole
series of concepts such as crises and coctions, in theories concerning
tumours and swellings, in the importance placed on diet, etc.; one may
well wonder if the notion of individual schools even had any relevance
for Cosian doctors. Furthermore, recent work has shown that there was
without doubt an internal progression in the School of Cnidus, which
could not be credited based on the outdated and impersonal style of
the Cnidian maxims.[27] This goes some way to explaining the complexity
of the problems posed by the *Hippocratic Collection*.

Thanks to the recent work done in the automatic indexing of the
books of the *Hippocratic Collection*,[28] we now have a solid foundation
for our understanding of the vocabulary of pain at our disposal. If we
disregard *algos*, which was replaced by *algèma*, and keep only *ponos* and
*odunè* as in the previous texts, we can get some idea of the distribution
of these words in the *Treatises*. In descending order, the texts where
*odunè* appears most often are *Diseases II* (73), *Diseases of Women II*
(67), *Epidemics VII* (54) and *Internal Ailments* (68). The *Ailments* and
*Nature of Women* both have the same number (49). The case of the
*Cosian Hypotheses* is interesting since it is the only one of the Collection
in which there is a substantial reversal of the relationship between *odunè*
and the other terms: *algèma* is used seventy times, *ponos* sixty-three
times and *odunè* only eighteen. However, this text, like the *Aphorisms*,
is made up of small sections taken from other texts in the Collection
and collated, perhaps to serve as a teaching manual or clinical hand-
book. These figures must be handled with care: first of all because the
frequency itself must be related to the overwhelming importance of
*odunè* in the whole work, as reflected by an overall comparison of the
respective number of appearances of the words in the *Hippocratic
Collection*: 772 times for *odunè* as against 14 for *algos*, albeit in competition with a much more recent word ἄλγημα *(algèma)* which appears
194 times. However, even if we include the use of *algos* as a verb *(algeo)*
(185 times), the superiority of *odunè* is clear. This depends on both the
subject and the length of the treatise, even though the most frequently
used words are not always the most instructive when viewed from the

perspective of a history of medical knowledge. At most, one could conclude that the so-called Cnidian treatises concentrate more on pain and that, when they do so, it is the word *odunè* that is used to convey it: this does not seem unreasonable when one takes account of the fact that traditionally the Cnidians, not the Cosians, have been considered as those most interested in the localisation of disorders, and that the contexts in which *odunè* is used show that the word is almost always used in a precise sense—either by qualifying, or by giving some clue as to the whereabouts of, the pain. Furthermore, a study of the contexts in which *ponos* and the verb *poneo* (to suffer) are used shows that the verb is used without any qualification to describe a general state of suffering or illness, and that when the localisation of pain is referred to, it is almost always approximate, involving the use of prepositions such as περί *(peri)* or ἐς *(es)*, i.e. "in the area of" or "about." It is more difficult to find distinguishing criteria for the word *algèma* which, in any case, is used less often. In any case, the figures quoted here allow us to identify more precisely the passages that deal with pain and which are reasonably numerous. Does not the text *Of Art,* which defines the doctor's duties, indicate that it is his job to alleviate suffering,[29] to know when to intervene and when there is no cure for the illness? The Hippocratic text itself created a link between specialised medical deliberations and lay deliberations. The author who wrote *De Prisca Medicina* stated: "In my opinion, he who wishes to discuss medical skill must above all make it his business to describe things in a simple way since the sole aim of the doctor's deliberations and research must be to deal with the illnesses that beset us all."[30] This in no way eliminates the idea of medicine as *technè,* since the patient knows neither how nor why his disorders come and go. The patient is able to describe his symptoms, however, and to recognise from his own experience and his "bodily sensations," what the doctors discover and try to explain. This is the specific value of the case history as it is the initial and critical step in coming to a diagnosis: "The only thing for them to remember whilst listening to the doctor, is what they have experienced. The physician who strays from their beliefs will not favour the right frame of mind and they will stray further from reality."[31] The book *Of Ailments* is also addressed to the simple individual so that he will be able to look after himself and eventually talk about what the doctors told him and prescribed. This common knowledge is very likely to be the condition for co-operation between doctor and patient, perhaps insuring a non-questioning attitude on the part of the former to the prescriptions of the latter. Judging from the numerous passages in which the doctor complains of shortcomings in regimens, this condition was not achieved. Thus, the way in which a patient is able to express the pain he feels will provide the doctor with some helpful information. The intellectual process which leads to a diagnosis in Hippocratic medicine requires a special relationship between physician and patient. In what we call "the Hippocratic triangle," the way in which the patient describes pain takes on a greater significance than in other concepts of medicine.[32]

Pain signifies, σημαίνει *(sèmainei)*, which is certainly not to be taken as an isolated symptom but rather as part of an overall picture of how the patient looks, what his behaviour is like compared to how he generally behaves, his stools, urine, sweat, etc.[33]

In a system of medicine which places great importance on prognosis because illness is seen as a process, the progress of which allows a definitive diagnosis to be made[34]—and in which the skill of the doctor is judged as much by how right he was in his forecasts as by the success of his healing—the moment pain begins in the pathological process, its duration and its relationship with any of various discharges (bleeding, suppuration, etc.) all serve as major clues in the understanding of diseases. According to *De Prisca Medicina,* in cases of ophthalmia, pain and inflammation last until the very moment when the congestion of the eyes begins to settle down thanks to the effect of the coction; in pleuritis, which is not called pleurisy since the pleura were not known of, pain in the side starts to calm down from the moment that expectoration, which dislodges phlegm from the infected side, takes place. Aphorism 47 in the second section states that "pains and fevers manifest themselves before rather than after the time at which pus develops."[35] Moreover, Aphorism 32 in the fourth section states: "In disorders caused by illness, if somewhere becomes painful, that is where any build-up will settle."[36] We can add to this interpretation of the timing of pain as a sign, a classification of its intensity: whether it is light or gentle, strong or sharp.[37] The exact location of the pain equally has a role to play in identifying the illness and in indicating what treatment should be followed. *Diseases I,* when referring to build-ups of pus in the abdomen, states: "It is particularly through pain and where it is felt, that these are found."[38] Furthermore, Aphorism 33 in the fourth section makes a direct link between the site where pain preceded all other signs of the ailment and the site of the ailment.[39] However, knowledge of whether the pain is above or below the diaphragm does give an indication of whether treatment should be instigated from above or from below. The quality of descriptions of pain is certainly less in the *Hippocratic Collection* than in the other texts studied and so too is its power of suggestion. This is perhaps due to the limited number of metaphors, which fits with a wish to provide precise accounts. On the other hand, this seems to improve the significance of pain from a medical point of view.

While the understanding of sensory pathways seems quite poor by our standards,[40] on the other hand, the aetiology of pain is not omitted. *Of Parts of Man* joins up the way in which pain is produced "through cold and heat, excess and want" with the cause of illness and in turn makes use of the principle of likes and the principle of opposites. Thus "it is brought on by heat in cold people, by cold in warm people, by humidity in those with a dry constitution."[41] This aetiology does not agree with the one put forward in *De Prisca Medicina,* which relates the origin of suffering to either the strength and quality of the humours or to the configuration of the organs. In the first case, it is the predominance of qualities such as bitterness, mildness, acidity, etc., rather than hot and cold, the lack of a moderating blend and their digestion, which brings on pain.

A variety of indications are given on the cures for pain, as well as for its spontaneous suppression at the onset of fever. Bleeding has a relatively conservative place in such therapeutics: it is only mentioned a few times, for example in *Epidemics II* with the servant of Stymangès, or in *Epidemics V.* Aphorism 38 in the fifth section describes the relief brought about in a person suffering from pain at the back of the head by opening the perpendicular vein. Incisions were used in pleuritis but most references deal with the use of heat in the form of affusions, fomentations or baths; cephalalgia, which cannot be translated into modern language because it is used to cover all sorts of pains in the head, was treated by abundant, hot affusions onto the head in such a way as to evoke a discharge of mucus;[42] similar observations were made in respect of otitis and pains in the hip other than sciatica. Sometimes cold is also used for cephalalgias as the incident described in *Diseases III.*[43] In this case, the author proceeds into an involved discussion of the advantages and disadvantages of hot and cold. When pain is erratic or long-lasting and unyielding, the physician may be persuaded to use cauterisation: "If the pain is concentrated in one particular area and remains there without being removed by medicaments, the painful spot wherever it might be, can be cauterised using a moxa of unbleached linen."[44] There is the same course of action for those suffering from gout. The sense in this practice must be viewed as a secondary manœuvre to be interpreted in the context of humoural medicine rather than from a practice of stimulation which was less established at that point in time. The Hippocratic work which, as we have seen, in certain sections places great emphasis on a therapy where opposites cure opposites, declares in two places that, where pain is concerned, there is good cause for curing like with like: in *Epidemics V* we are given the advice: "Match like with like, for example pain calms pain." This can be compared to Aphorism 46 in the first section: "From two pains occurring simultaneously but not in the same place, the strongest obscures the other."[45] We do not know which practices were inspired by such formulae for the relief of pain. However, from such remarks we have proof of the quality of the observations made by doctors in the Collection and of the amount of generalising and conceptualising which, although different from our own, goes beyond empiricism.[46] Furthermore, gynaecological treatises in particular note that the Hippocratic doctors knew how to use narcotic plants such as mandrake, henbane, nightshade and poppies.[47]

When scanning this quite lengthy period from Homer's time to that of the Hippocratic texts, because the participants throughout this period are not comparable, it would be dangerous to try to extract some constant factors or a uniformity of practice in the answers to pain in Greek society. It is possible to go back and discover attitudes which endured, however, such as the acceptance of pain as an inevitable fact of life for both the ill and healthy alike; indeed such findings must be taken into account without valorisation or occultation. The rationalisation of illness and pain in the Hippocratic work, which is made by a reinterpretation of symptoms, preserves both the word of the patient and the unique relationship he has with the physician. However, since the latter takes charge of the consultation and interprets the case history, he can also manipulate and distort his statements. Neither from the patient's, nor *a fortiori* from the doctor's point of view, is there any truly candid or straightforward exchange of information due to an underlying view of the causes, meaning and treatments of pain. This stems from a complex elaboration that is more or less shared by both parties and which includes a large measure of more archaic ideas—

which are closer to religious thought—than the tragic genre has allowed us to unravel.[48]

The widespread dissemination of Hippocratic ideas, whose original source had not been Athens, continued throughout the Greek cities of Asia Minor and, with Alexander's conquests, to the entire known world. Soon, new Hellenistic cultural centres grew up, particularly in Egypt where Alexander established the city which bears his name in 331 BC. Medicine, without disavowing Hippocratism, sought better anatomical foundations through the practice of dissection.

### The Influence of Alexandria During the 3rd Century BC: Herophilus and Erasistratus

Alexandria's fame lasted for a number of centuries due to several favourable factors: it enjoyed royal patronage, at least indirectly, and it also had a renowned library and was an intellectual and scientific learning centre where scholars from the entire Mediterranean Basin converged. Rather than just having a "school" in the strict sense, Alexandria was also characterised by its mental attitude, its intellectual curiosity, its taste for the arts and sciences, its tolerance and the diversity of its cultures. Nonetheless—despite the prestige of Egyptian medicine itself—the establishment of the Hellenistic culture on Egyptian soil seems to have encouraged few cultural exchanges, but served instead to enable its physicians to freely practice postmortem examinations, as H. Von Staden points up in his authoritative book devoted to Herophilus.[49] Herophilus, whose dates were approximately 330 or 320 BC to 260 or 250 BC, was even able to experiment at will on live criminals who were handed over to him by the authorities for his research. This practice of vivisection, originally permitted by Ptolemy I Soter and then by Ptolemy II Philadelphus, was also adopted by Erasistratus (304–245 BC), a contemporary of Herophilus, though perhaps slightly younger. The practice was soon condemned on ethical grounds and in the name of each physician's moral duty, but it held an unequalled fascination throughout the centuries.

The information Herophilus gained on the brain and the nerves represented real progress in relation to the work of his predecessors. While Aristotle had considered the brain to be a simple cooling organ—designed to regulate the heat generated by the heart's action and to induce sleep—when Herophilus became interested in the brain's anatomy, he was able to distinguish the ventricles, and he conferred a particularly important role on the cerebellum in controlling voluntary movements. He describes seven pairs of cranial nerves and located the origins of motor nerves in the spinal cord and the brain. Nonetheless, the term νεῦρον *(neuron)*, or nerve, did not yet correspond to a very well-defined anatomical reality[50] as is clear in this commentary by Rufus of Ephesus on Herophilus and Erasistratus: "The 'nerve' is a simple, solid body, the cause of voluntary motion, but difficult to perceive in dissection. According to Erasistratus and Herophilus, there are 'nerves' capable of sensation, but according to Asclepiades, not at all. According to Erasistratus there are two kinds of 'nerves,' sensory and motor; the beginnings of the sensory nerves, which are hollow, are to be found in the meninges, and those of the motor nerves in the cerebrum (enkephalos) and the cerebellum (parenkephalis). According to Herophilus, on the other hand, the *neura* that make voluntary motion possible have their origin in the cerebrum (enkephalos) and the spinal marrow, and some grow from bone to bone, others from muscle to muscle, and some also bind together joints."[51] Herophilus, who had studied vision, tried to explain the way in which nerves functioned through the concept of sensory *pneuma*, a term rather difficult to translate but which refers to a material element—though highly subtle and invisible—responsible for ensuring a number of bodily functions including assimilation as well as sensory and motor activity, and transmitted via the nerves, thus explaining why these were deemed to be "hollow." Ancient scholars in fact described several different types of *pneuma*, including natural, sensory or psychic depending upon the type of activity under consideration. The *pneuma* theory, originated by the Stoics and also adopted by Aristotle, played a major role in the history of the physiology of sensibility and later evolved under the concept of "animal spirits"; this theory was also accepted by Erasistratus.

Erasistratus, originally from Antioch in Syria, studied medicine in Athens and in Cos under the tutelage of Praxagoras before going on to Alexandria. He dissected the brain and studied the nervous system, confirming the distinction between sensory *(aisthetika)* nerves and motor *(kinetika)* nerves. He became interested in the cerebral convolutions, noting that they were more complex in human beings than in animals, and he described the cerebrum and the cerebellum as two separate organs. Contrary to Herophilus, he rejected the humoural theory, and sought to eliminate any unnatural causes from physiological and pathological explanations, developing a mechanistic, corpuscular philosophy and, to the notion of forces of attraction at work in the organs, he substituted the principle that nature cannot abide a vacuum.

Herophilus and Erasistratus explored a great many other aspects of medicine—including studying the pulse—and they used human dissection as the basis of their anatomical knowledge which thus enabled them to lay the foundations for an understanding of the nervous system. They prepared the way for Galen's work almost four centuries later though they themselves seem not to have enjoyed the glory of any direct posterity from what we can determine. Celsus' account of their method is particularly edifying: "As pain and illness may invade our inner organs, they cannot determine any way of restoring them to integrity without knowing their structure. It is therefore necessary to open up cadavers to scrutinise their viscera and entrails; and Herophilus and Erasistratus have gone even further as they have opened up live criminals which kings conveyed to them from the dungeons so they might capture in the raw what nature had kept hidden from them . . ."[52]

## Roman Medicine During the 1st Century AD: Celsus, Popularizer and Encyclopaedist

The work of Celsus, who lived during the reign of Tiberius in the 1st century AD, illustrates an entirely different aspect of medicine. His scientific view was much more closely allied to the *Corpus Hippocraticum*, as it regarded pain as an important clue for prognostic purposes, and gave detailed descriptions of the various possible remedies. It is not entirely certain that Celsus was actually a physician, and his eight books on medicine may have been part of a vaster work that was probably an encyclopaedia also covering agriculture, martial art, and rhetoric. He undoubtedly used several sources for his information but his results do not in any way give the impression of being a laborious compilation. The structural economy of *De Re Medicina*, one of the most precious documents we have on the medicine of Antiquity, clearly indicates that it was not his intention to give any lengthy explanations about pain. The first book is general enough to cover the early origins of medicine, its branches, and the different medical factions opposing one another; the second book presents general pathological conditions and therapeutic measures; the third and fourth books treat different types of illnesses and their cures; the fifth deals with medications; the sixth covers the disorders of specific parts of the body; and finally, the last two cover surgery and the disorders requiring surgery, such as fractures and dislocations.

In relation to the three schools he mentioned in his Preface[53]—the dogmatists, the empiricists, and the "methodists"—Celsus followed a careful, eclectic approach which is also evident in his commentaries on pain. The rational school of medicine, upheld by the dogmatists, sought "the hidden causes of diseases, then the apparent causes, followed by those of natural actions, and finally the make up of internal parts."[54] Theirs was an ambitious medical programme as they strove to uncover the basic principles governing the body, asserting that no treatment was possible unless the underlying causes of the disorder were known. However, in view of the extreme diversity of medical opinions and the fact that "all physicians have cured some patients," wouldn't it have seemed just as sensible to have ignored hidden causes and depended only upon experience? Such was the view defended by the empiricists. The methodist camp proposed to identify disorders[55] on the basis either of "constriction" *(strictum),* or "relaxation" *(laxum),* or a combination thereof, and their general approach relied on common traits evident among several disorders as well as the therapeutic indications provided by the phenomena themselves. Celsus was very critical of the methodists' practices, however, as he felt they were unable to distinguish the singularity of individual ailments or to use the appropriate remedies; he also thought they did not distinguish patients' individual characteristics or "idiosyncrasies" which only steady, careful observation could reveal.

In Celsus' approach to medicine, and in the general medical attitude prevalent during Antiquity as a whole, pain had no significance other than its value in announcing specific individual disorders and in providing prognoses; its meaning went no further than such indications and, in consequence, required nothing other than taking the appropriate measures to alleviate it. Pains had specific attributes such as their own "colours" as well as their seasons; they also corresponded to the different periods of life, to individual temperaments, and sex: winter brought on headaches, coughs, and sore throats, side twinges, and all the visceral disorders, though the seasons most to be feared were summer and fall. Continual pain, with or without inflammation, was a worrisome sign as was its sudden termination if there were no other favourable signs. However, pain's role was not solely limited to signifying the general evolution of the illness; it also foretold its subsequent stages more precisely: it served as an early indication of gout when located in the hands and feet, while sharp pains after giving birth—with no other negative signs—foretold of possible abscesses or nosebleeds, and digestive troubles could be expected when the hypochondrium was swollen and painful. Each pain, depending upon its location and the moment at which it occurred in the overall process, as well as the various other signs involved in the condition such as fever, sweating, delirium, and the patient's position in bed, etc., could immediately be interpreted on the basis of semiological rules which were largely of Hippocratic origin. As was the case in the *Corpus Hippocraticum,* the prognosis was helpful in determining the diagnosis, and the course of the illness made it possible to identify and better understand the disorder itself.[56]

Despite signs to the contrary, pain cannot be viewed as an alarm signal nor as a sentinel here; it was not a state which announced an illness to come nor was it even its prodrome, but was already the disorder itself. Pain had no conceivable positive value, thereby underlining a concept of health which totally excluded the least possible upset or suffering as being completely incompatible with fitness. Each form and manifestation of pain required the appropriate response: carefully graduated remedies seemed to provide the perfect answer to differing degrees of suffering and the physician never admitted to helplessness where pain was concerned. In case of "side pain" *(pleuritis),* for instance, an acute disorder characterised by pain, coughing, and expectoration, "The appropriate remedy for severe and recent pain is bloodletting; but if the case is either of a slighter or of a more chronic kind, then this remedy becomes either unnecessary, or belated; and recourse is to be had in cupping after incising the skin. It is also appropriate to apply vinegar and mustard upon the chest. . . . If the pain persists for a longer time, it may finally be dispersed by resin plaster. Food and drink should be taken hot, avoiding cold. Along with the above treatment, however, it is not unfitting to rub the lower limbs with oil and sulphur. If the cough has been relieved, the patient should read a little out loud, and now take both sharp food and undiluted wine."[57] This type of detailed account is given with regard to a number of painful illnesses, and gives a measure of optimism about the physician's ability to provide relief, particularly through the use of dietary regimens. Podagra[58] pains in the feet and hands were treated by exercise and by applying cooling remedies, as well as by rest if the pain increased;

however, once the ailment had persisted for a prolonged period, dietary precautions such as forgoing wine along with sexual abstinence became insufficient. In the paroxismal phases of pain, in addition to the hot sea-water baths and the marsh-mallow root cataplasms, topical ointments were applied locally made from poppy capsules mixed with fat, and fomentations composed of poppy and wild-cucumber decoctions were used to saturate sponges intended for rubbing over the painful areas.

The therapeutic principles consisted in treating a heat-producing disorder with its opposite and in stimulating evacuations to rid the body of any harmful substances contained therein. These principles not only affected the symptoms but the presumed causes of the disorder also and, if the physician's victory over the disease was not always ensured, he did not lack for potential correctives and was able to draw on a reasonably diversified pharmacopoeia as well as on dietary measures. Pliny and Dioscorides left a meticulously detailed inventory of every available pain-relieving remedy obtainable from the three natural kingdoms.

### An Exceptional Clinician: Aretaeus of Cappadocia

In contrast to the encyclopaedic character of Celsus' *De Re Medicina*, and also of Pliny's compilations, we might compare the work of Aretaeus of Cappadocia who lived in Rome during the 1st century AD, and whose *On the Causes and Indications of Acute and Chronic Disease*[59] Dioscorides mentions. In the form in which this work has reached us—it may have originally been intended as an instruction manual—it amounts to a kind of diptych with the causes and symptoms of disorders on one side and the treatments on the other; in each group, the distinction is also made between acute conditions and chronic ailments. There are few anatomical descriptions in the treatise, though at the beginning of a chapter Aretaeus sometimes recalled the nature of certain organs—the kidneys or the womb for instance—and explained their function; he also provided a few such indications at the beginning of the prescribed treatments. The essential body is devoted to identifying diseases, which he does with unequalled precision on the basis of their symptoms and of their "history," though his rigorous clinical observations did not exclude further research into their underlying causes.

A good example of his approach is evident in the case of headaches "caused by the cold and dry together." "If the headache is incidental and only lasts a short while, even if this amounts to several days, we term it 'cephalgia'; if, on the other hand, the disorder persists a long time and recurs periodically at close intervals, and if it is also increasingly painful and more and more difficult to cure, we call it 'cephalea.' This condition takes on an infinite variety of different forms: in some it is a continuous pain, though quite mild, but without intermission; in others, it returns periodically in similar fashion to a daily or double-tertian fever."[60] The pain might only affect one side of the head, in which case it was called "migraine." Aretaeus described migraine attacks admirably, the deep orbital pain which seemed to spread to the cranial membranes, with facial convulsions and glassy eyes staring fixedly ahead like "a mantle of horn," or convulsing in their sockets, with pain extending down into the neck, extreme sensitivity to light and an inability to remain upright, along with vomiting of *pituita*—one of the body's four humours, considered to have a cold temperament. When this disorder became chronic: "life is, in short, like a form of torture for them" (the sufferers).[61]

Aretaeus' ultimate remedy for cephalea was very long and painful in itself and did not ensure a cure. It was partly based on what was previously said about causes, in other words, the necessity of evacuating the influx of cold humours in the head by using countermeasures, though it is difficult to determine just how the "dryness" was opposed, if not indirectly, through the blood's properties. The treatment consisted for the most part in blood-letting, first by incising the arm, and then the forehead; next, the patient's head was shaved and washed with hot water on the basis, here again, of fighting a disorder with its contrary; then wet cupping was applied on deep scarifications and a branding iron was used for cauterisation; the head could, if desired, be rubbed with rubefacient plants to provoke sudation as well. While this treatment lasted, the dietary intake was very frugal and evacuation of *pituita* was encouraged by giving clysters; when the patient finally began to convalesce, gymnastics were added to his regimen, along with bathing and walking.[62] In desperation, the famous hellebore could be administered as it had always been a powerful remedy for every type of head ailment.

The work's arrangement—clinical descriptions preceding the treatments—was sometimes accompanied by reflections on the conditions inherent in the passage from a normal state of health to a pathological one. This was so in the case of arthritic and sciatic complaints which were studied as a single group because they were all considered to be articulation problems; sciatica was not deemed to be a disorder of the sciatic nerve but rather a disease of the hip.[63] Aretaeus expressed considerable surprise about the insensibility of the articulations and of those other parts (tendons, fascia?) when severed or broken, and of the pain they inflicted when affected by a pathological condition. His explanation—offered only as an assumption—involved both the texture of the parts whose tissues were generally too dense to be affected, and their temperature: "Those parts which are very dense, like the ones just mentioned, do not feel contact or wounds and are therefore not susceptible to pain because pain seems biting or sharp to the senses, and what is dense is not subject to being pricked or irritated; on the other hand, what is rare or slack may be pricked or irritated and can therefore feel pain. However, as dense parts subsist through natural heat, they may also become sensitive through the same heat. . . . But if this material detects a change in its degree of warmth, then a change in its sensibility occurs, and then the heat of this material becomes excited by the internal momentum of this sensibility faculty, thereby causing pain of a particular nature which is the result of an overabundance and an increase in this natural heat."[64] This density or rareness concept played a similar role to the hardness or tenderness of nerves described by Galen, and heat acted by modifying the qualities of living tissues by apparently diminishing their density, and thus increasing their porosity and receptivity to outside impressions. It is interesting to note that Aretaeus did not explain pain as the result of humoural causes but as the consequence of a process of internal deregulation inherent in solids. Aside from his qualities as a fine clinician, all highly admired by 19th century practitioners of Laennec's clinical method, Aretaeus was also very much committed to providing flexible theories to explain pain and disease.

## Galen of Pergamum: Pain, a Component of Touch

Conflictual or divergent trends were frequently brought together in Galen's work: it allied both medicine and philosophy, anatomy and physiology, a passion for logic and *disputatio*—an energetic form of debate often launched against his adversaries for polemical purposes—highly diversified and methodically evaluated medical practices, along with a careful study of the works of Hippocrates, whom he revered, as well as experimentation. Because of the coherence of his system, his powerful intellect, and his professional successes—as much in his native city of Pergamum where he was born in the 2nd century AD, as in Rome where he lived intermittently most of his life and was the physician of such emperors as Marcus Aurelius, Commodus, and Septimius Severus among others—Galen has a very special place in ancient medical history. His colossal works, including perhaps some five hundred published titles of which slightly less than one hundred are still extant, had been commented upon and translated for centuries from Greek into Arab, which contributed to its preservation, and into Latin. They constituted the sum total of Antiquity's medical knowledge and served as the authoritative medical references.

Galen placed considerable importance on pain in his writings, not only as a symptom, but also in his analysis of the mechanisms at work in sensibility and perception. Galen—who had a great deal of experience with dissection though probably only of animals (monkeys, pigs, cattle) rather than of human beings—adopted Erasistratus' distinction between motor and sensory nerves, but distinguished them on the basis of hard or tender nerves: "Each of the senses needs a tender nerve—a nerve because it is a sense organ; it needs a tender nerve because the sense must be inclined and affected a certain way by the external object for any sensation to be had. However, the tender is more apt to experience an impression and the hard more apt to act. For this reason, the tender nerves are necessary to the senses, and the hard nerves to all the other parts. . . . Moreover, the tender nerves and the hard nerves do not originate in the same parts of the encephalon and do not follow the same pathway to reach the senses."[65] It is clear in this passage that Galen had introduced a teleological perspective into his explanation of the structure and the use of parts. If the encephalon was generally soft, it was because this consistency allowed for the impression of sensations and because such mobility favoured the functions of imagination and intelligence centred therein; however, as motor nerves also originated there, it was necessary, according to the same providentialist logic, for the encephalon to have a dual nature, not in the sense that there were two hemispheres but because, based on such reasoning, its anterior soft portion would be in contrast to its other part, the hard "parencephalon," corresponding to the cerebellum.

The malleability and suppleness of the encephalic mass and sensory nerves only made up one of the conditions for sensation; for it to occur, the exterior impression had in addition to correspond specifically to the

sense upon which it was acting or, in short, there had to exist a form
of specificity. "Every sense is not modified by every perceivable object.
The shiny and luminous are affected by colours, the aerial by sounds,
the vaporous by odours; in a word, the similar recognises the similar.
Therefore, the sense related to air cannot be modified by colours."[66]
This perfect correspondence of the sense with its external stimulus did
not yet constitute perception, which required awareness in addition or,
in other words, the establishment of communication through the intermediary of a psychic *pneuma* between the individual's directive principle—located in the encephalon and the seat of all intellectual activity—with the sense organ. This psychic *pneuma*, a subtle, material
vapour which, in a sense, played the part of the "nervous liquor" of
later centuries, was a sensory projection of the encephalon.

The three conditions necessary for perception were clearly set out by
Galen: it required an organ to receive outside impressions, a connecting
passageway, and an organisational centre to transform the sensation into
a conscious perception. But to which particular sense did pain belong?
Galen associated it specifically with the general category of touch, the
sense responsible for tactile and thermal impressions when the external
impressions or irritations were of moderate intensity, and for painful impressions when these contained an element of violence: "The tactile sense
has a specific characteristic which distinguishes it from every other sense
in the way it functions: this is pain. It can follow different sensations
arising from other specific external sensibilities but, for its part, it not
only arises from exterior sensations but even more from sensibilities
within the body, and frequently with such violence that some are so
overcome by torment that they prefer to take their own lives."[67] Galen
then went on to describe how the eyes and ears could be the seat of
pain, and also how taste and smell could also experience disagreeable
sensations, but he once again emphasised that "the most intense pains
arise in the tactile sense." In order to understand the meaning of pain
as a variation in degree of intensity, it should be remembered that for
Galen every sensation supposed an internal change, which was the
requisite condition for stimulating the psychic *pneuma*. Without such
an alteration or transformation (*alloiosis*, ἀλλοίωσις, and *metabolai*,
μεταβολαι), no sensation was possible and this was the reason it was not
the continuation in an altered state which was painful in itself, but the
actual moment of transition from one state to another—made all the
more acute depending upon the speed of the shift.[68] Once the organism

had assimilated and adjusted to the alteration, the pain and even the awareness that there had been a change ceased.[69] That was also why the absence of pain could be misleading in cases when diseases had become established imperceptibly.

As a sign of internal changes in our organs, or as a sign of changes coming from the external world, these two causes of pain had an obvious and virtually immediate utility: pain served to warn and to protect every living being, and these functions accounted for the arrangement of the nerves and membranes which enveloped the brain, the pia mater and dura mater in particular, which Galen described: "Nature has had a triple motive in arranging the nerves: it provided for sensibility in the perception organs, movement in the locomotion organs, and the faculty to recognise any lesions sustained in all the others."[70]

Galen devoted several chapters of his work, *De Locis Affectis*, to analysing the diagnostic value of pain. The latter, in addition to other symptoms, was responsible for identifying any unhealthy organ or afflicted part. Each organ had its own specific function and any handicap to that function amounted to a specific symptom of that organ, a sort of pathognomonic indication pointing not so much to a specific disease but to a specific organ. Galen was responsible for classifying the different forms of pain which have been handed down virtually to modern times: "pulsific" or throbbing, "gravative" or weighty, "tensive" or stretching, and "pungitive" or lancinating. As is so often the case with him, he sets out his descriptions in the form of a pointed discussion with one of his rivals, Archigenes, who had also written a treatise on "The Diseased Parts" and who regarded pain characterised by a feeling of numbness as a specific manifestation of disorders of the nerves.[71] Galen did not consider the sensation of numbness to be a specific type of pain as it could be felt in any part of the body and, in fact, numbness actually countered pain to a certain degree. His comments on Hippocrates' aphorism stating that "a moderate numbness eliminates pain"[72] clearly suggests the practical applications of this state of insensibility which could be produced either by cold ambient air or by applying refrigerant medications locally. Perhaps this simple observation could be included within the theoretical framework of "anodyne" remedies in which opium served as a "cold substance," for instance; however, its origins were undoubtedly based on the method of representing health and disease through a system of "qualities" such as cold and heat, moisture and dryness, each corresponding to the four humours: blood, yellow bile or choler, phlegm, and black bile. What then could the sensation of pain particularly associated with the nerves have been? For Galen it was a feeling of tension because he envisioned nerves like the strings of a zither which would break when over-stretched, with the sensation extending from the point of origin to the extremity of the nerve.

In a humoural system of pathology such as Galen's, inflammation characterised by pain, heat, redness, and swelling (the four Latin "or" suffixes: *dolor, calor, rubor,* and *tumor*) played a major role. The "pul-sific" or throbbing pain of inflammation drew its name from the fact that arteries pulsate, though this is not ordinarily perceived. Only in cases of extreme inflammation can the rising motion of arteries be felt—or what amounted to the pulse for Galen—provoking a reciprocal compression of the arteries and surrounding tissue in the inflamed zone. This pulsating pain, characteristic of the arteries, then extended to the neighbouring areas insofar as they were sensitive. In the medical system handed down by Galen, throbbing pain became the specific sign of every inflammatory disorder. Various subtle gradations and degrees of intensity could be distinguished depending upon the severity of the disorder, its progression, and its rhythm: throbbing pain might become "pungitive" or lancinating when it extended beyond its initial territory as was the case with migraines or headaches. In a carefully structured system based on deductive reasoning such as Galen's, each sensory feeling was classified and categorised by order of importance to reach a better understanding of the afflicted part.

As an example, the explanation of inflammatory pain as the result of compression provided Galen with the basis for his differential diagnosis between "peripneumonia" and "pleurisy": "Pulsific pain therefore does not occur in peripneumonia but rather in the case of pleurisy due to the nature of these parts; the lungs are insensible, and pleurisy is a disorder of the membrane which lines the walls of the thorax. The part of this membrane in contact with the ribs is necessarily compressed, though the entire intermediary area escapes such pressure and only becomes painful due to the natural course of the inflammation."[73] Though it is always very difficult to know exactly to what pathological conditions the clinical descriptions of Antiquity actually correspond—despite the continued use of identical names which can be mislead-ing[74]—it does nonetheless seem that we have indeed been given a description of pleurisy in this passage in view of the reference to the membrane which enfolds the lungs rather than an account of simple *pleuritis* or side pains, as was the case in previous texts, because the word *pleuron* could not refer to the pleurae—unknown due to a lack of dissections—but only to the sides.

On a more general level, the entire anatomical vocabulary must be viewed with the utmost discrimination because the distinction made between veins and arteries did not have the same significance it has today as blood circulation was then unknown, for instance. The same is true of the ambiguous meanings of the word *neuron* that we translate as nerve and which we come across in both Herophilus as well as Galen, who explains: "Under this common denomination for the nervous category, I include the ligaments and tendons";[75] but he was also including the nerves in the modern sense, of course.

A dragging, oppressive pain of the type accompanied by a feeling of weightiness ("gravative" in French, from the Latin *gravis,* or heavy, weighty) might be felt in organs such as the kidneys, the liver, or the lungs "when they become afflicted by certain unnatural disorders of the tumoural type,"[76] but they were then also accompanied, more often than not, by an impression of tension (extension or distension) of the neighbouring areas. The rigorous observation of the different forms of pain was based on the conviction that there existed a link between each type and each "part" or type of part afflicted, characterised by its own specific nature or *physis* (φύσις). If correctly analysed, the symptom made it possible to identify the specific reality of the disorder. The actual seat of the disorder had a much more precise anatomical significance than in Hippocrates, who also drew indications from the localisation of pain, though it was of a more "regional" nature. The qualitative description of the different pains was closely linked to their presumed cause, and their meanings were part of a perfectly elaborated rational system of thought based on humours and qualities. This is clear in the next passage when Galen, having thoroughly analysed throbbing and weighty pains, expands on the subject: "Where other types of pain are concerned, they afflict certain parts but not others. It is therefore important to always keep them in mind and, knowing at the same time the nature of each part, recalling the type of pain associated with it, or not. . . . There is a type of pain which exists on its own in the afflicted part as a result of an abnormal change in the crasis [*i.e.* the combination of the humours], independently of any outside influences. Another which results from tension that is not found exclusively in the troubled part, but it is sometimes caused by the neighbouring parts. Another type is provoked through the contact of exterior objects with the suffering part when there is compression, contusion, or injury caused by a foreign body producing the pain."[77]

Just as there were two possible causes for disease, internal disruption and external aggression, the same was true about the causes of pain: one type resulted from the individual "temperament" or, in other words, stemmed from the way in which each individual had his own personal constitution composed of a blend within him of hot and cold, of dryness and moisture, whose different combinations could each predominate in turn except within the perfectly balanced temperament; the other type could tear, cut, and destroy. The Galenic definition of pain was adopted for many centuries and structured on the basis of these dual possibilities. "I have often mentioned the two basic types of pain in my other writings; these are the sudden and major change in the constitution of the temperament and the interruption of continuity. . . . Any part which has been distended, compressed, crushed, or injured suffers as a result from the break in its continuity. When an individual is injured by a pin prick, he endures the same pain as the one caused by the erosion produced by a sharp humour. In both cases, the continuity is compromised. The pain caused by a biting humour or by the overabundance of a humour is therefore not caused in the same manner."[78] The pungitive pain, the one that felt like a sharp point being thrust into the body, would specifically have been associated with the membranes while, going into greater detail, disorders of the colon would have been characterised by boring pains reminiscent of a screw penetrating deeply into the flesh.

Whether circumscribed to one area or diffuse, steady or shifting, acute or chronic, the multifacets of pain could nevertheless be deciphered. The success of this classification system was undoubtedly due to a number of reasons: it rested, for its definitions, on the solid logical foundations of an Aristotelian system of classification which made it possible, through a complex combination of attributes, to account for the extreme diversity of observable phenomena and to reach sensible conclusions. It also had the ambition, or aspiration of better understanding the reality of sickness through the help of this system, and of providing medicine with a rational foundation and basic certainties which had previously been the object of considerable disagreement. Finally, if we consider the limited possibilities of acquiring knowledge in Antiquity and even until the end of the 18th century or, in other words, essentially until the development of Laennec's clinical method, it should also be remembered that there were very few other diagnostic methods available and that this body of knowledge was far from negligible. The precise, painstaking and detailed knowledge of the meaning of different types of pain was ultimately relegated to a subordinate role—rather than actually invalidated—to be replaced by other more recent investigative methods and different, newer forms of analysis (tissues, cells, etc.). It may be that this body of knowledge about the semiology of pain, accumulated over the course of many centuries and carefully perfected through a multitude of observations—though frequently obscured by the successive strata of added commentaries and different interpretations—is worth saving from total oblivion despite the anatomical misconceptions which inevitably condemn whole sections of it.

## Ancient Philosophy and Pain: An Overview

Ancient medicine, in playing its instrumental role in uncovering knowledge about the body and in constituting a semiological classification of pain, came to no specific conclusions about its "utility," for all that. This obviously did not stem from any desire to divorce medicine entirely from philosophy, quite the contrary. Though Hippocrates has been credited with being the first to mark this division,[79] medical doctrines were for the most part all based on philosophical systems that explained Nature's origins and its composition. The moralist philosophers incessantly pointed up analogies between grief and sickness, both expressed by the term *aegritudo,* or the analogy between anger and madness. If, despite their basic opposition, the two rival ancient philosophies of Epicureanism and Stoicism bestowed no utility on pain, it was due to each of their concepts about the individual's relationship to his own body. This approach—which had consequences on the definition of health and, more generally, on the very nature of pleasure and pain as well—depended on what each system considered to be the ultimate end, the goal towards which each individual must strive, or the *to telos.*

For Epicureanism, the highest good was pleasure and therefore its doctrine only associated with *telos* what nature has ingrained in every living being. The Epicurean belief itself, however, was far removed from those caricature portrayals of it as a frenetical, pleasure-seeking philosophy of dissipation. It was based on a carefully calculated gratification system designed to ensure that the pleasures pursued were not followed by greater pain; it taught to rigorously work out the consequences of one's actions and to avoid getting caught up in only the present moment. If, when at full liberty to choose, "every pleasure must be enjoyed and every pain must be rejected," the practical rule might be expressed in the following way: "Certain pleasures must be rejected in order to savour greater pleasures later, while some pains must be patiently endured to escape more severe ones."[80] This evaluation, which promotes an austere lifestyle, is more the result of a diligent attempt to avoid any troubles and to reach a state of ataraxia than to a devoted quest for pleasures. For the wise, happiness was to be found in a tranquil ataractic existence rather than in the pursuit of "active pleasure" that titillated the senses.

The very absence of pain was therefore a pleasure, as Socrates explained at the beginning of *Phaedo* when he had just been released from his chains: "What a strange thing what we term pleasure seems, my friends! And what an odd relationship it has to pain which passes for its opposite! They refuse to be brought together in man; if we chase after one and catch it, however, we almost invariably also obtain the other as if, despite their duality, they were attached to one and the same head . . . that is the reason where one is present, the other soon follows behind. I believe this is happening in my case as well, because following the pain I felt in my leg due to the chain, I am now beginning to feel the pleasure that comes after it."[81]

While Socrates limited himself to asserting that the two states of pleasure and pain were inseparably linked, the Epicureans for their part clearly stated that, for them, no possible intermediary state existed between pleasure and pain: there was no room in their system for a sense of well-being or of discomfort being modulated in one direction or another. Thus, all absence of pain was a cause for rejoicing and all things which caused joy were therefore a pleasure; the Epicureans even added that "The pleasure we consider to be the supreme pleasure is the one we are conscious of when every possible type of pain has been eliminated."[82] Conversely, the slightest injury or the least sense of discomfort was immediately interpreted as pain—*i.e.* the ultimate adversity or evil—to be shunned at all costs. Despite the notion that pleasure and pain were not limited to the senses, and though the wise Epicurean very much enjoyed the pleasures of friendship, as well as recalling past pleasures with satisfaction, nothing worthwhile was possible without the senses. Such a doctrine was not necessarily in contradiction with virtue, but the latter was subordinate to pleasure: it was because virtue could lead to pleasure that the wise man pursued it, but not for its own sake. In short, no precept was more important than to live pleasantly, while its opposite, "The greatest adversity is certainly to live in pain," was equally true. This Epicurean doctrine did not triumph in Antiquity for reasons that were not entirely due to its definition of the greatest good but also because its political implications might easily have incited wise men to renounce any involvement in city affairs.

It was violently criticised by the Stoics who caricatured and satirised it as a voluptuous, sensualist, and debauched doctrine. At the time when monotheistic sects were sweeping through the Roman world, closely followed by Christianity, an objective alliance was forged between its Stoic critics and the new Christian morality opposed to such a philosophy based on materialistic principles and claiming to liberate mankind by restoring—thanks to Epicurus—personal dignity which was being trampled beneath the advancing footsteps of religious superstitions.[83] Such a philosophy inferred the closest attention to the self, an earnest self-centredness which was doubly condemned both by Stoicism and by Christianity.

Stoicism stretched over a number of centuries and it would be impossible in such a brief overview to present all its various facets. It therefore seems preferable to stay within the confines of the Stoic tenets, current during Roman times, promulgated in the works of Seneca or Plutarch. Like Epicureanism, Stoicism was founded on the basic tendencies of individuals, though it did not interpret them in the same way. It considered that prior to any notion of pleasure or pain, nature prompted every living being to stand firm in its own personal reality, to experience self-love. The ultimate purpose, which incited people to choose between different forms of behaviour, and also instructed them that they were not all equivalent, was whatever remained in harmony with nature; this criterion was what conferred value to all things (*axia*, ἀξία). Such a concept should not be viewed lightly, although appropriate behaviour was above all to preserve one's natural constitution and to do those things which were in harmony with nature. The good was not in nature itself, in fact, but in "the degree of correlation between what is appropriate and the choice made, then [in] the permanence of that choice which is ultimately in unfailing agreement with itself and in harmony with nature";[84] in other words, this was a reflexive process rather than an instinctive mode of behaviour. By appointing virtue as the highest good and it alone, Stoicism placed its emphasis on both the moral rectitude of the intention, rather than that of the action, and the inner liberty which made it possible to be free of outside constraints such as social upheavals, changes in fortune, slavery, or sickness.

The Stoics' attitude with regard to pain was generally one of encouraging endurance, "bear it and stay impassive" ('Ανεχοῦ καί ἀπεχοῦ), but the essential precept, backed by all sorts of practical suggestions such as being well prepared to face future ordeals, etc., rested on the much more fundamental conviction that "pain is not an evil." By this was meant that pain was given no place within the arena of moral good or evil and, in this sense, it had no value as far as the wise person was concerned and had no place among his preoccupations. To use a little sophistry, evil—or in this case pain—was the object of a much stauncher denial where disdain was only one consequence: "A happy life is recognisable by its morality which leads to the natural conclusion that only that which is moral may be considered good," which leads to "there is no evil but what is morally vile."[85] The reasoning was as follows: if pain were an evil, it would be impossible to be indifferent to it, to despise it and, finally, to overcome it, which would lead to the position of the Epicureans. It would have been impossible to find a brave man "if it were not well established that pain is not an evil." It was probably as a result of this highly-deliberate system of denial—devised to preserve individual liberty at times to the point of suicide—that the attitude of silence or disregard for pain became prevalent during the Middle Ages in the West, rather than through the influence of Christianity, though both became closely linked in unseating the body's pre-eminence.

### ⬦ PHARMACOLOGY IN ANTIQUITY ⬦

Though before Dioscorides many Greek physicians had described medicinal remedies obtained from plants, it was his *De Materia Medica* which became the authoritative reference in Graeco-Roman times, as well as during the Middle Ages, and virtually right up to the beginning of Modern times. This extraordinary success, to which Galen undoubtedly contributed by paying homage to his work, is due to the considerable increase in the number of remedies compiled by Dioscorides (1st century AD) compared to his predecessors, and the way in which they were presented. He expanded on the possibilities drawn from the three natural kingdoms—mineral, botanical, and animal—by providing a list of nearly 900 substances that could be used in the composition of medicines in contrast with only about 130 in the *Corpus Hippocraticum*. In the field of botany, in particular, which includes 600 entries, he devoted every effort to describing each plant in detail, indicating its original habitat, specifying the exact parts to use, indicating when to harvest it, how to prepare it, the methods of preserving it and how long it would keep. Along with his descriptions of their medicinal use, he also warned of any hazards or drawbacks for any given preparation. Though it seems likely that Dioscorides travelled considerably— it is said that he studied in Tarsus and in Alexandria before he went to Rome, and he himself also alludes to his life as a soldier— he probably did not himself observe all the plants he discusses but called on the writings of his predecessors, Sextus Niger, in particular, as did Pliny the elder to a great extent. This common reference undoubtedly explains the numerous resemblances between Pliny and Dioscorides.

The order in which the latter classified his information does not always seem entirely clear to us, and as early as the 6th century a manuscript of his medicinal substances appeared ordered alphabetically. The five books of his *De Materia Medica* originally contained: 1. aromatics, oils, syrups, and tree saps; 2. animals, cereals, and herbs; 3. roots, seeds, and certain herbs; 4. all the roots and all the herbs which had not been mentioned previously; 5. wines and minerals. It is thus obvious that these remedies were not classified according to the effects they supposedly produced, and the ones intended to relieve pain can be found in various places. Among them are included the opium poppy, the field poppy, lettuce, belladonna, henbane, and black nightshade.

This same classification problem, concerning the order in which information is given, also occurs in books XX through XXVI of Pliny's *Historia Naturalis* devoted to medicinal plants. This was not intended to be a medical treatise but an encyclopaedia in which the entries were arranged either by their morphological resemblances or their onomastical resemblances (in other words, based on their names, which complicated matters still further in cases when the resemblance concerned Greek names and not Latin ones), or based on their uses. In this way, a group whose leaves were used is listed, another gives plants with bulbs, and yet another section includes roots, etc.

The poppy, for its part, was listed among the aromatics. The unbearable smell of the black opium poppy's sap made it possible to discern any substitutions, as did its bright pure flame when alight and the fact that it floated in water. Pliny informs us that its use was already controversial in Antiquity as it had occasionally been used as a mortal poison, while Erasistratus defended its use particularly when instilled into the eyes. The latter practice seems to have gained little acceptance. The poppy's exuded juice was applied locally for pain of the joints, for instance, and instillations were applied in the ear in case of earaches, while pastilles could be taken in milk to induce sleep. Pliny carefully explains the way in which to prepare it: "We recommend that the capsules or the calyx be incised, and it is the only variety [he is referring to *Papavera nigrum*] which is thus incised. This juice, like that of any other plant, should be collected on wool or, if there is but little, on the thumb nail as is done for lettuce; however, the juice of the poppy is abundant and more can be collected the next day when, having dried, it is thicker; it can then be kneaded into small cakes which should be set to dry in the shade. It has not only soporific qualities, but it can even cause death during sleep if taken in too large amounts. We call it opium."[86]

Book XXV of his *Natural History,* which provides a panoramic sampling of all existing botanical works available for medical reference, ends with an enumeration of each part of the body and the remedies specific to it. Pliny also lists the most severe types of pain: among the thousands of different illnesses which can afflict mankind, "to try to distinguish which ailment is the most terrible seems almost like folly as each believes his present ill to be the most cruel of all. On this point, however, experience has shown that the disorders which cause the most atrocious torments are stones in the urethra, while pains in the stomach are in second place, and headaches in third; there are few other complaints which lead to suicide." And yet ancient medicine did not lack for therapies designed to fight pain. Though we may at times have trouble identifying all the plants mentioned, we nonetheless come away with the feeling that a thorough knowledge of the virtues of plants existed, as well as a deep commitment to their rational applications. If we examine the treatment for bladder and urinary conditions, we note that it included diuretics, responsible for encouraging the evacuation of calculus as well as the secretion of urine (plantain, perhaps as a decoction, taken with dried raisin wine, and woundwort leaves), and sedative remedies: extract of henbane from the Solanaceae family, "panaceas" or *panacis radix*—which probably included different plants such as lovage and laserpitium officinalis or L. panax (Umbelliferae), hyssop (Labiatae), and even hellebore for which, along with its purgative properties, Pliny also mentions the narcotic value. Each of the narcotic remedies were prescribed in highly diverse forms (frictions, applications in the form of salves, fomentations, fumigations, decoctions, potions, etc.) either for external use or to be taken internally in draught form. Pliny unfortunately does not always provide all the necessary indications for preparing these plants and, as his references were virtually always from written sources, he sometimes assumes that compound correctives were actually new types of plants, for instance!

This rich pharmacopoeia testifies to the constant efforts made to develop diversified new substances for use alone or in combination, though the results of this research were partly lost during the Middle Ages. It also reveals a definite attitude that sought to fight sickness and pain actively and energetically through every means available as the only choice was most certainly either relief, or suicide. It also clearly demonstrates that life was not viewed as sacred but that value was placed on living a full life in as healthy a body as was possible.

# 2

## The Middle Ages and Pain
### A World to Investigate

From late Antiquity to the dawning of the Renaissance, the Middle Ages certainly did not amount to a single monolithic whole, in terms either of its duration or of its geographic distribution. The West and East were at opposite poles as the result of the Roman Empire's internal chaos and the subsequent disintegration of its social life, with invasions on one hand, and the new intellectual centres arising in Syria and Persia exerting their growing influence, on the other. From its establishment in 330 to its fall in 1453, Constantinople—the Oriental capital of the Roman Empire—served as a centre of cultural exchanges between the different worlds and as the junction point between Graeco-Roman culture and Eastern civilisations. It would indeed have been most interesting to explore this universe further for its own rich complex history rather than in passing, as a simple transition, and to have taken the time to place greater emphasis on its individual development. The 7th and 8th centuries had little in common with the powerful feudal monarchies later established in the Western world, for instance, and any notion of Eurocentrism during that entire period is completely out of the question.

In view of the diversity of the cultural centres during that time, including Alexandria, Bagdad, Antioch, and Gunde-Shapur in Persia, and also because Arab science began its penetration of Europe starting in the 11th century by way of several routes, particularly through Italy and Spain, we will limit ourselves to a few broad outlines which characterised pain during that period. The most striking feature seems to have been the continued predominance of Galen's teachings, which held sway as a perfectly coherent system, providing answers for every problem encountered in medicine even beyond the Middle Ages. Another factor is linked to the importance of Arab medicine which produced so many renowned physicians, Avicenna and Rhazes among others. This medicine harvested a much richer medical and philosophical heritage from Greece during that era than any which managed to reach the then-slumbering Occident; the introduction of this Arab medicine in the 11th century acted as a catalyst in awakening Western thinking and subsequently its universities. Finally, Christian beliefs concerning pain certainly dominated the Middle Ages in the West, and left their enduring influence.

## Galenism in the Middle Ages

Galen was already well recognised as a medical and philosophical mentor during his lifetime, but it was virtually inevitable that his huge and diversified body of writings should lead to condensed versions and summaries of his ideas. Moreover, Galen's insistence that all parts of medicine were linked, both in theory and practice, anatomically and physiologically, made it very difficult to use only fragmentary portions of his doctrines. The discipline's own specific characteristics therefore led to a practice which may seem contradictory in that it adopted the use of commentaries and abridged canons. It was impossible to touch upon any medical matter without coming across Galen's ideas on the subject, and the taste in late Antiquity for vast compilations and encyclopaedias reinforced the tendency to list short citations accompanied by explanatory comments. This trend is clearly evident in the writings of Oreibasius. Modern scientific attitudes undoubtedly have quite a false appreciation of the duplication or collation value of what has been said by others on a given subject. Recopying Galen in the Middle Ages did not, for all that, signify abandoning a personal viewpoint; before any scholar allowed himself to enrich or expand upon some topic or another, wasn't the wisest course of action to recall what the recognised authorities had had to say? However, even if we accept this subtler outlook on references and their interpretation, there is no doubt that a number of aspects of Galen's work and what had served to nourish it were dropped along the way. Dissections were abandoned and there was an increasing disregard for anatomy, while theoretic discussions began to predominate over observations—along with deductive reasoning and *disputatio,* already prevalent in Galen—and then took over. What is perhaps more serious is that many physicians were introduced to Hippocrates through Galen. Thus Galenism has been handed down as a distorted medical doctrine, diminished as well as padded out in relation to Galen's own original writings.[1]

Various features have remained consistent throughout Galenism, however, such as the four elements (air, water, fire, and earth), the four humours (blood, yellow bile, black bile, and phlegm), and the conception of temperaments. The human body's functions were divided into vital, natural, and animal, while prevention and healing, as well as the causes of disorders, could be attributed to a careful breakdown of the "six non-natural agents" (food and drink, air, exercise and rest, sleep and wakefulness, etc.). The "humoural" framework of Galenic pathology is also perfectly obvious in the explanation of pain seen as the result of variations in the quality of the humours, though the "interruption in continuity" might be equally responsible or, in other words, a break in the parts.

It was under the dual influence of Avicenna and of Galen that Guy de Chauliac defined pain in his *Grande Chirurgie (Chirurga Magna),* published in 1363: "Pain, according to Avicenna, is a feeling of contradictory qualities. But along with these contradictory humours which might inflict pain, according to Galen, there may be alterations which break or cut, stretch or abrade: pain is therefore the result either of personally generated contrary qualities, or interruptions in continuity caused by accidents . . ."[2]

The treatment principles consisted in counteracting the disorders by their opposite, such as humidity with dryness, for instance, and heat to ward off cold, etc., which accounts for the use of "anodyne" remedies (to relieve pain); these were cooling or, generally, "of a contrary nature to a disorder, such as opium, mandragora root, nightshade, henbane, and poppy,"[3] and could be administered in the form of suppositories or as eyewashes, while substances such as saffron, myrrh, and castor oil could also be added. However, Guy de Chauliac conferred much more importance on evacuation or purges, including blood letting and temperament "alterations." In order to favour such evacuations, remedies that encouraged inflammations to "mature" or suppurate were also prescribed, prompting the emergence of a wide range of fats and oils, mixed with soft white bread along with eggs and made into plasters that diffused a certain heat. Guy de Chauliac also mentions another way of alleviating pain, which entailed "removing the sensibility of the part," presumably through the use of a ligature. If the general theoretical framework is Galenic, the sources are more diversified where remedies are concerned, however. This is undoubtedly due to the preservation of Dioscorides' *De Materia Medica* and Galen's writings *On Medicines,* along with the existence of many herbal and antidotal handbooks, sometimes only fragmentary but frequently completed by local lore. The medieval pharmacopoeia to relieve pain and cure various disorders included a certain number of popular remedies which had confirmed uses; there was less discrimination in determining the real virtues of plants, and preparation and administration methods were often closely akin to magical ritual, a practice which the Christian awe of miracles seemed to encourage.[4]

## The Transfer of Medical Knowledge from the Orient to the Occident

Galenism served not only as a model in the West but also further afield as Arab philosophers and physicians were instrumental in saving and disseminating his body of work, thanks largely to the translations undertaken at the urging of such enlightened caliphs as Al-Ma'mun.[5] A majority of the medical translations from Greek into Arab were done in Bagdad in the 9th century by a Nestorian Christian, Hunayn ibn Ishaq, thereby passing on the vast majority of Hellenistic learning to Arab scholars during the 10th century. This was the period in which the monumental Arab encyclopaedias were begun by Rhazes and Avicenna (980–1037). Avicenna was more of a philosopher, contrary to Rhazes, and in his *Canon of Medicine,* he assembled the sum of contemporary medical knowledge into a rigorously ordered system where deductive reasoning had an important place, especially where differential diagnoses were concerned. His *Poem of Medicine* amounts to an abridged version of his *Canon* and was probably intended for students. In it, pain had less importance as a useful diagnostic aid for identifying disorders than the pulse, urine samples, or evacuations; however, it was the harbinger of "crises,"[6] those salutary conclusions to illnesses proclaimed by an evacuation (sweat, hemorrhage, etc.), according to a concept that was as much Hippocratic as Galenic. Thus, painful formications in the sides with paroxysms announced a crisis as did pains in all the joints, and the exacerbation of all such symptoms were a favourable sign. Certain types of pain announced the end of the crisis, as did abdominal pains, though these should not be accompanied by headache. Pain played a major role in the prognostication of illnesses and also indicated the afflicted areas.[7] The sober and concise *Poem* provides a few indications on "sedative medications" which included two classes of antagonistic remedies, some which numbed like opium and others which reacted entirely differently: "Medications to suppress pain are warming and draining towards the exterior, diluent and emollient."[8]

The originality of Arab medicine remains difficult to assess in the absence of a detailed analysis of every available source. However, both as a result of its own considerable contributions to the general pharmacopoeia and the elements it had assimilated from Greek learning, its introduction to Europe—both through the Salerno school and through Spain—was a decisive factor in the awakening of Western medicine.

### Christianity and Pain in the Middle Ages

The growing importance of Christianity in Western Europe throughout the Middle Ages leads us to wonder about the attitude of early Christians in relation to medicine, and how they regarded pain. According to Georges Duby,[9] there are few accounts of the ways in which people dealt with pain at least until the 12th century when religious preoccupations underwent a shift; this became increasingly focused on Christ's Incarnation and his sufferings on the Cross and may have been responsible for the appearance of a greater concern with bodily suffering. Prior to that period, in a society largely ordered by men who were either powerful Church authorities or feudal lords constantly warring with one another, there seems to have been little room for pain whose expression would have been considered strictly a female prerogative. It is difficult to answer such questions categorically as what we may have assumed was a change in mores may simply be due to a different approach revealed by our references. Nevertheless, it seems probable that Christians viewed pain as both a form of Divine retribution, or as a sign of having been especially chosen and, as such, deserving of rewards in the hereafter—an attitude which may have encouraged a stoic indifference to pain. The monastic practices of mortifying the flesh, along with the performance of the basest menial tasks and the encouragement of a special order of flagellants for a time before it was banned, the fact also that Christianity originally presented itself as a religion offering salvation and healing (in the truest sense) by virtue of faith and prayer, all amounts to painting a landscape in which there seems to have been little room for intimate attention to the body; it seems that pain tended to be eclipsed whenever it was perceived and also that medicine and religion may have even been in competition with one another. There was nonetheless a limit to the indifference shown to the physical body in Christian thinking, in such instances when pain's overwhelming nature might lead to suicide. Moreover, the excessive agony in figurative scenes about the endurance of physical pain in "purgatory" originated around the twelfth century and the complacence with which suffering saints were depicted on stained-glass windows might also serve as useful clues about yet another relationship to pain.

What we know of other epochs teaches us that human beings are ready to try almost any remedy to relieve intense pain, including consulting non-official medical channels. Why should it have been any different in medieval times? Didn't the restricted use of "sedative sponges"—imbibed with a narcotic substance and applied to the area to be incised or held under the nose—nonetheless persist? Hadn't opium and various other sedatives been used continually since Antiquity? There is no question but that the Church's teachings tried to provide an acceptable meaning for pain, either as something to be endured as a divine gift or as a sacrificial offering which brought faithful believers closer to Christ, or as a means of redemption. However, what we really do not know beyond all these explanations is what people actually did in the Middle Ages when they were suffering.

# 3

## Pain in the Renaissance

The Renaissance, of which one could see signs in art and literature as early as the 14th century and which shone with all its splendour in the middle of the 16th century, was not however a golden age in which the love of luxury and lust, the love of art and culture and the love of life dominated without there being a darker or different side. Perhaps the group sculpture of Laocoon and his sons, discovered in Rome in 1506, could be regarded as a symbol of the contrasts of the Renaissance, by combining the beauty of ancient sculpture with the expression of suffering, the energy of the struggle with the despair of facing the deadly embrace of the snakes which were about to strangle the Trojan and his sons. Nor should one limit one's view of the period as there was a sharp distinction between the beginning of the 16th century which was relatively prosperous and successful, and the end of the century which was darkened by the wars of religion, famines, poverty and epidemics. The years 1520–35 were already perceived by contemporaries as being a precarious and menacing time, a "calami-tous" period when "the world was out of control."[1] Certainly, the last third of the century was marked by successive waves of the plague starting in 1563–64, with strong proliferations in 1578–79, a peak in 1582, apparently culminating in 1586–87 but then breaking out again with renewed vigour in 1596–98.[2] The importance of this scourge, made worse by shortages and famines which occurred at least every five years or more, was such that from 1561 the daily lot of men's lives was to endure misfortune and confront fear and the experience of pain. One is reminded of Montaigne who was obliged to flee from his house for six months due to the plague "in a very laborious search for a refuge for [his] family: a wandering family, frightening their friends and themselves and fearful wherever they tried to stay, having to change lodgings as soon as one of the group was beginning to feel the slightest twinge in the end of a finger. All illnesses were taken to be the plague: there was no time to identify them. And the way things stood, according to the rules of medical art, when any danger occurred there were forty days of mortal terror with one's imagination playing with one's mind and sorely trying one's health."[3] This first-hand testimony to the agonies of quarantine for those who were fearful that the evil would manifest itself any minute, is enough to make one wonder at what point pain becomes self-perpetuating and is amplified by the fear of death. In this phase of his meditation on death it seemed to Montaigne that the fear of death was aggravated by fears of the pangs of death, and despite his assertions about the plague that "it is a death that does not seem to me to be of the worst type; it is generally quick, producing dizziness but not pain and with the consolation that it is a common condition and therefore without ceremony, mourning or mourners,"[4] the death of the plague-stricken victim was not such a sudden demise as Montaigne had assumed. The surgeon Ambroise Paré in his *Traité de la peste*, published in 1568, describes the ills which befell the victims of the plague—he enumerates the "head and kidney pains, rashes and pimples affecting the skin, abscesses, charbon boils, dysentery and an infinite number of others: and let us begin with pain in the head which is so common in this illness."[5]

To these calamities which were not easily controlled and were interpreted as manifestations of the wrath of God, one may add the evils which resulted directly from the fury of men: incessant wars of conquest between France and Spain, as well as wars of religion which ravaged the whole country between 1562 and 1598. The actual conditions of war worsened: battles were of a longer duration, armies were larger, populations were heavily requisitioned by soldiers living at their expense, there was looting and plundering, armed bands of mercenaries were constantly changing sides while long sieges starved and decimated whole towns. War was everywhere and the invention of gunpowder which was used in France around the time of the reign of Charles IX, modified the nature of wounds for the worse. The massacres of Saint Bartholomew (1572) were the best known outbreak of the religious violence which erupted and then lasted all the longer in that neither side was able to gain just recognition nor lasting victory. No matter whether the outbreak of Catholic violence, the amplitude and cruelty of which exceeded that of the Protestants, was considered as ritual violence heralding divine retribution, or as a premeditated and rational campaign of destruction,[6] it made a lasting impression on the conscience and served, in the collective imagination, to equate the heretics as a body with the disfigured and broken body of the executed criminal. The poet Ronsard, who was himself a Catholic, wrote in his *Discours des misères de ce temps* of 1562 about "the cruel storm [which] threatens the French with such a pitiful shipwreck"[7] and the disasters of this "monstrous chronicle" in which the country had become its own executioner. However, that did not prevent him from arguing violently and intolerantly against the Protestants. A few years later, Agrippa d'Aubigné, a Protestant poet and companion of the King of Navarre, painted the horror of the carnage using the allegory of "France, afflicted mother, burdened by the two babes in her care."[8]

The violence of the images described by the poet to emphasise the shame of a civil war fought in the name of religion and his vivid rhetoric, aimed at rousing indignation, are perhaps no more than a glimpse of this period of atrocities:

> Oh poor France! Oh bloody land
> Not only earth but womb! Oh mother if it is a mother
> Who betrays her children by the softness of her breast
> And when they lie hurt, shakes them by the hand!
> You give them life and under your breast
> The bloody quarrel is envenomed by the obstinate![9]

There are few texts from this period quoted more than those of the Apocalypse, while Baroque poetry, which still flourished at the beginning of the 17th century, delighted in drawing attention to human fickleness and to the "perpetually changing world." In so doing, it multiplied the metaphors of fleeting shapes and of the precariousness of the times, like soft wax which could be moulded into any shape as is illustrated in the work of the poet Du Bartas. Winds, dreams and smoke usher the spectacle of a world which suddenly turns all the known laws upside down, creating a breach for the immense chaos: a world where contrasts are intermingled, *e.g.* fire and water, pain and joy, where hierarchies are overturned, where man faced with pain and death experiences a feeling of abandonment and identifies with the martyrdom and agony of the Cross, as this passage from La Ceppède suggests:

> Here is man, oh my eyes, what a deplorable sight!
> What shame to look at him, the lack of food,
> The pain and blood lost so extensively
> Have so crippled him he is no longer desirable.[10]

It is a period when paintings of the Descent from the Cross proliferated, when sculpture as well as painting portrayed the Mater dolorosa and when Piétas seemed to take on all the sufferings of the world and soothed them with their compassion.

The deepening intensity of their religious fervour allowed Protestants and Catholics to make sense of this spectacle of death and see within it God's punishment and his denunciation when faced with these

> cities drunk with blood and still corrupt
> Which are thirsty for blood and drunk on blood
> You will feel the terrifying hand of God.[11]

However, at the very heart of these atrocities emerged other attitudes which gradually broke free from any religious interpretations, refusing to see the evils of the world in collective terms or as an endorsement of original sin, but rather focusing attention on the individual, strictly human, experience of pain and illness.

## The Individual and the Experience of Pain

This "creation of the individual" during the Renaissance which we know was also the result of a long process, however, is inseparable from the new way of looking at that object which became a favourite theme for description and representation: "the observed body, the dissected body, the loved and also hated body."[12] It is often asked whether the Reformation had initiated a new relationship with the body, or more exactly whether there was a specifically Protestant outlook within a more general framework of the relationship between man and God without any intermediaries. Since the relationship with God occurred mainly within the privacy of hearts, where acts of devotion were tolerated as concessions to the weakness of men, but divine images rejected as idolatrous, the body and the use of the senses hardly formed a feasible basis for mediation with God. The constant repetition that man could be saved only by divine grace and not by his own deeds, and that everything he had at his disposal was a sign of God's favour, undoubtedly had incalculable consequences and contributed to the dedramatisation of the relationship between the sinner and his body. Calvin's catechism[13] is very careful to emphasise this infinite distance between God and man, who had no hope of obtaining salvation by his own personal effort (which did not mean that he might exempt himself from it). By emphasising the decisive role of grace necessary for salvation, and repeating that without divine grace, human "deeds" themselves were not enough to merit salvation, Protestantism, whether or not it stressed the dogma of predestination, freed the body from the grip of the Church.[14] In this respect, however, it may be necessary to introduce distinctions between Lutherans and Calvinists.

There is a great difference between the general outlook and the practices proposed for use in prayer by St. Ignatius Loyola in his *Spiritual Exercises*. The fervour of the founder of the Society of Jesus is based on a more profound understanding of man and calls upon both his memory and his imagination, as well as his understanding and his will. The fact that the preliminaries to each contemplation on the life of Christ should be a "composition of place" where the believer must try to envisage tangibly the circumstances where the scene takes place, the fact that in the "three ways of praying," St. Ignatius Loyola sets out principles on how to pray, for example by only uttering one word per breath in order to meditate better, the fact again that the senses are involved either in the meditation of hell—hearing the cries of the guilty, imagining breathing sulphur or touching the flames—or, on the contrary, in experiencing the sweetness of heavenly life, all reflect that Loyola's exercises require the total commitment of the person praying: "By this term 'spiritual exercises' are meant all kinds of ways of examining one's conscience, of meditating, of contemplating, of praying both out loud and mentally, and other spiritual methods that we will deal with now. Indeed, just as strolling, walking and running are bodily activities, likewise the different ways of preparing and making the soul ready to rid itself of all its disturbing afflictions, and after casting them off, searching out and determining the will of God for regulating one's life for the salvation of one's soul, are called spiritual exercises."[15] In this commitment to the service of God by the body and soul unreservedly, the experience of pain occupies a special place: the pain of sin is like the first necessary step on the path to penitence and is in fact at the very heart of the exercises of the "First Week." But, as emphasised by the "additional things to observe in order to perform the exercises better," it serves to bring out this pain of sin perceptibly through external penitence which is a result of the former and which follows precise rules regarding nourishment, rest and the body: "It consists of making him suffer tangible pain by wearing a hair-shirt, ropes and chains that bite into the flesh; by accepting these disciplines or by injuring oneself and by carrying out other austere practices, the most appropriate and least dangerous method is that the pain should be felt only in the flesh but not go so far as the bones; so that penitence causes pain and some infirmity. It therefore seems more apt to use disciplines which employ small ropes that cause pain externally, thus avoiding everything that could cause any serious internal infirmity."[16] Certainly, this treatment was reserved for those who wished to follow the path of spirituality, but it says a great deal about the relationship to the body, which had to be noted if it were to be subdued, "in order to compel the sense to obey reason" and, by so doing, show that it is possible to triumph over the sinner. To these aims, which involved encouraging indifference to earthly joys, was added the ever present model provided by *Imitation of Christ*: this anonymous work written in the previous century, which was one of the most important sources for Catholicism, has also been the object of numerous interpretations; but the Passion of Christ who, according to Christian dogma, chose to suffer and die in order to free men from their sins, sanctifies pain, makes the pain felt by the sinner acceptable and, in the case of *The Spiritual Exercises*, is also desirable; it serves as a means of getting closer to God by achieving an understanding of the sufferings endured by the "God-made-Man." Whether the apprenticeship of pain be a test on the road to salvation in the hope of eternal life or an opportunity to offer one's suffering to God as proof of one's love, as was the case for mystic writers such as St. John of the Cross or St. Teresa of Avila who were contemporaries, the relationship between Christian faith and pain is a spiritualised and sublimated one which deliberately turns its back on innate spontaneity and the voice of nature. For the non-believer, this is without any doubt one of the most unacceptable points in Christian dogma.

In contrast, Montaigne's attitude in this respect is still more remarkable. In the last quarter of the century, and Montaigne again provides an excellent example of this, the debate on the body ceased to be simply one involving doctors or specialists interested in the hygiene and physical exercising (gymnastics, games) of the body and turned to analysing the individual self. By asserting that "I myself am the subject matter of my book," Montaigne introduced the body in all its states to the dignity of literature and the value of introspection. By so doing, he made the writing of the *Essais* the ideal medium for exercising his critical faculties and for determining the boundaries of his "innermost heart" which also implied an unprecedented confrontation with his own body, including all its pleasurable experiences and sufferings. The individual's increasingly secular background and the adoption of a personal code of intimacy that suggested the following maxim, "My opinion is that one should lend oneself to others but give oneself only to oneself,"[17] were irrevocably linked to a new relationship with the body; a body no longer seen as a simple carnal envelope imprisoning the soul, checking its yearning to soar and dragging it down into the earthly mire, according to a Platonic philosophy that was widely adapted to Christian thought; a body which was neither transfigured by the sufferings undergone to imitate Jesus Christ, nor glorified by the pleasure and beauty that it revealed, but a body accepted for the fundamental reality of its feelings, with its mixture of pain and joy, its daily humility and mediocrity. Without doubt, this is the significance of the physical portrait which Montaigne painted of himself and of his character in this unprecedented venture in self-examination, the object of which was neither the confession of sins nor heroic exaltation but the understanding of oneself. It was at such a price that the experience of pain was integrated into the writings of one to whom self-study was intimately linked with the art of enjoying life and of "spending it completely at ease." Having, through the premeditation of death, adopted a Stoic attitude which, by the habitual thought of death, was destined to help him learn how to withstand and calm his fears, Montaigne drew the conclusion "that we spoil life by the concern about death and death by the concern about life" and added: "To the number of several other capacities included in this general and principal chapter on how to live, is this article on knowing how to die; and one of the least difficult if our fears did not add weight to it."[18] In a pun he writes that death is really "le bout, non pourtant le but de la vie" ("the end, but not the goal of life").

This attitude involved a personal and moral obligation to undertake self-study and it agreed with fundamental humanist principles of free examination and independence in relation to tradition and authority, as well as the practice of examining one's conscience as advocated by the Reformers. However, it was not a generalised attitude but rather constituted major evidence of the emergence of a new sensibility. Moreover, even when misfortune and pain were seen as signs of God's displeasure, as was still the case in the work of Paré who saw "the conversion and amendment of our lives"[19] as their main antidote, these by no means led to an enhancement of the value of pain nor to a resigned acceptance of it. For a surgeon such as Paré, if the illness was sent by God, "the means and the assistance are similarly given to us by him to use as instruments for his glory, looking for remedies in our ills, even in creatures to whom he has given certain properties and virtues for the soothing of poor suffering people; and wishes that we would make use of secondary and natural causes as instruments of benediction: otherwise we would be ungrateful and would scorn his active kindness."[20]

## The Revival of Anatomy

Ambroise Paré, perhaps more than Rabelais, epitomises that need for action and knowledge which is at the very root of medicine. The Renaissance doctor and surgeon alike were faced with a whole set of new experiences which modified the study of diseases. Thanks in part to the discovery of the New World and the enrichment of the pharmacopoeia, and in part to the discovery of numerous texts from Ancient times, he had quite diverse therapeutic methods at his disposal. This situation which evolved gradually at the end of the Middle Ages, allowing for example access to Hippocrates' books on surgery (notably his treatise on fractures and dislocations), without doubt gave a considerable boost to medical studies. Furthermore, Italy, and notably the universities of Padua and Bologna, was the place *par excellence* for these renewed discoveries with a much more complex and richer ancient period than scholastic tradition had given it credit for; thus, the Arab world ceased to provide the only access to certain scientific texts which it had helped to keep alive. One form of this critical spirit towards tradition consisted in confronting the teachings of Galen with anatomical observations. Little by little respect for authority gave way to the need to "see to know," with the slow return to the dissection of cadavers. *De Fabrica Corporis Humani* by Andreas Vesalius, published in Latin in 1543,[21] constituted a milestone in a long-used process which, in previous centuries, could have been associated with the names of Mundino or Carpi. The term *fabrica* which means not only the structure of the human body but also the ultimate result of a providential creation, suggested that the prerequisite for understanding the body was to take the human machine to pieces bit by bit. It is significant enough that Vesalius opened his "Fabrica" with a critical statement on the separation which had taken place between the three previously connected methods of treating the body: diet, medication and surgical art. The disdain for the manual arts, as opposed to the liberal arts, led to entrusting the surgical art to the care of barbers whereas, as Vesalius says, "it is the most important and oldest branch of medicine, that which (and I doubt that there is another), in the highest degree, depends on the observation of nature."[22] Surgical techniques and anatomical knowledge needed to unite in order to force the body to divulge the secrets of its organisation. This outlook aimed to put an end to the division mentioned above and also involved modifying the hierarchy between "internal medicine" (which was a domain reserved for doctors who were in charge of diagnosis and treatment, but who left the preparation of remedies to apothecaries-druggists, who in turn had long been confused with grocers) and "external medicine" which encompassed the sphere of treatment carried out by the surgeon-barber. In the end it would take two centuries to leave behind this division of duties and to restore the surgeon's dignity. However, one must weigh this up cautiously in order to understand that the fear of carrying out surgical operations involved the fear of being confused with the barbers and not only the apprehension about doing something to the body. Relegated to subordinate tasks, these individuals had little opportunity to carry out important operations in their practice in town. Part of the revival of anatomy in the 16th century resulted from the surgery of war, though those who had taught anatomy in the previous century were physicians (*i.e.* not surgeons), a paradox which should be emphasised. This situation where specialisation led to reciprocal ignorance is clearly described by Vesalius: "Since physicians judged that only the treatment of internal ailments was in their domain and thought that their knowledge of the viscera was quite sufficient, they neglected the structure of bones, muscles, nerves, veins and arteries which supply the bones and muscles as if it did not concern them. Add to this that having abandoned all practical work to barbers meant not only that doctors lost all real knowledge of viscera, but also all dexterity in dissection, to the extent that they no longer attempted it themselves. However, the barbers to whom they had abandoned the technique were so ignorant that they were incapable of understanding the writings of teachers of dissection."[23] The reinstatement of anatomy in relation to Antiquity did not occur only through a greater practice of dissection but also by associating this practice with the possibility of discursive reasoning on the use of parts, in which the order of anatomical description was important. Without doubt, it was a condition for daring to proceed with major surgical operations. Knowledge of the structure of the human body must therefore have involved both a critical re-appraisal of ancient learning and, at the same time, provided clear information that would be accessible to those who were not learned but who nevertheless had the greatest need for this anatomical knowledge. It was for this reason that the use of vernacular languages had such importance. However, Vesalius, still writing in Latin, emphasised the need to distance oneself from Galen each time that his animal dissections had misled him. This was demonstrated by direct observation of human cadavers but did not stop him from also dedicating a chapter of his "Fabrica" to animal vivisections. "Everybody gives such total credit to Galen that it would be impossible to find a doctor who would admit that even the slightest error had ever been found in his books on anatomy, and even less possibility of one of them being found. In the interim [let us note that Galen often corrected himself, rectifying mistakes he had made in a previous book in the light of experience and thus providing contradictory theses at short intervals], the revived practices of dissection, the critical interpretation of Galen's works and the justified amendments made to several passages, show us clearly that he himself had never dissected a fresh human cadaver. Misled by his dissections of the monkey . . . he was frequently led to unduly accuse ancient physicians who had practised human dissections of being wrong."[24]

This relative liberation with regard to the anatomical knowledge of Galen is also conveyed by the order chosen to reveal the make-up of the human body: after a study of the skeleton (bones and cartilages) then of the ligaments and muscles, Vesalius dedicated his third book to veins and arteries, the fourth to nerves, the fifth to organs of nutrition, the sixth to the heart, and the seventh to the brain and sensory organs. In spite of what Vesalius had to say about it, this sequence was a little different from the Galenic method which, being dictated by the principle of human activity as it occurs in the physical world, began its anatomo-physiological descriptions in the treatise *On the Uses of the Parts of the Body of Man* with the hand, arm, foot and leg, before going on to the trunk and the three great cavities of the body (abdomen, thorax and cranium) and finished with the "conduits," *i.e.* the tubes which constitute the veins, arteries, and nerves. Vesalius' architectonic vision of the body undoubtedly explains why he begins with its structural foundations which ensure stability and the possibility of movement, then subsequently takes an interest in the way in which the parts are connected and finally in the organs capable of impulsion and attractions. This Vesalian method was not completely distinct from the Galenic one, and for this reason permitted a certain amount of confusion to occur, as Juan Valverde de Hamusco stated in his *Historia de la composicion del cuerpo humano,* published in Spanish in 1556. This Spanish doctor who wrote in this language "considering the great need which our nation has for men who understand anatomy," which was considered as "an ugly thing to the Spanish,"[25] had himself studied in Italy—in Pisa under Realdo Colombo, in Padua and in Rome. He placed more emphasis on the joints of the body and the reproductive organs which he ranked in third place. However, in other respects he followed Vesalius as much in the general order he presents as in the description of each part for which he studied the number, place, form, substance, connections, use and function, according to the Aristotelian categories.

### Ambroise Paré: Innovations of War Surgery

We must place the work of Ambroise Paré (1509–1590), "surgeon to four kings of France,"[26] in a context which is triply expounded by the revival of anatomy, the architectonic conception of the body and the rehabilitation of *technè* or surgical skill as an instrument of knowledge.

However, one cannot emphasise enough the permanence of Galenism as a physiological and pathological model. All the great treatises on medicine of the period testify to this, such as that of Jean Fernel, for instance. In his *Pathology,* like Galen, he made pain a symptom of touch: exactly in the same way as the sense of vision is harmed by a light that is too bright and hearing by sound which is too loud, the sense of touch, "which is not a simple and unique instrument," is damaged by pain provoked by an excess in the qualities of hot and cold, humidity and dryness, softness and hardness, bitterness and sweetness, the heavy and the light. In contrast, the balanced mixture of these properties, their "temperament," does not elicit any sensation; "I have shown in my *Physiology* that this sense is not at all affected by things which, having been mixed, are similar to it, and that it does not perceive them but that, on the contrary, it suffered in various ways due to things which were dissimilar and opposite, and could not feel them without injury or lesion."[27] Fernel, who returned to the list of signs liable to be distinguished from pain, emphasised that it was not the perception of noxious qualities which constituted pain but the affection which consequently arose from such a situation, exactly in the same way as anger or sadness arise from the sight of something hostile. He rejected as sophistry the discussion of whether the "interruption in continuity" (*i.e.* the rupture or tearing of the tissues) was the cause of pain or if it was a noxious quality: "Galen, commenting on a passage by Hippocrates which states that any illness is a wound, maintains that all pain is due to a violent separation, and to a division within that which was once whole. Averroes, criticising Galen, is of the opinion that when there has been an 'interruption,' pain is provoked by a depraved and noxious quality which insinuates itself within."[28] This debate, in which Fernel refused to take part, pitted the two explanations against each other—one which was more concerned with the qualities of the humours while the other was more concerned with the state of solids. For Fernel, as for Paré, the two explanations of pain had to be related.

The development of anatomy was not translated into an immediate gain in the understanding of the mechanisms and treatment of pain: although, without doubt, a more precise description of the structure of bones and ligaments of the joints had been acquired during this period, the Galenic heritage continued to reign and carry weight since it constituted a rational and coherent medical system, capable of providing doctors with a global frame of reference.

The political events of this period without doubt involuntarily contributed to a profound change in the status of surgeons, epitomised in France by the career of Ambroise Paré, who progressed from the position of barber-surgeon to that of a fully-fledged surgeon: we must remember here not so much that the wars, notably between France and Spain, were incessant but that the ways of waging war had changed. Most particularly, from 1523 onwards there had been the use of firearms which coincided with the outbreak of the wars of religion—civil wars which endowed the traumas of war, the wounds and haemorrhages, with an inevitability and a burden of reality from which no one could escape. The *Dix Livres de chirurgie,* published by Ambroise Paré in 1564 (his first treatise on this question dates from 1545), provides a clear picture of the devastating role of the hackbuts which, like a bruise produced by flying stones, provoked "an aggravating pain" which discoloured the flesh and quickly made it gangrenous. It is because it was realised somewhat accidentally at the time of the campaign of Perpignan that the wounded whose injuries did not receive boiling oil or cauterisation healed better than did the others, that Paré came to understand that this treatment had to be abandoned. He showed similarly, along with other Italian surgeons such as Maggi, that bullets and musket-balls were not dangerous as a result of the combustion they caused or because of their venom, but because they bruised, lacerated and tore the flesh. Consequently there was no reason to use supposed antidotes to the poison caused by bullets (carried by the powder and the air). Rather, it was better to extract them, having placed the wounded person in the same position he had been in when injured in order to determine the position of the object and its trajectory more accurately, and to provide rational treatment by removing the debris of clothing, equipment and splinters found in the wound in particular. The seriousness of such wounds was certainly linked to their position in the body, but also to the mass and speed of the projectile itself. This concept induced Paré to use a much more conservative type of surgery which won him great popularity in the armies.

We have evidence of the fears caused by these pains in a dedication written by Paré, at the beginning of his *Dix Livres de chirurgie:* "To those who being pleased to recover wish to suffer no ills," in which he observes that the society of his time shuddered at the very mention of pain, "ever since Epicurus": since both the mad and the wise

> believe all pain to be of an extreme rage
> causing them to turn pallid and to lose heart
> when the mere word of pain is pronounced.
> From there have come a thousand wanton ways
> a thousand joyful discourses and mawkish slippery practices
> appropriate for exempting this soft tender body
> from that which some pain could well teach it.[29]

This fear of suffering led many of the sick or wounded to label the surgeon as an executioner and to avoid incisions, cautery or burning at all costs. However, the labour of the surgeon also involved provoking useful pain and thus he had to ward off the impostors and the "trickeries" of the swindlers, sorcerers and quacks who promised to be able to avoid it. The story of the Duke de Guise, Henri le Balafré (Henri "The Scarred") who allowed a lance that had pierced his face through from one cheek to the other to be removed and contented himself with only an "Ah!" as a manifestation of his suffering, is often cited as an example of courage. In the 16th century, testimonies illustrate both the capacities of endurance and the reluctance of the sick to suffer ablation, amputation or cauterisation of their flesh. Many preferred to await death rather than accept such tearing of the flesh. Despite this permanent spectre of pain, the atrocities of war and of epidemics, it is without doubt pointless to ask whether the pain tolerance threshold was different in the past from today. Nothing indicates that men then were more hardened than those of today: this is seen in the account of that manservant described by Paré who avowed in relation to a terrible toothache that "for the extreme toothache that he had, if he had not been afraid of being damned, he would have thrown himself into the moat from a window and would have been drowned in order to escape from his pain."[30]

What is probable is that pain in the 16th century, due to life expectancy, more often took the form of a sudden violence suffered by the body than a chronic pain stretching over years, although the experience of pain due to the stone endured by Montaigne from the age of forty until his death is certainly a counter-example. For Paré, "pain is thus a sad and adverse feeling, as a result of a sudden change or from an interruption in continuity."[31] This definition was taken from Galen.

This breaking of the tissues of flesh by bruising, incision, etc., or change in the qualities of the body were still seen, according to the Galenic representation, as a mixture of humours and qualities which implied, in order to be felt as painful, the sensibility of that part of the body and the awareness of the pain; it was necessary "that he be wary of the said alteration or interruption: otherwise, if no cause at all can be seen for the pain, notwithstanding the sensibility of the part, there will be no pain."[32] This definition of pain was not new, since these conditions dated back to Antiquity. However, we must see in it the reaffirmation of the necessity of individual awareness of one's body in order to perceive pain. In a context where the trials for witchcraft and debates on the subject were very much in mind, this reaffirmation could implicitly signify a rational interpretation of the phenomena of ecstasy as a separation of the body and the soul, and might provide an explanation for the real or imagined insensitivity of witches. This is the interpretation that was provided by Hippocrates' aphorism, according to which, when two pains are present, the strongest (which holds our attention) "obscures" the other.

## The "Anodyne" Remedies: From the Ligature to "Sweet Vitriol"

The works of Paré transgressed the usual separation of the surgical and medical domains; this actually caused him some difficulty with the Faculté de Médecine in Paris and obliged him, in certain editions of his works, to suppress, transfer or merge certain medical observations into special "surgical" sections which included tumours, for instance. This was the case with pain associated with fevers, as in principle he was not allowed to treat the subject. The humoral aetiology of pain led to a hierarchical order in which pains were categorised according to their causes and to therapies partly based on countering the illness by its opposite: "From hot and cold very strong pain is made; from dryness, moderate; from humidity, almost none or dormant."[33] This explains the abundance of designed remedies to refresh as much as to calm: the use of "anodynes" (in the true sense, medicines meant to suppress pain) fits well with this logic of curing each illness with its opposite. Certain "anodynes" were primarily characterised by their "coldness" which stopped the animal spirits from going as far as the painful part and, as a result, suppressed its sensibility: into this category fell the "stupefacients or narcotics" which were often applied topically to one part. Amongst many other examples, one can cite a formulation of hyoscyamus (henbane), hemlock, solanum, mandrake and opium.[34] It is this very logic of desensitisation of a part by interruption of the circulation which encouraged Paré to use ligatures not only for haemostatic purposes after surgical intervention instead of boiling oil or cauterisation but before amputations: "Extreme ligatures and compression also remove the feeling of a part, as when it is necessary to amputate a limb: in this way they will be added to the number of improper anodynes."[35]

This new technique, which replaced the sleep-inducing sponges which had fallen into disuse in the course of the Middle Ages, was one of the only methods available since the discovery of ether (so-called *spiritus aethereus* by Frobenius in the following century), the "sweet vitriol" of Valerius Cordus, was not immediately made use of as its anaesthetising powers were not yet understood. In Valerius Cordus' *De Artificiosis Extractionibus Liber,* published after his death by Gessner in 1561, he explains in minute detail the way to prepare and distil "the oil of vitriol" which he distinguished into either bitter or sweet: "The oil of vitriol, that some call the oil of life or artificial melancholy, is nothing more than an aluminous quality and a substance extracted, by art, from vitriol and mixed with a small quantity of sulphur . . . and, assuredly, this oil can be made into two types, one bitter and the other sweet: the bitter oil is made from a mixture with large amounts of alum, and a little sulphur, whilst the sweet is simply made up of sulphur. And it is nothing else than a sulphurous liquid extracted from the bitter oil."[36] After having described the long operations of separating the two "oils" and having described the appropriate vessel for distillation, Valerius Cordus indicated the practical virtues of this "oil" which was much more effective than sulphur because it penetrated easily throughout the body and could be given with all confidence: "It acts strongly in the body against all putrefactions, and above all against plague, and to reduce thick and viscous humours in pleurisy, peripneumonia and rebellious coughs. It stops calculi from forming in the kidneys and the bladder and heals ulcerations of the bladder."[37] It is clear in view of the slowness and the precautions taken with the preparation, that this sweet oil of vitriol could not have been in current usage, and the passage from *De Artificiosis Extractionibus* certainly testifies that Cordus did not even dream of its potential uses for suspending the consciousness and numbing the senses. However, in the *Paradoxes* by Paracelsus, there is a curious passage in which he mentions the potency of this sweet vitriol, but one is not sure whether it is exactly the same substance since Paracelsus insists on its stability, in contrast with other preparations extracted from vitriol: "In relation to this sulphuric acid, it has to be said that of all the things extracted from vitriol, it is the most remarkable, since it is stable. Besides, it is so mild that it can even be taken by chickens and they fall asleep for some time because of it, but wake up a little later unharmed. On this sulphuric acid, no judgement need be expressed, except to say that in illnesses which should be treated by anodynes, it calms all suffering without causing damage, relieves all pain, extinguishes all fevers and prevents complications in all illnesses."[38] Undoubtedly, the suspicion which weighed heavily against Paracelsus—

in part because of his contempt for Galenism as well as due to the confusion between chemistry and alchemy, and later the criticism of iatrochemistry—contributed to his text being forgotten and to the discrediting of the trials he alludes to and which may or may not have been repeated. Furthermore, there existed all sorts of medicinal preparations aimed at calming and easing the suffering part or perception of pain: decoctions with a base of barley, lettuce sap, camomile flowers, sweet-clover, periwinkle or water lilies, fomentations, ointments and liniments. In the formulations, the most often used ingredients were rose water, plantain, nightshade and betony; one also finds endive sap, wild chicory, sorrel and oxyrhodium which was frequently used in lotion form. Spices, notably cinnamon, were part of a rich pharmaco-poeia to which purgatives, bloodlettings, scarifications and vesicatories should always be added. Their function was to draw the boiling fumes which were trapped in the brain to the exterior, by providing a way out. However, Paré recognised that in the most violent and the most stubborn pains such as the colic, nothing was as good as a "syrup" with a base of poppies. The *Livre Neuvième de la Chirurgie,* dedicated to the "stone" (renal lithiasis), reveals enough about the way in which the surgeon of the time might approach the problem of pain, dissociating the illness from the operation: "One will know the stone in the bladder by these signs: the patient feels a weight . . . at the seat and perineum, with jectigative [repeated shooting] and pungitive pain, which extends as far as the end of the penis, so much so that he pulls and is always rubbing it, so it becomes excessively elongated and relaxed, and though he does his best to empty it, for the pain that he suffers causes a great need to urinate, he cannot do so very freely and sometimes manages only drip by drip, and in urinating feels extreme pain, crossing his legs and sitting on the ground crying, moaning and gripping strongly, because the stone is an unnatural thing."[39] These concrete descriptions of the continual pain suffered by the patient, together with the explo-ration of the bladder with a probe, allowed the diagnosis of the stone to be confirmed, and made it necessary to have an operation to remove it. Paré listed the instrument (hooks, probes, duck-billed forceps) and the technique to be used in such operations in order to avoid injuring other parts and in order to leave a wound which could be easily sutured and would heal quickly. Of what the patient actually endured during the operation we have no indication other than that it was necessary to tie his ankles and legs by passing a rope around his neck and tying his hands against his knees, and that, in order to assure that the patient was adequately immobile for the accomplishment of the surgery, he had to be held securely by four men, "strong, not fearful nor shy."[40] The awareness of the inflicted pain was not absent but merely set aside for the duration of the operation. Ambroise Paré often observed that "the primary intention of the surgeon must be to ease the pain" for reasons related to the intimate association of pain and inflammation or afflux of humours into a part of the body, with the risks of fever, boils (*i.e.* abscesses) and gangrene which they brought. He also wished to receive some co-operation from the patient, a form of acceptance of the surgery: "The indications as to the state of the patient's valour and strength must take precedence above all else: for if he is failing or is in a weakened state, it is necessary to forsake all other things in order to be helpful to him; as when necessity forces us to amputate a limb or to make some large incisions, or other similar things. Nonetheless, because the patient does not have enough valour and strength to tolerate the pain, such operations need to be postponed (if possible), for long enough for nature to be restored and strength regained by good diet and rest."[41] This ethical consideration respected the free-will of the subject every bit as much as the practical considerations; the appeal to the "valour" of the patient was not only a way of soliciting his courage, but also the resources of his nature and "energy." This co-operation was deemed indispensable for achieving a cure, in a medical system where the state of mind deeply influenced the evolution of morbid conditions.

For Montaigne the philosopher, the question in effect was really to ask what Nature could do in order to "manage" pain: but this question was bound up with a more substantial consideration of the influence of one's view of pain on the reality of what one experiences. This issue is situated at the intersection of the subjective and the objective and involves the experience making up each individual in which both the body and the soul play their parts. Montaigne was brought up on Stoic precepts, having learned celebrated examples of determination and steadfastness: from Mucius Scaevola enduring the burning of his arm to prove to Porsenna that he did not fear death, to the little boy from Lacedaemon preferring to tolerate in silence the fox gnawing at his abdomen rather than having his pilfering discovered.[42] Accordingly, in the first strata of the continuous writings that make up the *Essais* in which nothing is suppressed and where additions and successive corrections were made in the course of the trials which he underwent, testifying to his changing beliefs, Montaigne seemed inclined to think that the fantasies inspired by pain and its proximity to death served to emphasise the fear we have of it: "It is easy to see that that which sharpens our pain and heightens our sensual pleasure results from our brimming imagination. In animals, this is in check and allows their body to experience feelings which are free and naive and consequently unadulterated in almost every species, as we can see in their similar ways of proceeding. If we did not trouble the spontaneous movement of our limbs, we would feel better for it as Nature has given them a just and moderate temperament with regard to pleasure and pain. This cannot fail to be just, being equal and common. But since we are freed from such rules, and may abandon ourselves to the wandering freedom of our fantasies, let us at least temper them towards the most agreeable side."[43]

It is not enough to repeat that death puts an end to even our greatest pains, as was already emphasised by Cicero in his *De Natura Deorum*, nor to denounce the futility of views which exaggerate pain: from the time he first started to write his *Essais*, Montaigne isolated the problem of pain from that of death: "Thus, let us concern ourselves only with pain which, I grant you willingly, is the worst misfortune to befall us and, of any man in the world, I am certainly the one to wish it the identical ill and to do my utmost to avoid it so that, up to the present, I have not had, thanks be to God, much to do with it. But it is in us, if not to eradicate it, at least to decrease it by perseverance; but should the body yet be affected by it, nevertheless it is necessary to keep the soul and mind steadfast."[44]

Thus, in agreeing that pain is the supreme ill of the human condition and that "extreme pleasure does not affect us as does a light pain,"[45] Montaigne adopted a philosophical attitude which is perhaps closer to Epicureanism than to Stoicism. This involved seeking well-being in the absence of pain, in that state of ataraxy which he called "indifference" rather than in active joyfulness: "Having no ills at all is the finest possession that a man could hope for," he said in numerous passages, but it is clear that his reflections on pain bore no relation to the Christian doctrine of sin. This meditation on pain, this face to face with the inherent self, is authentic and is a form of introspection made within a philosophical framework devoid of transcendency. It quite simply had no place for the theology whose answers, one might suspect, had no more weight for Montaigne than those of the Stoic philosophers; in circumstances where the individual is confronted with the experience of pain and illness in solitude, this outlook could not be satisfied by long-suffering principles of fortitude and endurance. When he became intimately acquainted with pain, he relegated the effort of putting on a bold front when facing the strain of sickness and the tortures of stones or colic to the theatre of rhetoric and actors. In so doing, this lover of life contented himself with doing his best not to wish for death and distanced himself considerably from heroic examples he had once admired. For Montaigne, there was no shame in vocal or physical expressions of pain and he established a philosophy which allowed the manifestation of "these exacting groans along with sighs, sobs, palpitations, growing pale, that Nature has put out of our control. Providing that the heart is free of terror and speaks without despair, it should be satisfied! What does it matter if we contort our arms, providing that we don't contort our thoughts."[46] When experiencing pain, the only thing that Montaigne forced himself to retain was his lucidity of thought—not because he thoroughly dissociated the body from the soul, but precisely because he knew to what extent the task was difficult and how much it depended on one's philosophical outlook. This requirement, based on our natural ability to withstand pain and the limits of our resistance, should be measured in the light of the real sufferings endured by Montaigne, which he described in his *Essais* or in his *Journal du voyage en Italie*.

## Conclusion

The belief in an intimate, innermost self as described in the autobiographical meditations of the *Essais*—and generally considered to be co-responsible for what has been termed the "birth of the individual"—may well have had a conflicting impact on pain's expression. Such a conviction is based on a clear demarcation between what belongs to one's fellow man, *i.e.* that which may be "shared out" or "revealed," and that which belongs strictly to the individual self. This distinction, illustrated in the shift in autobiographical styles, ultimately relegated personal physical pain—a subject which Montaigne had thoroughly explored—to the private inner world; it became concealed all the more insofar as it was exposed to public scrutiny for a time, however briefly and cautiously. The period in which the *Essais* were written therefore appears a rather unique time in which the straightforward expression of pain, without pretense or camouflage, was presented in the brutal light of harsh reality and seemed overwhelmingly intolerable as a result. Later autobiographies tended towards a greater introspection as well as the analysis of spiritual suffering rather than the candid recollection of physical pain; there are, of course, a number of exceptions. Reading between the lines of what is said or left unsaid, it seems that the pain associated with an illness or wound was much more distressing for the one afflicted than the actual disorder itself, and more disturbing for his or her immediate entourage than the evident pain which could be witnessed publicly during such "spectacles" as floggings or executions. This proximity to pain, which the public authorities sought to restrict, was to become the sole province of the physician or surgeon, and was perhaps ultimately a contributing factor in the professionalism and specialisation of medicine, along with a number of other considerations. The means of alleviating pain at the physician's disposal were both empirical, thanks to the remedies handed down from Antiquity, and based on a comprehensive body of knowledge largely influenced by Galen's works. Despite the renewed interest in anatomy which had begun to tear down this edifice, its humoural foundations retained their firm foothold though iatrochemistry also had begun to contest its inroads. It was not until another scientific model with altogether new requirements came along that a shift in the basic foundations of knowledge occurred.

# 4

## *Pain in the Classical Age*

In the 17th century, mankind's knowledge took a great leap forward thanks to Harvey's discovery of the circulation of the blood, which he described in his *Exercitatio anatomica de motu cordis et sanguinis in anima,* published in 1628. Indeed, this work was a turning point in the history of medicine, because it would allow physiologists and doctors alike to gradually break free from the legacy of Galen. They would become free to develop new methods of investigation and to dare uphold that their observations and deductions could be more accurate than those handed down from Antiquity. However, it should be added that this analytical attitude, which had begun to take form in the previous century, also benefitted from the success of the revolution in physics and astronomy which came about as a result of new explanations of the universe formulated in terms of mathematical principles. The condemnation of Galileo by the Church in 1633 certainly led Descartes to postpone the publication of his *Traité du monde;* however, it also shows that the science which was gaining in strength was one which was concerned with the natural laws governing the motion of bodies, their modes of transmission and velocity and that, in order to progress, all it required were concepts explaining space, gravity, dynamics, etc. At the same time, it was also tending to reject, once and for all, Galenic "faculties," all types of occultism and nonsensical explications such as the belief that "nature cannot abide a vacuum." In this context where Mechanism was triumphing in the natural sciences, medicine itself tried to envisage the human body as a complex machine which could be compared to an ensemble of ropes, levers, and pulleys. It tried to reason in a "geometric fashion," *i.e.* by rigorously stringing together all its propositions and accepting only that which could be proven. These new tendencies which sometimes have been summarised as constituting a "first biological revolution,"[1] had indirect but important consequences both on the way sensory mechanisms were depicted and, more particularly, on our understanding of pain.

Though there were certain common views within the "scientific community" of this period, the way science was evolving in France was very different from the research being carried out simultaneously in England or even in the different Italian states. What we refer to as "The Oxford School of Physiology,"[2] which included not only the work of Harvey but also, slightly later on, that of Lower, Mayow and, notably, Willis, placed much more emphasis on experiments, was much more active in the rebirth of chemistry and, at the same time, was much less involved in religious debate than were its French counterparts. It is also probable that there were significant differences in the interpretation of Aristotle's work and that Cartesian dualism and the theory of animal-machines were viewed in terms of the critical analysis of the vegetative and sensitive soul. Furthermore, the physiologists from Oxford—foremost among whom was Robert Boyle—defended an atomical, corpuscular concept of matter which was more complex than that put forward by Descartes. By reducing matter only to its spacial extensity, considering that movement was not one of its inherent characteristics, Descartes proposed a mechanical model whose specifications could be very attractive to the life sciences, but which came up against a number of difficulties. However, his role in developing the concept of the reflex, his interest in involuntary movement which was partly linked to the debate about the souls of animals, and his efforts to make progress in cerebral localisations by proposing that the soul was situated in the epiphysis or pineal gland, were important developments prior to the much more precise and exact work of Willis. Indeed, Willis carried out numerous dissections and was always intent on linking his anatomical observations to his clinical work.

## The Cartesian Machine and the Theory of Sensation

In the fourth part of the *Dioptric,* which was published after the *Discourse on Method* and was dedicated to the senses in general, Descartes dismissed the idea that there were two sorts of nerves, sensory and motor, as well as the hypothesis of a localisation for "the faculty of feeling in the skin or membranes" of the nerve, *i.e.* its sheath, and of the capacity to move "within the internal substance of nerves." For him, *animal spirits*, a sort of wind or subtle flame emanating from the blood, provided a way of explaining movement whereas "the fine threads of which the internal substance of nerves are made subserve the senses."[3] One can see a legacy of the Galenic "pneuma" in the animal spirits which also have their role to play in these threads enclosed in tubes that are more or less swollen with spirits. They make it possible to grasp the almost immediate communication of a contact made from any point of the peripheral nervous system to the brain in exactly the same way the bell in a church tower rings when we pull on the rope attached to it. This image of the rope being pulled is used several times by Descartes, in fact as early as 1637 in the *Dioptric*: "If we but lightly touch and move the position of these limbs where one of them is attached, at the same time we also move the part of the brain that it comes from, in the same way as by pulling on one end of a taut rope we also make the other end move."[4] But it is especially in his *Principles of Philosophy* of 1644 that Descartes goes into greater depth with his idea of sensations, and pain in particular, as a way of understanding the union between the soul and the body. He analyses the problem of pain in phantom limbs by citing the case of a young girl who had her hand and forearm amputated. Drawing on the conclusion already reached by Ambroise Paré, he saw in the persistence of pains that seemed to come from severed extremities, proof that "pain in the hand is not felt in the soul as being in the hand but as being in the brain."[5] He tried to provide a rational explanation for this by arguing that the enduring "agitation" of the nerves from the amputated hand, which now terminated at the elbow, produced similar sensations to those felt had the hand still been there. This exposition by Descartes was altogether remarkable in that he didn't regard the pain from an amputated limb as being imaginary, but rather as being so real that there could be a precise explanation. More particularly, it was his way of explaining that pain was a perception of the soul. He believed that the experience of pain confirmed the existence not only of the body to which the soul is joined but also the reality of external bodies. However, according to Descartes, the senses allow us only to become acquainted with what is useful or noxious but never with the precise nature of pain.

The perception of pain was a perception of the soul which could be produced either by the action of objects external to the body or from the body itself. In the first case, pain was linked to "touch," one of the five external senses as opposed to the "internal senses" which comprised natural appetites (hunger, thirst, etc.) and feelings such as joy, sadness, love, anger, etc. Descartes did not view pain as being a specific sensation, but rather as being one mode of action of the animal spirits involving the nerves of touch. Thus contact with external objects enabled the soul to perceive such qualities as hardness, weight and warmth since there were a multitude of ways in which nerves were activated or, conversely, prevented from reacting; for Descartes, however, the perception of pain was not an additional quality that could be equated with hardness or warmth, or their opposites. Rather, it was a particular modality of these various sensations as was pleasure: "Beyond that [the feelings of hardness, weight, etc.], when these nerves are solicited a little harder than normal albeit in such a way that our body suffers no damage, the soul feels a tickling which it also perceives as a confusing impression; and this impression is naturally agreeable especially since it acknowledges the strength of the body to which it is joined in that it can endure the action which causes this feeling without being injured. But if this same action has a little more force to such an extent that it injures the body in some way, this makes our soul feel pain. Thus we can see why sensual pleasure and pain are two completely different feelings to the soul, not withstanding that one often follows the other and that their causes are very similar."[6] As for the perceptions that come directly from our bodies such as the heat of the hand, or pain, these are not different in any way from those produced by external objects. It is only through the order of succession of sensations that we are able to determine whether the sensation is already within us or whether it arises from an external source.

In explaining the transmission of a sensation, whatever its source, Descartes wanted to consider only the movement, form, situation and size of the body. The image of the taut rope could illustrate two characteristics of the transmission: the speed of passage from the receptor to the brain and the necessity for continuity between the two; evidently, however, it was inadequate in its detail since it resolved neither the localisation of the sensations in the brain, nor the exact nature of what had caused them. With respect to the first problem, Descartes chose to situate the *sensorium commune* (or "convergent point for all sensations in the brain") in the pineal gland because, he said, it is the only organ which is not replicated in the brain: although we have two eyes, ears, and hands, only one sensation is felt by the soul, and there must, therefore, be only one place where the sensations come together and which permits the nature of the sensation to be well defined. Descartes added to these reasons derived from an approximate anatomy (the surgeon La Peyronie thought that the corpus callosum could also have answered the criterion of uniqueness) some arguments linked to the position of the pineal gland in the brain; he judged the gland to be sufficiently central to receive even the smallest inputs arising from the peripheral organs and, in turn, to have some effects upon them. From this strategic position, the soul was like the spider at the centre of its web, controlling even the finest nerve threads whose extreme thinness ensured their extreme sensibility as well as their flexibility and suppleness which enabled them to avoid being damaged. This metaphor was used in the following century, notably by La Mettrie and then by Diderot, in making claims about the soul's substantiality. By situating the site of the soul in a particular place in the brain, Descartes opened the way to subsequent research on the localisation of cerebral functions, such as that carried out by Willis, by Gall and Spurzheim, by Broca, by Ferrier and by others. He was not concerned in any way with determining a particular centre in the brain for pain itself nor in singling it out as a specific sensation. However, as a result of his localisation, he made materialistic interpretations which he himself did not accept as possible. As he emphasised in his *Passions of the Soul* published in 1649, "the soul is linked with every part of the body all at once" and "its nature is such that it has no relationship with the extent, dimensions or other characteristics which make up the body."[7] The pineal gland was quite simply the privileged location harbouring the soul's functions, its "organ"; curiously Descartes' main argument for the soul being united with *all* other parts of the body was the single, indivisible nature of *the body,* whose parts were interdependent. The Cartesian system may have appeared all the more suspect to the religious authorities as the soul he thus localised in the pineal gland was indeed the spiritual and reasonable soul—the only one recognised by Descartes—and not a soul "of lower rank" such as the sensitive soul of the Aristotelian school.

With respect to the second problem, the nature of what was transmitted, Descartes went into details throughout his work about the production of animal spirits which were rarefied and propelled towards the brain by the warmth of the heart in which, prior to Willis, he imagined there being a sort of internal fermentation. The spirits went to the brain by the shortest, straightest route: these continuously moving bodies—of which only the most subtle and agitated parts could follow such a path after getting past a series of screens and very narrow openings—were the real means of nervous transmission. Descartes extended Harvey's model of circulation to the movement of spirits, and in particular the concept of valves which acted as so many little doors opening under certain circumstances to let blood or spirits through and also preventing reflux: the pores which Descartes thought of as being in the internal substance of the brain were like a number of doors which did not open for all the animal spirits, but only for those most appropriate for transmitting sensations and eliciting movements. How then was it possible to explain why just any object could not excite just any nerve? According to Descartes, six conditions played a role in activating nerves, all of which depended on geometry and mechanics: the place from which the action originated, the strength and other qualities of this action, the disposition of the threads which made up the brain's interior, the unequal actions of the spirits, the variations in the external limbs and finally the concomitance of several actions. In the example where a flame was placed close to a foot, the fire or the heat opened the pores which controlled the tubes which drove the spirits to the brain and it was they which, in response, determined the spirits' transmission "partly in those muscles which are used to make the eyes and head work in order to see, and partly in those that are used to make the hands move and bend the entire body to protect it."[8] Behind this explanatory model, we can see the principle of artificial fountains and in particular the machine at Marly, or that of organ pipes, as well as the construction of automatons: the machine becomes the model used to understand the living. In the Cartesian model, variations in sensations and increases in their intensities, *e.g.* when the heat of a flame becomes a burn, can equally be produced by a lengthening of courses taken by the animal spirits, or by the opening of new pathways; thus the sensation of burning reaching the brain provokes a transfer of spirits not only to all the external limbs in order to prompt a flight response, but also towards nerves "whose purpose is to cause internal emotions resembling those we experience following pain: for example, those which pinch the heart, or set the liver aflutter and such others."[9] This also explained why passions appeared to arise from the heart, whereas they came about through the movement of animal spirits from the pineal gland. However, it is in the *Passions of the Soul,* and not in the *Traité de l'homme* (where these examples of the reactions to burning were found) that Descartes used the expression of "reflected spirits" which, as G. Canguilhem pointed out,[10] have been improperly deemed as a basis for the concept of reflexes. The passage where Descartes describes the "reflected spirits" of the image formed in the pineal gland does not deal at all with reflex movements but, on the contrary, concerns a well-adapted response in line with the affective experience and the individual history of the subject who has already felt the pain and fears it.

Descartes tried to dispel the confusion which was often made between pain and sadness, arousal of the senses which for him was a sign of sensuality and joy. By means of a type of association or automatic succession, sadness always followed pain because the soul recognised the weakness of the body and its inability to resist the injuries that afflicted it. But the proof that a distinction could be made between them was the capacity to "suffer pain joyfully, and to accept unwelcome titillation".[11] it is in this way that it was possible to explain how a member of the audience at the theatre could enjoy being moved by suffering which he was not party to, or how the hope of something good coming out of something bad could transform its nature.

Thus pain, of all human experiences, was one of those where the question of the relationship between body and soul was asked most pointedly. If pain required a judgement by the soul, as was believed by Descartes, then it is legitimate to ask whether, in his system, there is any place for considering pain in animals.

## The Animal-Machine and Pain

Are animals capable of feeling pain? The answer to this question should be obvious, since everyone has heard a dog moan or cry out in pain when it is hurt or has been hit. However, this question which had been the subject of debate throughout the 17th century had repercussions the following century and also engaged both scholars and theologians of various persuasions.[12]

The Cartesian answer to the professed common sense evidence that animals suffer in the same way as man, is closely linked to both Descartes' analysis of involuntary movements and his study of reflexes. Thus, when a friend approaches us with his hand in the air as if to hit us, although we know that such could not be his intention, we involuntarily close our eyes and try to protect ourselves: this example given by Descartes in the *Passions of the Soul* leads him to a more general consideration of a group of movements, produced "by objects of feelings and by the spirits without the co-operation of the soul":[13] "Every involuntary movement that we make (as when we breathe, walk, eat and finally do all the other activities which are common to man and beast) depends solely on the formation of our limbs and of the course which the spirits, excited by the warmth of the heart, follow naturally in the brain, nerves and muscles, in the same way as the movements of a watch are produced by the single force of its spring and the configuration of its wheels."[14] If the distinction made between voluntary and involuntary movement is not totally new, what is novel is the interpretation of animal pain as a series of instinctive movements and reflex movements which have all the outward signs of being the same as those man associates with pain, though these are but an illusion of the senses. The Cartesian thesis of the animal-machine, which was first set out in the fifth part of the *Discourse on Method* and then in the second *Meditation* and the *Réponses aux Objections* as well as in numerous letters,[15] could be summarised the following way: the animal doesn't suffer since he doesn't think that he is suffering.

In order to understand what seems not only at odds with our basic intuitions but also as a challenge to common sense, we must put ourselves in the context of the debates of that era: schematically, the pain experienced by the animal is at odds with the Augustinian belief that "nobody suffers pointlessly"; if Christian theology states that all suffering has a meaning in the perspective of eternal life, one might ask why God has allowed innocent creatures to suffer, since we cannot credit them with a soul that thinks nor, consequently, with having free will. Thus, it is not possible to interpret such sufferings as either a punishment or a trial. In the following century, philosophers adopted this argument as a highly convincing defence against providentialism. Scholasticism tried to answer this problem by envisaging the soul of animals not as a substance but as being incorporeal and mortal. However, as Descartes pointed out in his *Sixièmes Réponses* to the objections raised by his *Méditations métaphysiques*, this solution carries the risk of simply introducing a difference in degree between the soul of man and that of animals, and it would no longer be possible to make a case for conferring immortality on the former if it were denied the latter: "In addition, this distinction, which is only one of greater or lesser degree, in no way changes the nature of things, though perhaps they do not make animals as reasonable as man, they might nevertheless occasion the belief that there are in them spirits of the same nature as others."[16] On the other hand, if animals calculate as we do (for example, when a dog remembers that he was beaten for having done wrong or tries to avenge the death of his master), and in particular are capable of showing manifestations of pain, joy, etc., though they are built like automatons, there seems therefore to be no reason why man should be accredited with a spiritual and reasonable soul and why thought itself should not be conceived of as a mechanical activity: these two problems meant that there was hardly any possible solution other than to refer back to either the Aristotelians' sensitive soul, or Gassendi's point of view. In the analogy between man and animals, the pain which the latter could experience either raised it to the level of man or else reduced man down to the level of the animal: in both cases, man lost his position in nature at the centre and head of creation. However, if we take account of the vitality of the mechanical model in the life sciences at that time, the problem of pain in animals also had consequences for the differences between an organism and a machine.[17]

As was pointed out somewhat ironically in the "Rorarius" article of Pierre Bayle's *Dictionnaire historique et critique,* "It is a pity that M. Descartes' opinion is so difficult to defend and so far removed from plausible reality; for it is also very useful to the true faith and this is the only reason that prevents some people from renouncing it";[18] in note C relating to this passage, Bayle, in retracing the origin of the debate on the soul of animals, recalled how efforts were made to involve religion in this discussion in order to destroy Cartesian theories and how, in actual fact, it was the Cartesian theory that had seemed to be the most orthodox as the possibility of animals feeling pain brings into question the justice of God and forces us to reconsider the way we treat them: "The animal's soul has never sinned, and yet it is subjected to pain and misery. . . . How do we treat animals? We induce them to tear each other to pieces merely for our pleasure, we slaughter them for our nourishment; we delve into their entrails while they are alive, so as to satisfy our curiosity. . . . Is it not cruel and unjust to submit an innocent soul to so much suffering? We may liberate ourselves from all these difficulties by accepting the dogma of M. Descartes."[19] Another note in the *Dictionnaire* examined the theories of Leibniz who was critical of the Cartesian machinery and gave animals a soul,[20] provoking a clash of views between him and Bayle.

In the 17th century, this debate was carried out simultaneously on both theological and metaphysical grounds. It also had practical consequences for the legitimacy of animal experiments, notably vivisection, which Descartes was not reluctant to practise. In the background of all the subsequent debates on pain, one can always find the lasting implications inherent in the controversy concerning the souls of animals, at least on the continent. In Great Britain, for very complex reasons based in part on a different philosophical tradition—as represented by Bacon's *Novum Organum* which endeavoured to advance science through inductive reasoning—as well as a less reductive mechanistic philosophy than that of Descartes' associating the notion of force to the definition of matter, and also due to a different relationship between science and religion, the problem of pain was approached in a totally different atmosphere, as is apparent in Willis' work.


## Thomas Willis: Pain and Reflex Movement

Thanks to Thomas Willis, knowledge of the anatomy of the brain took a great step forward and one could say that his *Cerebri Anatome,* published in 1664, was "the very foundation of the anatomy of the central nervous system and the neuro-vegetative system."[21] The precision of the anatomical descriptions, for example his classification of the cranial nerves or the precise way in which he distinguished the different parts of the brain (particularly the difference between grey and white matter, striated bodies, etc.), were always accompanied in his work by an interest in use and function. His anatomical work involved observing pathological cases and equally dissecting various animals which allowed him to make comparisons with human anatomy. Although, traditionally, animal spirits found in the cerebro-spinal fluid (simply described as fluid bathing the brain) were ascribed intellectual functions (reason, imagination, memory), Willis gave details of a much more complex system whose distinctive characteristic was the localisation and specialisation of functions according to the different regions of the cerebrum and cerebellum. "As the office or use of the cerebel in general, nothing of it occurs, spoken by the Ancients, worthy its fabrick, or agreeable to its structure. Some affirm this to be another Brain and to perform the same action with it; . . . Others place the memory in this part."[22] For Willis, the cerebellum, which was the source of the animal spirits of sensitive souls, had to be distinguished carefully from the brain: "The office of the cerebel seems to be for the animal spirits to supply some nerves, by which involuntary actions (such as are the beating of the heart, easie Respiration, the Concoction of the Aliment, the protrusion of the Chyle and many others) which are made after a constant manner unknown to us, or whether we will or no, are performed."[23] Willis provided any number of expressions to describe these involuntary actions which take place without our being conscious of them or noticing them at all *(nobis insciis aut invitis)*, and it was within his studies of reflex movements that he addressed the problem of pain. This sensation, which warns the organism that it is in danger, *in return* automatically elicits a movement aimed at saving the body from aggression, a movement of turning away, of protection, or of flight. Normally, when the brain is calm, the animal spirits "run" in a continuous flux which is regular and fans out from there to all the organs. This can be expressed by a metaphor of radiance *(radiatio)* together with an image of a series of waves which expand as they move away from their point of origin. In spite of the terminology being the same, Willis' animal spirits are quite different from those of Descartes, since they ensure both centripetal and centrifugal functions. What in Descartes' work were mere metaphors to invoke the animal spirits, namely the wind or subtle flame, are systematically used by Willis who, with the aid of a pyrotechnic model, illustrates the means of transmission of sensations: "The animal spirit is the light before it becomes a fire. Its transfer is determined by the ignition and its effect is of the nature of a deflagration. In this physiology, nerves are no longer ropes or tubes, but are like fuses."[24]

Each time that the involuntary part of the sensitive soul is influenced and agitated by unusual and noxious impulses, it is forced to modify the route of the animal spirits: they reflux or are reflected, provoking contractions or slackness; and at that time, the sensitive soul becomes subject to passions and is made to behave in a disordered "animal" fashion.[25] This response to an unusual sensation therefore consists of a reflex movement which is dependent upon the cerebellum: the "intercostal" and "vagus" nerves, which correspond to the sympathetic and parasympathetic systems, play a role in producing these reflex movements which were analyzed in depth by Willis. However, we can see that one of the original aspects of this system was the attempt to determine a relationship between conscious cerebral activity and the automatic regulation of involuntary movements by the cerebellum. The experience of pain is one example of the interaction between the two "systems": the annoyingly disagreeable sensation or external perturbation which reaches the corpus striatum in the brain is immediately communicated to the cerebellum whose troubled animal spirits, quite unbeknownst to us, bring about modifications of the pulse and respiration, as well as spasms of the viscera, convulsive movements, etc.[26] In his *Pathologiae cerebri et nervosi generis specimen* published in 1667, Willis studied convulsive or nervous illnesses. He placed great emphasis on the way in which muscular contractions are produced, involving a sort of explosion at the very moment when the two sorts of chemicals found respectively in the blood (nitro-sulphurous particles) and in spinal fluid (salty and spirituous particles) merge.

Willis, like Descartes, was looking for physical bases for sensations and movements when he gave a decisive role to the animal spirits made in the blood. However, he also introduced a chemical dimension which the French scholar objected to as unfounded fantasy, and gave much more detailed anatomical bases for the localisation of functions. Contrary to Descartes, Willis upheld the distinction between a spiritual and immortal soul and a bodily soul shared by man and animals which he described throughout his *De Anima Brutorum:* this bodily, material soul was itself made up of two parts: one, found in the blood, was in charge of vital or natural functions; the other, in the nervous system, dealt with sensations and movements. Here again, we come across, at least approximately, the Aristotelian tripartite approach to the functions of the soul. However, there is a totally new anatomical base and a more elaborate physiological explanation for reflex movements than is found in Descartes. In Great Britain, the renaissance in medicine did not involve such a dramatic breaking away from the teachings of Aristotle, even if it implied an abandonment of Aristotelianism as defined by scholastic tradition.

## Sydenham's Pragmatic Medical Approach: The Excellence of Laudanum

Nicknamed the English Hippocrates in recognition of the value of his medical observations, Sydenham's concern for analysing "epidemic con-

stitutions" (as in the treatise *Of Air, Water and Places*) and his attention
to the "course of nature" in illnesses have more than earned him the
right to be included in a history of pain. What brought him fame was
the preparation of laudanum which he named and used not only to
alleviate pain and induce slumber but also to treat dysentery epidemics,
hysteria, nervous illnesses, attacks of gout, etc. "The laudanum tincture
which I have mentioned as being given in daily draughts is quite simply
prepared in the following manner: sherry wine, one pint; opium, two
ounces; saffron, one ounce; a cinnamon stick and a clove, both pow-
dered. Mix and simmer over a vapour bath for two or three days until
the tincture has the proper consistency, strain and lay by for use." The
tincture form was easier to use and, above all, could be administered
in accurate doses, which meant that it could also be given to children,
as pointed out by Sydenham;[27] his intention in publishing this recipe
was to undermine the quacks who claimed that the success of their
remedies was due not to the inclusion of opium but rather to some
secret, ingenious means of preparation. England was more willing than
France to use this remedy in the treatment of pain although, as we have
already noted, its use was extolled a century earlier by Ambroise Paré. The
isolation of Hecquet, who was alone in Paris in pleading the case for
opium against all the others, can perhaps be explained by Galen's
reluctance to use it, the strength of Galenism in a highly conservative
medical establishment, especially in Paris (as illustrated by the criticism
of Harvey's discovery), the renewed condemnation of opium by G. E.
Stahl who brought it to the attention of the authorities, and finally to
the bad reputation of iatrochemists and Paracelsians in France, who
themselves were very much in favour of using opium.[28] Thus, in 17th-
century Europe, there was a first "Opium War" in which most doctors
were participants. Sydenham championed it unreservedly: "And here I
cannot but break out in praise of the great God, the giver of all good
things who hath granted to the human race, as a comfort in their
afflictions, no medicine of the value of opium; either in regard to the
number of diseases that it can control, or its efficiency in extirpating
them. . . . So necessary an instrument is opium in the hand of a skilful
man that medicine would be a cripple without it; and whoever under-
stands it well, will do more with it alone than he could well hope to
do from any single medicine."[29] In the following century, it was a
different story as the use of opium no longer came up against such
resistance on the part of French physicians. It is probable that its
therapeutic successes and the fact that the great Dutch doctor Boer-

haave had stated that he was in favour of this remedy, played a major part in this change of attitude. Sydenham cited experience in order to justify the use of a remedy without which, as has already been said, medicine would have been ineffective and inadequate: a clinical study of hysteria had taught him that recurring pain could be eased in patients only by taking laudanum over a period of years, progressively increasing the dose in order to overcome the effects of habituation; however, despite this extensive use, he found no evidence that it had noxious effects on either the brain or nerves. Suffering from gout himself, he gave a wonderful description of it which allows us to distinguish this affliction from rheumatism. He had thus personally experienced pain which became exacerbated to its very limit and knew that, in a similar situation, a patient could not bear even a cover to be brushed over his foot: "The pain which is at first moderate becomes more intense. With its intensity, the chills and shivers increase. After a time this comes to its height, accommodating itself to the bones and ligaments of the tarsus and metatarsus. Now it is a violent stretching and tearing of the ligaments. Now it is a gnawing pain, and now a pressure and tightening."[30] Having himself experienced the long and agonizing torment of gout and the ebb and flow of the pain, he knew that in this case only opium could provide some relief. A century later, for similar reasons, Brown's attitude would be the same.

If, as regards practical medicine, Sydenham's contribution is important, as regards the theory of sensations and movement he kept to an explanation which involved a disorganised afflux of animal spirits into a part of the body which became painful if it were innervated. The disorder felt by the patient came as much from the accumulation of the animal spirits in one part of the body as from their lack in another part. Thus, in this respect, he took a special interest in the aetiology of nervous disorders such as hysteria and hypochondria and, for this reason, became involved more closely with the relationship between body and soul. Any violent movements of the body and violent agitations of the soul could equally produce this perturbation in the distribution of the animal spirits in the body, and be the origin of these nervous disorders which are often characterised by erratic pain and an hysterical lump *(globus hystericus)*. Their origin was all the more difficult to uncover, as these maladjustments or this "ataxia" of the spirits, often occurred without the patient being aware of it. It was only the failure of established remedies that led the doctor, by deduction, to guess what the trouble was. These observations led Sydenham to conclude that there existed an "internal man." This idea would blossom in the next century and was taken up by Locke, Buffon, Van Swieten (in his paper on Boerhaave's *Aphorisms of Surgery*), Cabanis and Bichat, until the concept of cenaesthesia was developed at the beginning of the 19th century through the work of the German physiologist, J. C. Reil. In contrast with the external senses which inform us about our relationship to the outside world, the internal sense, which is accessible to us only through reasoning, regulates the movement of the animal spirits: "Just as the outer man is built up as a framework of parts belonging to the outward sense, there is also an inner man similarly constituted of parts, though consisting of the due and proper arrangements of the spirits, an arrangement cognisable only to the eye of reason; an arrangement, too, which is so united and intimately combined with the temper of the body, that it stands or falls according to the firmness of the constituent principles."[31] Sydenham superimposed another duality upon the classical one of the soul and the body. This was a prelude to the distinction which was to be made by Bichat between the two lives—the internal or organic life and the life of relation, albeit that he did not base it on any anatomical difference between the various sense organs.

Thus, in the 17th century, the majority of doctors or scholars who were interested in sensory physiology and the physiology of movement, were confronted by a class of phenomena which took place subconsciously, and in which pain played two roles: by bringing about reflex movements for retraction and flight, pain increased knowledge of neuromuscular actions; by passing a certain threshold above which a sensitive feeling was rightly perceived as painful, it allowed the internal life of the body to reach the consciousness when otherwise it would have remained hidden to both the patient and the doctor.

## Pain and Religion in 17th-Century France: A Few Questions

The problem of pain has some evident theological implications since it harks back to the meaning of Evil in the world: a number of insights into this subject were provided when considering the question of the souls of animals, where pain was a problem because original sin could not be attributed to animals, and because at the same time this posed the problem of the suffering of innocents, particularly of children. The definition of God, which implied goodness and omnipotence, clashed with this searching question: did God want his creatures to suffer?

An examination of the attitude of doctors in the 17th century prompts two preliminary remarks: no matter what the religious convictions of one person or another might be, pain, at least in the writings of the medical elite, produced a single reaction only shared by the physician and patient alike, namely the search for relief. In the debate on the use of opium, the arguments put forward were not of a theological nature on the necessity or value of pain; rather, they were based on different scientific concepts. If one compares France and England on this issue, the controversy depended on the respective importance of iatrochemistry in each of these countries, or to the manner in which freedom from the ideas of Galen came about. In any case, the physician's vocation is to relieve the pain even when he cannot provide a cure; the patient asks instinctively for such help. This of course does not signify that the religious concepts which then dominated society, modelled education, and ruled the principal acts of social life were absent from the modes of thought, gestures and attitudes of doctors and patients. However, everything took place as if a clearly established role division completely separated the domains of the doctor and the priest, and it would, in truth, be difficult to uncover the origins of this separation. Religion tried to give a meaning to suffering; its teachings and ceremony accompanied or organised illness and death; it tried to console and explain. It doesn't actually seem to have dictated the behaviour of the physician insofar as his goal was to rationalise his practice; however, it does not seem improbable that a scientific ideology, which in effect relegated the calming of suffering to second place, could well, perhaps unconsciously, have been influenced by religious ideology. We must, no doubt, qualify the scope of these remarks which concern the medical elite rather than the majority of practitioners; we must certainly emphasise that for the overwhelming majority of the population, coping with pain was not a choice but an inescapable plight forced on them by poverty, the physical remoteness of the rural physician, as well as the cultural traditions which designated the priest or the lord or lady of the land as mediators between the sick and the doctor or surgeon until he intervened; in the wretchedness of want and the surrender to pain, religion served as a palliative to the impotence or absence of medical help, without erasing the hope in the heart of every living being that they should not suffer. No doubt, too, we should stress the extent of the imbalance between men and women in humanity's struggle against pain. On the subject of "female nature," everything—including the most contradictory remarks—was said: at times, that woman, being more sensitive, impressionable or weaker than man, had a lower pain threshold than he and that, consequently, little notice should be taken of her cries and tears; at other times, it was said that, because she was more sensitive, she was also more flexible and could put up with pain better; or again, as she was more used to suffering, be it simply because of the experience of childbirth, she was ultimately more resistant.[32] Moreover, all of these theories were successively presented in the somewhat mitigated scientific guise as an alibi for the modes of representation and social organisation which had been established to benefit the powerful of this world. There was, however, one specific point on which the doctor's attitude was in complete agreement with the Church's directives, namely at the moment of birth, when the child's life was routinely given priority over that of the mother. A variety of attitudes about the pain and outcome of childbirth arose as a result of this position, to which the doctrine of the Church had largely contributed.

Secondly, the Church's point of view is certainly not uniform concerning the problem of pain: the rifts between Catholicism and Protestantism are, on the whole, not so much due to two different ways of viewing this passage on earth and the values of this life, within these two theological currents, but rather on the unavoidable compromises that must be made with the values of the world in which one lives. Jansenism in France exerted a strong influence in the world of literature from Madame de La Fayette to Racine, and contributed to a contemptuous view of the body while certain texts such as *Prière pour demander à Dieu le bon usage des maladies,* written by Pascal around 1659, are very representative of one possible attitude regarding pain, a somewhat mystical attitude legitimised and sanctified by the Christian doctrine, even though it was certainly not the easiest to preach or to have accepted. Whatever discretions the Church used, it was always clear that if it were accepted that Man's ultimate purpose on earth was to both serve and love God, then health was no more important in itself than any other earthly possession. It also followed that illness and pain could be viewed as benefits. This is Pascal's argument as set out in his *Prière:* "If my heart was full of affection for the world when it had some strength, deny this strength for my salvation, and make me incapable of enjoying the world, be it by feebleness of the body, be it by zest for charity, in order to be exalted by you alone."[33] In a form that is perhaps meant only for the chosen ones, this prayer reflects the very essence of Christian doctrine on pain and illness: an apprenticeship in detachment from the world through pain and illness, providential sign, pain-punishment, pain which brings one closer to God through empathy. "Touch my heart so I may repent my faults, since without this internal pain, the external ills which you inflict on my body will provide me with new chance to sin. Make me truly understand that the ills of the body are nothing more than punishment and the complete manifestation of the ills of the soul. But also make them the remedy, by making me consider, in the pain that I feel, that which I did not feel in my soul albeit it was quite ill and covered in ulcers. . . . All that I am is odious to you, and I find nothing in me that could please you. I see nothing there, Lord, apart from my pain that in some way resembles yours."[34] In what, for the non-believer, seems a sort of frenzied denial of the world and a self-renunciation, Pascal then goes on to ask of pain: "O that I should not feel these pains without consolation; but that I should feel pains and comfort together, to the point where I feel only your comforting without any pain."[35]

The fact that this prayer conforms to basic Christian dogma and that we can recognise within it the principal elements which have been handed down for centuries, does not effectively inform us to what extent these ideas had spread throughout society. When Bossuet, from the height of his pulpit, was citing the words of Ecclesiastes in his funeral orations, La Rochefoucauld, a great observer of the world, explained how self-esteem, or the love of oneself, was the mainspring of all our passions and all social life. Was it possible that it was this self-love which trapped one into cherishing pain and priding oneself on enduring it? Who could possibly hate himself so much as to wish to be in pain? Which doctor could betray his calling to such a point that he did nothing to alleviate it? The Classical Age debated passionately about the meaning of pain. Whatever weapons were available in a given society to fight pain, the underlying metaphysical significance of pain remained unresolved but, confronted with the personal experience of pain, ordinary doctors and patients appear to have been convinced of the urgency to further knowledge in order to put an end to what, for the individual, is always the unacceptable and unnameable.

# 5

## *Pain in the Age of Enlightenment*

In the course of the 18th century and most particularly in its latter half, a shift was becoming clear in the perception and definition of pain. This coincided with well recognised changes in long-established lines of thought—changes which were linked to the dechristianisation of society, the secularisation of thought and the separation between science and metaphysics, which was one of the sources of rationalisa tion. It is increasingly accepted that this "Enlightened thought" took on different forms depending on the cultural area and that it cannot possibly be reduced down to or simplified by the title "The Rational Century." It can be said without paradox that, from the physician's point of view, it was characterised by the definition and the measure-ment of sensibility, that is by the research into the properties of the living fibre. Sensibility was a physiological concept before it was a psychological or aesthetic one and its characteristics had been defined through the observation of "subtle anatomy" by the physicians of the previous century, particularly the English and Italian ones, and through animal experiments. The concept of sensibility constituted the frame-work within which the problem of pain could be examined.

Medical reflection on the problem of sensibility was preceded and doubtlessly made possible by the "Philosophical Revolution" which had been set in motion by Locke in England and relayed with modifications by Condillac to France. Locke's empiricism and Condillac's sensualism without any doubt focused attention on the physiological conditions for sensations. They both did this by a critical examination and analysis of the functioning of the human spirit, by refuting the existence of innate ideas which preceded all experiences and by making the sensation the starting point for all knowledge. Moreover, if in the absence of sensation, the individual can neither know nor act, and if he needs to be spurred on by desire, *i.e.* from lack and need, or even anxiety ("unquietude" or, in the etymological sense, lack of rest), one can understand why it became indispensable to understand how sensations were produced and transmitted. It also becomes clear why there was such an interest in the cognitive faculties and the psychology of those in whom a sense was missing, such as the person who had been blind from birth and who might eventually recover his sight, and why there were dreams of man being transformed as a result of a supplementary sense. Within the framework of natural philosophy, the problem of sensations and sensibility became a major stake.

One can easily demonstrate this departure from the field of theology and even metaphysics, provided one remembers that the strong controversial rejection of an attitude can perhaps also be interpreted as a sign of its existence. The theodicies of the previous century were severely criticised because evil, the massacre of innocents, the blameless victims of wars or earthquakes, the existence of monstrosity or illness, seemed like countless scandals, denials of Providence and Divine Justice. One remembers Voltaire's reactions at the time of the Lisbon earthquake, or the attack which Diderot makes Samderson, who had been blind from birth, utter when he puts the question to the Protestant vicar: "Observe me well, Mr. Holmes, I have no eyes at all. What have we done to God, you and I, for one of us to have this organ and for the other to be denied it?"[1] Certainly, the struggle against systematic optimism in the best of all possible worlds or the prudent withdrawal which anticipated there being a principle of compensation within the universe that might ultimately re-establish a balance between afflictions and blessings constitute attitudes which are more or less critical of the finalistic and providential themes of the beginning of the century. These were the basis of the whole Christian apologetics and were perfectly set out in Abbé Pluche's *Le Spectacle de la nature*.[2] But essentially, the main change occurred elsewhere, and it would be hardly of any use to return to these problems in a study of medical ideas about pain. This change lay precisely in the fact that for the physician or the physiologist, the problematical question of pain could be placed aside from the problem of sin, evil and punishment. This does not signify an indifference to the way in which society apprehended the problem of pain, quite the contrary: in an age when the physician-philosopher was preoccupied with morality and happiness, when he was interested in a special way in the connections between the physical and the moral, his thoughts on pain and pleasure came up against the problem of how society was organised. What was true for the physician was perhaps more so for the physiologist: even a religious or indeed devout figure such as Haller could approach the question of pain without introducing religious obsessions; it is true that this was easier for someone whose work involved experimenting on animals, rather than being a physician. With Haller and the beginnings of the experimental method, the definition of sensibility and the respective functions of the nerves and the muscles found themselves based on more scientific foundations.

We must qualify this overall picture of the secularisation of pain in the medical conscience by mentioning the influence of animistic positions which we will examine further. We should also add the importance of careful medical observation and examination, followed by a faithful and exact description of nature, which was a major characteristic of this period and for which Hippocrates had provided an unrivalled model. Even before clinics had become the rule in institutions and had been developed systematically within the hospital system thanks to the Revolution, physicians concentrated on investigating the symptoms and the "history of the disease." This medicine of observation was resolutely urged as opposed to the systems in use the century before and was reflected in the interest for nosologies which henceforth had to be based on symptomatology and not on a classification of supposed causes. The *Encyclopédie,* with the articles "Observer" and "Observation" by Ménuret de Chambaud, one of the vitalist physicians from the Montpellier school, provides a remarkable testimony to that medicine of the Enlightenment, whose objective was to be a practical medicine and a medicine based on observation.

## The Value and Forms of Pain: Towards a Semiology of Pain

### *The Value or Usefulness of Pain?*

Even before any classifications of its different forms, pain was seen as a warning or alarm signal which confers the feeling of our own existence and alerts us to the dangers which are challenging our bodies. Well-meaning and beneficial, it diverts us from a harmful lifestyle, warns us of the approach of illness and invites us to change before it is too late. One could also possibly transfer the Delphic maxim "Know thyself" onto the exclusive province of the body. Furthermore, the person who took the trouble of examining himself would perhaps find indications or warning signs of afflictions to come, in a hypersensitivity or in feelings which were on the boundary between pleasure and pain. The individual could then proceed to an auto-anamnestic history before coming to a medical diagnosis: "This bitter fruit of nature hides the seed of a great blessing; it is a beneficial effort, a cry of sensitivity through which our intelligence is warned of the danger menacing us; it is the thunder which rumbles before crashing."[3] The theme of the usefulness of pain, which is so frequent in the medical texts of the period, signifies firstly that pain is a sort of sixth sense, a vigilant internal sense, which may sometimes even indicate to the physician how he should proceed: "Sincere friend, it wounds us in order to serve us and each day medicine successfully imitates beneficial irritations; associated to the spasm, it diminishes the plethora, dissolves the swellings, chases away the heterogeneous humours, found on the head, it produces a haemorrhage or beneficial vomiting. . . . When, under the name of gout, it comes to plague old age, it protects against all other infirmity and promises a long life."[4] The pain from gangrene prompts amputation in order to conserve the rest of the body.

By having a preventative role which set aside self-respect for the sick part or body, pain was sometimes taken to be the voice of nature, its "crisis." As such, it was to be emulated by the physician. The frequent identification between pain and fever had its counterpart: this was because either one of them could seem like a reaction of nature and thus be considered as something which was beneficial and which must be permitted to express itself. Was it not sometimes said that operations which entailed great pain had a greater chance of success than did others? "In childbirth, the pains, . . . although very strong, are absolutely necessary; and far from calming them, one looks to provoke them when they are too weak. The pain which follows the majority of surgical operations indicates that a process or reaction on the part of nature is underway, and also becomes itself one of the means of healing; in general, those operations in which sedatives have been used with the aim of sparing the sick some pain have been less successful."[5] However, this point of view which was still held in 1805, and which perfectly summed up a whole set of very old medical ideas, was far from being a unanimous one in the Age of Enlightenment: quite the contrary, it seemed like a backwards and, on the whole, controversial opinion. An abyss separated the few supporters who upheld the "usefulness of pain," a phrase which actually encompasses many different concepts, from those who publicly endorsed its expression and leaving it to run its course. Cries had their use and an excess of courage could be detrimental.

Hoffmann, for example, was in favour of a very old philosophy whereby one had to express openly that the body was ill, whereas the prolonged effort of keeping things to oneself would always result in an aggravation of the problem. The notion that a sensation should be expressed by cries and moans followed the same model as that governing unwholesome humours and tainted, retained blood; it involved the same logic that led to excising a wound, to encouraging suppuration and to giving vent to pain. Furthermore, traditional psychological analysis involved the dialectic representation of expansion and contraction: joy was the result of a type of self-exteriorisation or self-release, a blossoming or opening out, a dilation beneficial to excretion, respiration, etc.; by contrast, pain and sadness stemmed from retreating or retracting movements which suppressed sweat and transpiration and led to a movement from the periphery of the body towards the centre. Cabanis would take up this idea at the end of the century and attach a great deal of importance to that through which the living come into contact with their environment, especially their skin. However, although the image of the beneficial crisis in which deliverance was all the better the worse the attack had been present in the medical imagination for a long time, the physician of the Enlightenment carefully distinguished those cases where pain accompanied a crisis from those where it existed without any positive significance for the prognosis of the patient. This framework of thought rested on the Hippocratic aphorisms which demarcated the value of pain as a sign. However, the position is not ambiguous: "One could in general say that since nothing which causes pain is beneficial, it must always be regarded as detrimental in its own right, be it alone, or linked to another illness, because it depletes strength, upsets the functions, it stops the digestion of 'morbific' humours, [and] depending upon its intensity, it always produces some of the detrimental effects mentioned above."[6]

The majority of physicians advised that one should always pay attention to reports of pain, above all at a time when a large part of the population only took to their beds and went to the physician when pain prevented them from working. In gout or rheumatism—medical complaints which do not fall exactly into our modern nosological categories—the first priority was to calm the pain and only then did one think about treating the illness. The ethics which were usual for physicians at the end of the 18th and beginning of the 19th century can be summed up well enough by the words of Voulonne, a physician who was also a revolutionary: "Let us not forget that amongst the motives which must suspend or slow down progress in the art, is that humanity wants a great deal of importance to be attached to pain, which is often inseparable from the destructive action of the 'morbific' principle. Do not let our like perish in order to spare him temporary suffering; but before resolving to make him suffer, let us at least wait to be forced to do so: in that way, he need never reproach us [for having] either a deadly compassion or a barbaric haste."[7] These principles testify to the fact that the day-to-day practice of medicine involved the infliction of pain (through cauterisation, moxibustion, surgical intervention, etc.), and show that a calculated logic or, to be more precise, an evaluation of the price of life by comparison with the extent of suffering, took place. It was an attitude that we will find again in connection with the debate on the use of anaesthesia.

### Types of Pain

One of the most apparent functions of pain is to facilitate the physician's diagnosis, but this requires great precision in the description of pain and definitions upon which everyone can agree. The classifications of the different types of pain have been characterised by a great stability from Ancient times onwards in spite of the abundance of vocabularies used and the many metaphors incorporated based on the real experiences of individuals. The physician of the 18th century, whose diagnostic technique was built on a semiotic and nosological approach to illness, had to appreciate the value of pain in relation to other signs which he was in the habit of monitoring: pulse, urine, respiration, tongue, face, etc. Without doubt, from the outset, the coding of descriptions of different types of pain fulfilled the desire on the part of the physician to localise the pain for the sake of clarity.

Overall, the classification of pain into four principal types still holds: tensive pain, *i.e.* accompanied by a feeling of distension of the fibres, for example that endured by criminals and others who have been tortured,[8] or by someone who has had a dislocation rectified; gravative pain which is typified by an impression of weight, as is characteristic of situations in which fluids accumulate abnormally within a cavity (in dropsy for example) or when a foreign body is present (for example, a stillborn foetus in the womb, calculus in the kidneys, etc.); pulsating pain, which has a rhythm corresponding to the pulsations of the arteries, and is found particularly in well innervated areas during inflammation; finally, pungitive pain which "is accompanied by a sharp feeling as produced by a hard, pointed body which has penetrated the suffering part."[9] The last two are liable to have different intensities which can produce, respectively, lancinating pains (when painful pulsations give the impression that the suffering parts are going to break), and terebrant pains which conjure up an image of a drill which pierces the flesh, cutting and wounding it (as in the case of boils and abscesses). Pain caused by pins and needles and pruritic pain are not defined in their own right but as varieties of pungitive pain; however, a scale of intensity exists, from an everyday itch to the unbearable itching of prickly, abrasive pain. One can find variations and refinements of these four categories which have constituted the descriptive framework of pain for a long time and have turned out to be useful in making a diagnosis. We can find them with a few minor modifications in the article "Douleur" ("Pain") in the *Dictionnaire des sciences médicales* of 1814. The author, Dr. Renauldin, in noting that "it is easy to prove that the four kinds do not include all known pains,"[10] adds to them a special category for the itching pain of scabies and exfoliative dermatitis, a category in which pleasure is mingled with pain and which attracted attention at that time, perhaps as a result of Alibert's work on skin disorders. Dr. Renauldin also wanted to create separate categories for: *(i)* the "burning pain" which occurs in cases of skin inflammation or desquamation, in gangrenous anthrax or the buboes of the plague, in contrast to *(ii)* the "cold pain" characterised by shivering at the onset of a fever; *(iii)* the "contusive or crushing" pains of rheumatism, nervous ailments, etc.; and *(iv)* the most atrocious pains such as those from ulcers and cancers. It was in relation to the latter that he noted, without any disapproval, a case of suicide. This typing was somewhat more complex than that which was in vogue, as it attempted to systematise the link between types of pains and illnesses, using the nosological method as developed by Pinel around the same time. However, it remained within the general framework of the original categories. One finds it again in an almost identical form in the *Sémiotique* by Landré-Beauvais (1813) who similarly proposed a separate category for pruritic pain.

Could these efforts to categorise and characterise such painful symptoms have invalidated the relevance of the description of a class of "painful illnesses," as was made by Boissier de Sauvages and other nosologists such as Sagar, Vogel, etc.? Cullen, a Scottish physician of the 18th century, and subsequently Pinel, vehemently denounced this class, arguing that since pain was a quasi-universal symptom of all illnesses, there was no point in making a specific class for it. The argument had less weight than it seems when one reads the theory of the seventh class of painful illnesses, as detailed by Sauvages: within this class he maintained a topographical order *a capite ad calcem, i.e.* from head to toe (there were pains in the head, chest, lower abdomen, limbs and extremities). The common characteristic of this class was that pain was always the principal symptom, without there being any immediate evident organic cause for it: thus, category I was made up of "pains or anxieties often widespread or cutaneous" and should not be included among the phlegmasias, whereas category III consisted of "pains in the chest or throat but not breathlessness or anything like asthma," etc. It seems very close to the idea of a pain-illness, which would be defined much later by Leriche, rather than to being a description of a simple symptom: "All disagreeable sensations are not illnesses, since there is no illness which causes such a sensation. It is necessary, in order to make it such, that a constant or remarkable pain be so overwhelming that it obscured all the other symptoms."[11]

### The Localisation of Pain and the Seat of the Disease: From the Organ to the Tissue

The minutiae of the descriptions are explained by the fact that "the mode of the pain indicates, up to a certain point, the judgment which should be made about the illness";[12] the idea of judgment simultaneously implies both the determination of the types of illness involved and the perception one has of its evolution and outcome. Inflammations of the chest provided a good example of a differential diagnosis based on the nature of different pains: "The pain is tearing and superficial and increases with movements of the arm and trunk in pleurodynia or inflammation of the muscles of the chest wall. Pain is lancinating in pleurisy; it is deeper and often gravative in peripneumonia; it is more general, more widely prevalent and duller in catarrh."[13] In appreciating the signs provided by different pains, physicians placed the greatest importance not only on how much territory it covered but also on its temporal dimension (acute or chronic, permanent and immovable as in most neuralgias, or vague and erratic). The prognosis and diagnosis were based at least as much on the rhythms of the pain and the extent of the painful zone as on its exact location. In the example of cephalalgia, which was called *cephalea* when the pain was permanent and continuous, *cephalalgia* when it was intermittent and migraine when it affected only one part, there were two known causes, according to Vandenesse in the *Encyclopédie:* the involvement of the vessels found in the membranes (meninges) of the brain (especially the pia mater and dura mater), and the shedding of an acrid lymph either on the membranes or directly onto the brain;[14] the treatment, "about which it is not possible to give general rules," had to be adapted to these differing causes. Sciatic pain, which stemmed from the hip, had a mixed origin that was half humoural and half nervous (the depositing of bitter tartarous matter back on the nerves);[15] the distressing straining in tenesmus could have very diverse causes: irritations, ulcers, excoriations, etc. A change in name related to the affected area did not influence the treatment which depended on the supposed aetiology. Repeated *ad infinitum*, these aetiologies hardly varied for different parts of the body and were most commonly attributed to the combination of irritations of the nervous system and humoural alterations (sourness, bitterness).

However, whether pain was the symptom or the effect of the illness, the essential question for the physician was to know whether a particular type of pain could allow one to determine that a given part of the body was affected and/or that a particular illness had befallen the patient. As we have seen, this problem was already posed by Galen in his treatise *De Locis Affectis.* The classification of pains as a function of their types and intensities had prevailed over a topographical classification system but it still came up against the problem of localising the pain: cardialgia (stomach pain), cephalalgia, colic or intestinal pain, gout, rheumatism, sciatica, tenesmus pains. The question of the affected area or the damaged organ gradually assumed an increasing importance with the development of pathological anatomy and then of clinical anatomy studies which compared the symptom observed in the living with the post-mortem lesion. However, pain often left no trace in the tissues or the organs insofar as it was a typical "functional sign" and, consequently, pathological anatomy could not always shed retrospective light on the original diagnosis.

At the beginning of the 19th century, research clearly focused on lesions of the tissue rather than of the affected organ, particularly under the impetus of Bichat in France. He concentrated on distinguishing the pain affecting one tissue rather than another and by so doing used it as a powerful sign for diagnosis. The pain which was located directly in the nerve was of quite a different type than that which arose from the skin or mucosa; where the nerve itself was concerned, it was not its sheath (which Bichat called the neurilemma) that was most sensitive, but the medullary substance contained within it. Moreover, Bichat believed it was necessary to penetrate deeply into the brain for pain to be felt, although he did not manage to explain why; he therefore supposed that the integrity of the brain was damaged and no longer allowed perception, in contrast to what occurred when the nerve alone was affected.

In his *Anatomie générale,* Bichat recounted a deeply moving anecdote, which gave him the idea of basing a differential diagnosis on pain: "What above all else fixed my attention on the diversity of the pains which are found within each system, was the question from a man with tremendous spirit and composure, whose thigh had been amputated by Desault. He asked me why the pain that he felt the moment his skin was cut was completely different from the awful feeling that he experienced when his flesh was sectioned—where the nerves which were scattered here and there were injured by the instrument—and why this last feeling differed again from that which took place when his marrow was sectioned."[16] For Bichat, the only possible explanation was that there was a sensitivity specific to each tissue, in both the natural state and in the morbid state. However, this anecdote, which turned up again in several works, notably in the *Dissertation sur la douleur* by Hippolyte Bilon,[17] contributed to a rather mad and disturbing quest for knowledge in which it was the modulations in the patient's cries that allowed one to determine which tissues were affected. This was the extreme limit to which the tendency to regard pain not only as an alarm signal but also as an instrument of study for localising a disorder ultimately led. Such an attitude resulted, for example, in patients being left under observation for some time without treatment in order to avoid the risk of masking the symptoms. The coding of symptoms in relation to the location of the disorder, established in Ancient times, was clearly shifted by Bichat and his followers' approach, since they were concerned with discerning not only the diseased organ but also the affected tissue. They knew that each organ was made up of several tissues, that one of them could be affected whilst its neighbour remained healthy, and that a given tissue could be diseased and produce its specific pain in several parts of the organism, indeed wheresoever it was present within the organism; such a distribution of a given tissue was what Bichat called a "system" and had its specific pain. The "tissue" problem in pathology gave rise to the co-existence of two different attitudes which were to crop up again throughout the centuries: either a special interest for the organ and the affected tissue, as was the case for Bilon who above all was preoccupied with pinpointing and eventually treating the localised lesions of the mucous, serous and fibrous tissues, etc., or a more global approach to the suffering subject, in which the knowledge of tissue pathology constituted only one stage: this was Bichat's point of view.

### Knowing How to Evaluate Pain: Questioning the Patient

Exclusively qualitative and measureless notations concerning pain appear both in the questioning of the patient by the physician and in a patient's explanations of how he feels. As opposed to other signs, pain requires the spoken word and the patients' explanations. The development of clinical medicine revived certain questions: up to what point could one have confidence in what a patient said about his pain and at what point was it necessary to try to find objective signs like those provided by the development of pathological anatomy? The semiotic value of the pain did not do away with the need for the anamnesis and even the most modern questionnaires retain a large number of those categories which had an unrivalled level of precision, even if they are now worded differently. At the dawn of the 19th century, physicians were looking for a pure sign which would remove the ambiguities inherent in symptoms. They wished to find a sign, the meaning of which would be as certain as that provided by the lesion found at dissection. However, they were to be confronted not only with the multiple signs fundamental to pain, but also by that special exchange between physician and patient in which, whether consciously or not, the latter adopts a distinctive attitude in relating the details of his painful symptoms, partly as a game and partly for negotiating purposes. For the physician, a semiologist such as F. J. Double or Landré-Beauvais, it was not only a matter of eluding the deception of simulated pain, but rather one of becoming conscious that the patient's account, far from being necessarily a true description of the pain, was all part of the established pattern of communication between physician and patient, between he who knows and he who does not. It was also about the social hierarchy which separated them and was burdened and obscured by the image which the patient had of his illness, the reply which he believed the physician expected and the fear that one reply rather than another would lead to a diagnosis of a serious illness or a gloomy prognosis.

Clinical medicine, at the very moment when it was becoming standard for all physicians, also discovered its major obstacles: although the patient's account was always sought since many ailments are initially nothing more than a perception of pain, his word was also continually viewed with suspicion. The practice of medicine took place under the strain of these two different attitudes which led the physician along equally precarious paths: on the one hand, he sought to make independent observations aside from the patient's description, to decipher body-language by watching out for flinching when a painful area was either touched or pressed, or for involuntary reflex reactions, such as withdrawing or retreating; on the other hand, he used his recapitulative account, which summarised all the indications provided by the patient, to control the truth of what was initially said, and sometimes even to correct it. The physician might in addition deliberately introduce an error as a means of testing the patient's reply—in short, he might state falsehoods to learn the truth. Chomel, in his analysis of the difficulties of diagnosis, admirably described the role of pain in diagnosis and the strategy which the physician had to use: "The first question to pose is whether the patient has any pain. If he replies yes, one should enquire as to its exact location. In order to avoid any errors induced by a verbal reply, the patient is invited to place his hand on the site of the discomfort and to circumscribe it or indicate its projection."[18]

The purpose of the physician's account was to be objective and to clarify the patient's subjective, confusing, uncertain and contradictory description. This artificial repetition was also to allow reinterpretation. However, this mistrust, which also arose because of the extreme variability of individuals' reactions, did not necessarily result in an underestimation of the intensity of the pain described by the patient: "It seems, at the very outset, that in order to appreciate the strength of the pain one has only the sensation of those concerned and the descriptions they give of it. But, from experience, one knows how often the complaints of the sick bear little relation to their true pains. After a more meticulous examination, one finds that the physician may also ascertain the pain by the ravages it has inflicted on the constitution; and despite the fact that physical pains like mental pains produce different effects depending on patients' individual dispositions, one may, by taking this consideration into account, estimate the intensities of the pains by the intensities of their effects up to a certain point."[19] Landré-Beauvais, like Double, was thoroughly aware that the evaluations which physicians had to make varied due to idiosyncrasies: "Pain is a very frequent phenomenon in illnesses: in order to deduce signs from it, one must take into account the age, the temperament, the degree of irritability and sensibility, which parts are affected, the underlying causes, the type and the duration of the disease."[20] There was hardly a more difficult or controversial problem to solve than that of knowing what value to place on a patient's complaints: "One must not always judge the degree of pain by the complaints of the sick, nor the danger of the illness from the degree of pain. People who are in the habit of enjoying their own voices too much and of complaining, are vociferous about slight pains; whereas others patiently put up with the most violent pains."[21]

However, the physician's experience and acumen gave him a fairly accurate idea of the patient's degree of suffering at a glance. The dialogue which involved the patient talking about his illness and the physician listening to him was, for both of them, a personal experience which, from the physician's point of view, could not become meaningless due to force of habit nor because of his knowledge or the responsibilities of his work.

## The Different Medical Schools and the Problem of Pain

The three principal medical philosophies which were favoured by physicians of the 18th century did not give the same explanations or employ the same treatments. The mechanists, who wanted to return to the notion of the human body functioning as a simple machine and who held the upper hand in medical circles up until almost the middle of the century, continued to explain pain in terms of the distension or separation of the fibres. However, it was the vitalists, whose ideas appeared to be dominant at the end of the century and beyond, who adopted the concept of sensibility playing a role which was simultaneously physiological and psychological. The belief in animism, which had been represented by G. E. Stahl the previous century, persisted though it was certainly a minority trend; it held an ambiguous position with respect to pain: on the one hand, this ideology considered matter to be essentially passive and in effect accepted purely mechanical explanations; on the other hand, however, in the sense that it considered the soul to be directly responsible for all organic functions, it made pain a particularly important sign in illness and the sign of internal strife.

### Mechanism and Pain

When it comes to the work of Hermann Boerhaave, it is hardly possible to speak of pain as being a specific problem: it was a sign in a large number of illnesses, notably in inflammation which had been characterised since the time of Celsus through the four "or" suffixes (*tumor, dolor, calor, color*—tumour, pain, heat, colour); it was a standard part of any clinical picture and did not require a special explanation since all aetiologies of illnesses were reduced to alterations due to excess or lack—be it due to the condition of the solid fibres which were supposed to make up the organism, or to humours which were to a greater or lesser extent bitter, thick and pungent—as was recalled in the *Encyclopédie* or, more specifically, in the article "Douleur" which was presented as an extract from Boerhaave and Van Swieten:

"When pain arises from material obstructing any vessel whatsoever, the parts are over-distended; one must learn how to put an end to this cause, by achieving the resolution or suppuration of the obstructing material. It is no less necessary to reduce the movement of the humours by rest and the above mentioned means, when bitter substances have affected the suffering parts and are the cause of pain, because the action of irritants on the nerves is proportional to the strength they bring to bear on the sensitive parts and to their reaction in retaliation against them; . . . one must also be sure exactly what the dominant type of acrimony is, so that it can be corrected by specifics, just as when it is acid, one opposes it with alkaline or earthy absorbents."[22]

The supposed aetiology of the pain itself dictated the remedies which were the most appropriate to use, and which were the opposites of the source of the ailment. Relaxation was used to counter tension, acid against base, and obstructions were cleared through all forms of flux (blood-letting, purges, different evacuents).

Within the context of a humoural pathology combined with "solidism," it was not possible to blame pain on material causes other than those of stasis, obstruction, engorgement or distension of the fibres. The nervous system itself, in the absence of microscopic studies or observations of tissues, which did not yet exist, was seen as a set of tubes filled with animal spirits—"subtle matter" which came from rarefaction of the blood. This idea contrasted with that of the nerves being like ropes stretched throughout the whole body. Thus, Boerhaave's concept could be perfectly satisfied by a humoural aetiology which, overall, did not treat the nervous fluid at all differently to any other; according to a Galenic and later to a Cartesian model, explanations for phenomena as diverse as the heightened sensitivity of an inflamed part, the speed of propagation of sensations and the absence of sensitivity in a limb below a ligature, lay in whether the animal spirits were rarefied or too dense and on whether their pathways were interrupted or they circulated freely. Even mental pathologies (melancholy, hypochondria) fell back on explanations of a circulatory type and such models appeared to be as dominant in France as in England, for example, during a large part of the 18th century.

However, material explanations did not suppress the psychological dimension: in his *Aphorisms of Surgery*, Boerhaave dedicated a section to pain and noted that curious phenomenon whereby we can recollect that we suffered, but in no way remember the particular quality of the pain which we suffered.[23] He was sceptical about the power of the soul to be totally isolated from pain even when an individual was deep in meditation. For this reason, he had recommended the use of a red-hot iron in cases of collective hysteria, convulsions, etc., such as those he had had to treat at the convent in Haarlem; he was convinced that this threat would be enough to suppress all such symptoms of disorder. This rationalist point of view was generally shared by the physicians of the Enlightenment. It was necessary to explain mental phenomena such as the hallucinations of madness or the pains of amputees, which were lumped together since in neither case could one blame an external cause that would have been liable to affect the nerves. In so doing, the "mechanist" physicians conceded that "one cause, whatever it might be, would produce the same change in the brain as would have happened had there been a nerve fibre . . . so disposed that its dissolution could have been a consequence."[24] Van Swieten, commenting on Boerhaave's *Aphorisms of surgery,* explained these phenomena by using Sydenham's hypothesis on the existence, within us, of an "internal man." Man had a dual nature—he was linked to the external world through his senses while the internal man, accessible only through reason, generally remained silent: the heart, the lung or the liver can be painfully affected without our realising it, if the affected parts are devoid of "feeling": they are then as if abolished in our consciousness, they are like our body's "blind spots."[25] This commentary by Van Swieten constituted one stage in establishing the concept of cenaesthesia and opened the way for what Bichat would call "organic sensibility," a little later.

There was a somewhat different explanatory model of pain proposed by Friedrich Hoffmann, the other great 18th-century physician who exerted considerable influence on his epoch early in the century. By stating unequivocally that his principles were based on those of mechanics and that he wished to follow the rigorous discipline of geometry, this "mechanistic" physician, as they were called, adopted the old theory of the Latin "methodists" and thereby held that spasms and atonicity of the fibres were the principal causes of illness. But these two possibilities could simultaneously be interpreted as either falling strictly within a mechanistic model linked to the fibres' elasticity or as a model in which spasms and atonicity were linked to the general condition of the nervous system. One underlying concept included in the second hypothesis conferred great importance on the irritation of the nervous system and, by so doing, provided a transitional framework in which to develop a more complex and differentiated approach to pain. While retaining its function as a signal, pain was all the more worthy of interest insofar as it was part of a system of pathology which tended to put everything down to spasm—*In tantum laeditur in quantum convellitur*—and could be studied for its own qualities and above all for its own rhythm, with its inversions, paroxysms and remissions. As with fever which was often its companion, pain obeyed its own laws with respect to its manifestation and disappearance. It also raised the problem of those disorders which were called "matterless," *i.e.* those which were not characterised by a lasting and irreversible alteration of the solids and fluids, and which therefore would not subsequently provide any organic lesion that might be discovered by means of pathological anatomy. One can see to what extent these perceptions initially remained bound up in a representation of the organism as a balance of qualities and humours (*e.g.* "temperament"), but gradually began to encompass the concept of irritation with its ensuing share of tension, tingling, convulsions and spasm, and then irritability and sensibility.

With Hoffmann, the model of the body as a hydraulic machine became complicated since illness and pains were no longer simply explained by disparities between vessel diameters and the speed or nature of the fluids coursing through them. It was viewed as being activated by a lack or excess within the universal movements of diastole and systole in the heart, arteries and all elastic fibres that were capable of inherent and not simply passive movements, and called into play principally by the blood's most subtle fluid component. A logic such as Hoffmann's was as circular as it was "circulatory" and designated the place held by illness in the very definition of life: "It is a circulatory movement of the blood and the other fluids produced in systole and diastole by the heart and the arteries, or to put it more accurately, by all the vessels and all the fibres, maintained by the interplay occasioned by blood and the spirits."[26] Life was not only the result of circulation but also a responsive movement of fibres to a stimulus provided by the blood in its most fluid form. Thus, we move away from a model involving the adjustment of parts to other parts, to a model based on action and reaction, or of mutual correspondence between parts, rather than a simple mechanical transmission system. Therefore pain appears in the exact situation where the circulation of the blood and the nervous system are found to be connected and interacting, and this provides the explanation for illnesses and their symptoms: "It has, as a matter of fact, been noted that the stagnation or stasis of the blood in some viscera or the vessels of some part immediately unsettles the nervous variety and causes convulsions in it; this state of affairs gives rise to pains, to fevers, to haemorrhages and to the suppression of excretions."[27]

Spasms became the most frequent type of illness, even though their origin could be humoural and the spastic prolegomena in a given part of the body could be accompanied by atonicity or relaxation elsewhere. Because of this, Hoffmann's form of medical practice came to be especially interested in the "sympathies"—nervous correspondences, by which distant organs communicated within one another, but which were capable of transmitting painful signals from one side of the body to the opposite side, from the top of the body to the bottom or vice versa. However, this theory of the sympathies only took on its full importance with the vitalists' theories on sensibility.

### Animism and the Psychological Interpretation of Pain

Animism, such as it had been formulated by G. E. Stahl at the end of the previous century and as it was adopted by Dr. Boissier de Sauvages from Montpellier, was based on the concept of a substance which was so passive in its own right, even when alive, that it was incapable of any function without the intervention of the soul. In this respect, as its vitalist critics of the 18th century perceived well, animism held a symmetrical, rather than a truly different, position to that of the mechanists; the animists quite naturally accepted the mechanical theory of pain, "brought about by distraction, erosion, tingling, or other similar lesions of the nervous parts."[28] However, they superimposed a psychological theory on it, which viewed the soul as intervening directly in the functions of the organs (digestion, secretion, etc.). Bodily pain was interpreted as a sign that the soul was suffering and as an attempt to unburden itself of it: "The psychological theory of pain acquaints us with the motives which account for the soul's role in these disorders, its ultimate aims and the means it uses to relieve them."[29] Thus, the link between body and soul was established in this way and provided a vast potential for explaining madness. According to Boissier de Sauvages, for instance, the pain endured by hysterical individuals, which was a somatisation of mental suffering, did not arise from a derangement of the brain's fibres or to a spasmodic irritation of the womb acting in sympathy with the intellectual functions; it arose rather out of a conflict between the free or voluntary pursuits of the soul and the impulses produced by natural appetites. Consequently this meant madness and illnesses had a dual origin which involved the patient's responsibility for his own pain and which, for Boissier de Sauvages, was also a repercussion of the Fall: "In this imperfect state [of man], there is all the less harmony between the free-will and desires of men the more brutish they are and lacking in sufficient reason, philosophy or religion to bring tranquillity to their souls. Under these circumstances, the intellect finds adverse that which appears agreeable to the senses and the instinct regards as adverse that which reason had judged to be good. This situation accounts for both mental and physical ailments."[30]

This concept of pain being a punishment for original sin was to undergo curious transformations in the medical deliberations of the vitalists and Ideologues. Their interpretation became totally secular since pain as well as illness were seen as nature's punishment for omissions in one's regimen, while mental illness was perceived as a sign of conflict between the demands of each individual character and the constraints of the social order; this interpretation called for a fundamental social reorganisation when its standards (chastity in particular) went against nature. This explains why, as a leitmotiv, the physicians of the Enlightenment maintained that in order to be a good moralist, one must first be a good physician, thus reversing the traditional relationship between medicine and morality.

Thus, for the animists and many others, pain itself was detrimental if only because of the attention paid to it by the soul which might keep it from properly accomplishing its other functions. Since the soul was continuously occupied with the conservation of the body, it was not surprising that it should be affected by pain, and that it should suffer and neglect less urgent needs; it was this attention it paid the ailing part of the body that caused the insomnia which was inseparable from pain, along with the other associated phenomena (paleness, fainting, anorexia, weakness, etc.). However, this psychological dimension did not introduce any confusion between physical pain and mental anguish. Though both were the province of the soul, the former arose specifically from the body, whereas the latter arose from an error in judgement on the part of the soul or from an hallucination. They were both real, although it was necessary to introduce a distinction between "sensitive" pains, *i.e.* those produced by external stimuli, and "imagined pains," *i.e.* those without an external cause but just as real to the ailing person. In this latter category could be included the pains that the amputee still feels in his hand even after amputation, for instance, and all those stemming from the brain's ability to produce sensations independently of the presence of external objects, such as auditory or visual hallucinations which, again, may arise partly from mental illness. Animism paid particular attention to these interactions between the body and the soul, in which pain had a privileged place.

### Pain's Physiological Conditions: The Discussion Concerning Sensibility

By emphasising that pain was a disagreeable perception of the soul, physicians clearly indicated that it involved awareness and that three conditions, which had actually been recognised for a long time, were required in order for it to manifest itself: the existence of sensitive innervated parts which were capable of receiving information, a reliable means of transmission without interruption from the periphery to the brain and, finally, the integrity of the *sensorium commune*, a universal organ of feeling found in the brain, although no one wished to re-open the delicate question of its localisation, in view of the famous problem involving the "site of the soul" which had led to Descartes' indictment and difficulties.

In the second half of the 18th century, physiologists' efforts were aimed at delimitating the body's sensitive parts from its irritable parts. Knowledge of the brain's structure still remained quite sparse, however, despite the work of Vicq d'Azyr. It was only at the turn of the century that the anatomy of the brain would really take off.

### Haller, and the Experimental Determination of Irritable and Sensitive Parts

The idea that the living fibre would respond to all forms of external stimuli had been formulated by the iatro-mechanics from the Italian school at the end of the 17th century, by Borelli and Baglivi in particular, as well as by Francis Glisson, an English physician who was a contemporary of Newton.[31] Glisson spoke of the irritability of the fibre in order to signify this internal capacity for action and reaction. Although the model used to explain the movements of the fibre were still largely those taken from mechanics and especially the image of the spring, the important point in common among these learned individuals was a belief that the fibre, instead of being a passive instrument, was endowed with its own inherent power, a sort of *vis insita,* which could be brought into action in different guises, by external stimuli. This force was conceived along the same lines as Newton's power of attraction, not as an occult property, but as an unknown cause which had to be postulated in order to explain the fibre's reactions. These reactions were the fundamental sign of life, but despite close microscopic observations, such as those carried out by Baglivi,[32] all Haller was able to state in the middle of the 18th century was that the fibre was to the physiologist what the line was to geometry: it was certainly an indispensable concept for the physiologist, although the most he could do was observe bundles of these fibres. The study of the properties of the fibre was to be conducted not by microscopic observation alone, but by experimental investigations of the reactions of different fibres in living animals and eventually in pathological cases which were brought to the investigators' attention. And indeed this was the main question of concern to the great physiologist, Albrecht von Haller, who sought to distinguish the different properties or reactions of fibres depending on their nature. While the concept of irritability has been an all-embracing one, describing the faculty to react in response to an irritation no matter what its nature, Haller, for his part, wanted to establish rigorous experimental distinctions between the irritability of the muscle fibre, *i.e.* that which we term contractility, and the excitability of the nerve fibre, which he called sensibility. Physicians had known for a long time that there were stimulants which produced violent reactions in one part of the body but were without effect on another part: this was the case of emetics, for instance, which acted vigorously on the stomach but produced no irritation at all on the cornea; however, was this specificity due to the stimulus itself, or due to the part, *i.e.* the fibres affected by the stimulus?

It was in order to resolve this problem that, in his experiments, Haller multiplied and diversified the types of reagent and means used to stimulate a given part, using a process of elimination: thus, he successively applied thermal stimulants, mechanical stimulants (tearing, cuts, etc.) and chemical stimulants (oil of vitriol, spirit of nitrate) to each part. Electricity, and particularly galvanism when it was discovered, also provided a means of measuring the irritability of the parts and their residual vitality after death. The entire body was thoroughly investigated from head to toe: membranes, cellular tissue, tendons and aponeuroses, bones and cartilages, muscles, glands, nerves, etc.

The repetition of the experiments was an indispensable condition to arrive at definitive conclusions: "What I call the irritable part of the human body is that which become shorter when some foreign body touches it with a certain force. If we suppose that the external touch is the same, the irritability of the fibre is greater if it shortens to a greater degree. . . . The fibre I call sensitive in man is that which, on being touched, transmits to the soul the impression of this contact; in animals, about whose soul(s) we have no certainty at all, the fibre which will be called sensitive is the one that being irritated brings about obvious signs of pain and indisposition. On the other hand, what I call insensitive is that which on being burnt, cut, stung or contused to the extent of being completely destroyed, gives no sign of pain, nor convulsion, nor any change at all in the situation of the body."[33] This experimentation produced a general chart of all the irritable and sensitive parts of the body.

For Haller, the skin provided the yardstick of all sensibility and it was in comparison with it that all other reactions could be evaluated: "In whatsoever way one irritates it, the animal cries, is agitated and gives all the signs of pain of which it is capable. This great sensibility of the skin prompted me to use it as a constant in measuring sensibility; and I established as the most sensitive those parts which can be irritated without altering the tranquillity of the animal, whereas it does give signs of pain if the adjacent skin is irritated."[34]

In Haller's work, the animal's pain became an instrument of physiological investigation which allowed him to establish that only the nerves and the innervated parts are sensitive, whilst only muscle fibres are irritable. On the one hand, the irritability of, or power to shorten muscle fibres, is certainly a property of living beings, distinct from elasticity as it is irreparably destroyed by dehydration, and depends upon the age and the vitality of the subject; irritability decreases with old age. On the other hand, this property is equally distinct from sensibility, since irritability is maintained for a certain time after an animal's death or the separation of a limb from the rest of the body, or in one part deprived of all feeling due to the nerves being ligated.

Haller's conclusions, built solidly on repeated experiments, opened the way for a rigorous definition of sensibility. However, these conclusions were violently contested by both the vitalists and the animists. For the animist Robert Whytt, the soul, which intervened in not only intellectual functions but also organic ones, was present throughout the body.[35] Decentralised, in that it was not situated exclusively in the brain, it was responsible for contractile movements as well as the transmission of sensations: the maintenance of irritability in a recently deceased body only proved that though one believed that death was complete, in reality it was only apparent and actually came about later than one would have thought, which led on to a closer examination of the signs of death. During the Revolution, this position would be in the background of the debate, elicited by the guillotine, on the maintenance of some type of consciousness and particularly of pain after decapitation, which opposed Cabanis, Marc-Antoine Petit and others, against Soemmering, Oelsner and Sue the younger.[36]

The famous episode of the head of Charlotte Corday flushing when slapped by the executioner provided new substance to an old debate, and to often confirmed facts such as the maintained movements of reptiles after decapitation, or of chickens and ducks being able to shuffle several metres in the same circumstances; this idea, "widespread these days, that the feeling of pain is prolonged after the apparent extinction of life,"[37] and for which "Soemmering in Germany, [and] Sue the younger in France sought to find proof by numerous experiments; . . . they believed that in a head separated from trunk this ability to feel pain remained right up until its natural warmth ebbed; thus the most awful, the most painful of tortures was that of decapitation. But if pain requires a form of judgement, how could it come about in a severed head? How could a head have an awareness of pain, no longer having the integrity of its organisation? . . . In actual fact, one can observe movements in the eyes, lips, and eyelids of a severed head, one can see the cheeks colour momentarily: but these are animal movements, phenomena of irritability, and not the product of a pain felt and judged; they no more prove that the severed head is expressing anger than an amputated hand demonstrates a desire to hit out, when it closes into a fist when the muscles in the arm are prodded."[38]

As part of this debate in which Soemmering's views were perhaps not devoid of political motivations, Marc-Antoine Petit refuted the thesis of pain continuing after death by pointing out the confusion which lay between the concepts of sensibility and irritability, in other words between consciousness and involuntary or reflex movements. Although Sue, in his *Recherches sur la vitalité*, tried by experiment to measure the degree and duration of the various types of vitality found in different parts of the body after death,[39] he did not manage to remove the ambiguity of the phenomena which he was analyzing though he approached the problem from a conceptual framework which viewed death as being "parcelled" (*i.e.* occurring bit by bit or organ by organ) and envisaged it as a process which differed, depending on which parts and tissues were involved. The outlines of this concept had been delineated by Buffon, and Bichat had recently contributed an experimental basis substantiating it. The common ground between Whytt and these much later studies was the conviction that there existed involuntary and unconscious movements and actions, albeit that these were "dependent on the soul." On the other hand, for Haller, this was a contradiction in terms: what was in question in this dispute between Whytt and Haller was whether one could conceive of the possibility of a vegetative life, as distinct from the life of relations (which was still called animal life). Such a vegetative life had a specific anatomical basis and its own nervous system (the sympathetic and parasympathetic systems). Haller stuck rigidly to his belief in a strict dichotomy between sensibility which was associated with consciousness, and irritability which was independent of consciousness.

Where the vitalists were concerned, the debate covered several different issues simultaneously: firstly, they contested the experimental results produced by Haller, arguing about the weakness of the criterion for pain which he used. In the *Encyclopédie*, the article entitled "Sensibilité" ("Sensibility"), which was written by Fouquet from Montpellier, mentioned that the experimental conditions under which vivisection was carried out modified the animal's reactions and behaviour.[40] In passing, he denounced the "philosophical martyrdom" to which Haller had subjected dogs, though the essence of the argument was not his criticism of vivisection on ethical grounds. It was necessary to wait until later, most likely the very beginning of the Revolution, for a link to be established between the treatment of animals and the treatment of men: it is well known that vivisection was highly controversial in England and in France at the time of Magendie and thereafter.[41]

In the second half of the 18th century, the opposition of the vitalists to the Hallerian ideas also stemmed from a preference for observation since this did not alter the natural conditions in which living beings found themselves, in contrast to experimentation which changed or destroyed the very object of study. Haller was not oblivious to this argument since, in his *Mémoires sur la nature sensible et irritable du corps animal,* he pointed out what precautions should be taken in order for the criterion for pain to remain clear-cut: "One must try to provide the animal with the least painful situation, by tightening its bonds only as much as is necessary, without letting them cut into its flesh, and one would do well to cover its head and eyes. The mere approach of a man who has inflicted a cruel wound on the animal can bring back its cries."[42]

In the same way, Fouquet had recalled that the pains provoked by incision of the skin would, according to the Hippocratic formula, risk "obscuring" the level of sensibility or insensibility of the parts being studied—dura mater, tendons, membranes, etc.; these were in fact the parts about which there was the most intense controversy, particularly in the case of the dura mater, the sensibility of which had been discussed in the theses of Baglivi. No doubt Haller was aware of the risks of confusion since he wrote: "One must allow the animal to have time to calm down completely from the pains of the incision. One can easily identify the timing of this tranquillity, by observing the rest, the silence and a relaxation in the suffering countenance of the animal. . . . It is then that one can touch and feel the dura mater and the tendon, and one will have had good reason to wait for the confirmation of this very tranquillity, as one would otherwise be sure of being mistaken if one dealt with an animal in pain."[43]

Because there was no reliable criterion for pain in their eyes, the characterisation of parts of the body as being irritable, sensitive, both at once or neither had to be re-examined. To the lively debates already underway concerning the question of the sensibility of the tendons were added discussions concerning the glands and cellular tissue: this time it was Dr. Bordeu from Montpellier who was the main adversary. He had defined the actions of the glands as being the result not of a compression, but of a reaction to a type of stimulation or irritation— each gland had a sort of individual sensibility or life of its own, in exactly the same way as any other part.[44] For Haller, on the other hand, the glands were very sparsely innervated and the pathological phenomena which were found therein (tumours, scirrhus growths) would be totally painless. Behind this debate concerning the glands, which was identical to that concerning tendons, membranes, etc., the crux of the matter was to determine whether there could be parts which were deprived of sensibility in a living entity. For Bordeu, as for all the vitalists, every-thing that lived was endowed with sensibility; each part, each fibre possessed a particular life of its own and, in that way, contributed to the overall sensibility of the organism.

In addition the vitalists promoted the concept of an indissoluble, inextricable relationship between muscles and nerves in which the extreme finesse of the latter made it difficult to determine whether they were absent or not; as a result, it seemed very difficult for them to separate parts into those which were irritable and those which were sensitive. However, to these debates which were taking place largely on experimental grounds, a misunderstanding also arose in the definition of sensibility and its role in the organism. For Haller, sensibility was, in effect, no more than a property of nerve fibres, whereas, for the vitalists, it was a general property of living beings, present throughout and evidence of life itself, capable of assuming very different forms and magnitudes. Within the organism, movement and feeling were the "principal phenomena of life, likely to be reducible to a single primitive phenomenon; therein, one sees even before life begins, or a little after it ends, a unique property—the origin of movement and feeling— which is tied up with the organic nature of the principles which make up the body . . ., and which Glisson was the first to discover; he called it irritability but, in reality, it is but one mode of sensibility."[45] In that way, it could assume physiological as well as intellectual functions, and it did not necessarily have to involve the consciousness. In short, the vitalists were ready to recognise that the muscle fibre had reactions which were different from those of the nerve fibre, but to both they attributed the general property which they called sensibility and postulated that it was to be found in the smallest living fibre.

## The Theory of Sensibility Among the Vitalists and the Ideologues
### *Cabanis: Towards a Psycho-Physiology of Pain*

Sensibility, being common to all living beings, had been defined throughout the entries of the *Encyclopédie* by the vitalists from Montpellier, notably Fouquet and Ménuret de Chambaud, as being a property which was exclusive to organic bodies. For example Fouquet, in a sort of hymn to sensibility, "the basis and the conserving agent of life, the animal principle *par excellence,*"[46] labelled sensibility as the faculty through which one could understand the organic functions (digestion, etc.), just as one could understand feelings. However, although there were "different quantities" of sensibility in even the most minute parts of the body, this faculty was mainly concentrated in three centres or "seats of sensibility" (an expression that Cabanis would take up): the head, the heart or praecordial region, and the stomach or epigastric region together were considered as the tripod or triumvirate of life. In this monist and materialistic perception of man, sensibility provided an explanation for intellectual ("animal"), "vital" and "natural" functions: "The workings of the soul are no less related to sensibility. Pleasure, grief, all passions seem to be represented in this remarkable centre consisting of the epigastric region, through its abundant nerve plexuses; and certainly, there is no difficult combination, no strong attention, no effort on the part of memory without the stomach and the whole epigastric region being first oppressed by a feeling of queasiness which indicates the action(s) of these organs."[47]

By making sensibility the sign of life itself and basing psycho-physical unity of man upon it, the vitalists certainly had to attribute a great deal of importance to its various affective aspects. For the Ideologue Cabanis, sensibility could not be defined outside the realms of pleasure and pain since, in the true sense, what affects us can never be indifferent to us:

"Psychologists and physiologists have, as if in concert, organised sense impressions on the basis of their general effects in the sensitive organ under two headings which effectively encompass them all: pleasure and pain. I will not concentrate on proving that both of them contribute equally to the conservation of the animal; that they result from the same cause and always manage to balance each other out. It is enough to remark that one cannot conceive of animal nature without the notion of pleasure or pain as their phenomena are essential for sensibility . . ."[48]

In concluding his history of sensations, he criticised the model whereby useful pleasure was set in contrast to detrimental pain, on the basis that it could be beneficial as a reaction that provided a boost of energy, rather than on moral grounds, in the usual sense of that term: "Sensations of pleasure are those which nature invites us to look for: it equally invites us to flee from those of pain. However, it must not be believed that the former are always useful and the latter always detrimental. The habit of pleasure, even when it does not go as far as directly lowering one's strength, makes one incapable of supporting the abrupt changes which the hazards of life can bring. For its part, pain does not only provide useful lessons. it also contributes at times to strengthening the whole body; it instills more stability, balance, and equilibrium to the nervous and muscular system. But in order for this to occur, it must always be followed by a proportionate reaction; nature needs to recover energetically under its influence."[49]

With the work of Cabanis, as with that of Hufeland in Germany, one really comes to understand how applying the concept of energy to a physiological idea could be interpreted in a romantic way. It is only insofar as pain awakens or stimulates weak or dormant vital forces that one may refer to the usefulness of pain: in a philosophical context where pain was indeed what nature—restored to its rightful place by the medical outlook of the 18th century—urged us to avoid, pain was worth only as much as it was able to contribute to the development of life. One sees the extent to which it would be wrong to interpret the medical debate on the usefulness of pain as one which was indulgent on the subject of pain, or which encouraged enduring it with resignation. The theme of "the usefulness of pain" as an effective means to be employed in the struggle for life—which led to the therapeutic techniques using shock and stimulation—was part of a judicious calculation concerning the benefits which the patient could hope to expect from such treatments.

However, it was not really in this specific area that Cabanis was innovative; he was only adopting and expanding on existing themes which were scattered through the medical literature of the vitalists. On the other hand, in relation to the problem of an "internal sense," he further developed the interpretation on the internal man which Sydenham had conceived in order to explain the production of sensations not provoked by the outside world. In addition to the existence of visceral sensations, which had already been envisaged in 1755 by another physician, Louis Lacaze, in his *Idée de l'homme physique et moral,* Cabanis also thought that sensations could be born spontaneously in the brain and provoke pains therein which were very real, albeit they were strictly the product of cerebral activity, imagination or memory. The suffering endured in a nightmare was placed in the same category as that produced by dislocations or wounds: "Thus we clearly observed three ways in which sensibility works . . .: the first is related to the sensory organs; the second to the internal parts, and particularly to the viscera of the cavities of the chest and lower abdomen (in which we also include the reproductive organs); the third to the cerebral organ itself, if we discount the impressions transmitted to it by its sensitive extremities, be they internal or external."[50]

As far as Cabanis was concerned, the pains of hypochondria—which had nothing to do with humoural pathology—could be explained in this way. One of his principal concerns was to show that pain was not a purely physiological reaction to a stimulus which was too strong, too intense or wounding, but that it required the intellectual participation and mental activity of the subject. Extreme sensibility, which could be called hyperaesthesia, could be understood only by considering that there was an excess of attention permitting the individual to perceive even the internal functions of the viscera;[51] conversely, a lack of attention provided a rational explanation for the lack of sensibility or at least for the indifference to pain which was observed among the convulsionaries of Saint-Médard who, between approximately 1729 and 1732, had come to public attention because, not satisfied with their convulsive fits, the followers of François de Pâris were beaten with logs and tortured, without appearing to feel anything.[52] The physicians of the Enlightenment had interpreted these convulsions as hysterical paroxysms. Such a mental disorder no doubt had a physiological cause since, according to Cabanis, sensibility—that general power to react—could not take up position in one part of the body without deserting other parts, rather like a given quantity of fluid: for him, excesses of activity in the brain and the organs were inextricably linked through a process of sympathy and could only express themselves at the expense of other parts. However, he was no less certain that an obvious proof of the relationship between the physical and the mental was that, to a greater or lesser extent, each individual could contain or refuse to be overwhelmed by such outpourings. At the end of the century, the physicians who were least hostile to Mesmerism[53] considered that the beneficial effects of the "magnetic tub" were, in the same way, produced not by some mysterious fluid, but indeed by the psychological action which the practitioner of Mesmerism exerted on his patients by turning their attentions away from their ailments; they thus concluded that Mesmer could be successful only in cases of mental or nervous illnesses.

Thus, the work of Cabanis was produced within the framework of monistic thinking in which sensibility was the cornerstone of life and pain provided the ideal experience from which to study the relationships between the physical and the mental. His questions about the psycho-physiological conditions necessary for pain to reach the consciousness led him to view the perception of pain as being a complex, chronologically staged process, during the course of which any given sensation at any given time could be "absorbed" by another sensation which, so to speak, competed with it to reach consciousness. If we are not conscious of the working of our digestive and circulatory organs most of the time, it is usually because other, stronger sensations have come along and taken their place; this is true unless a pathological disorder of the organs has imposed their presence to the extent that they become troublesome. Thus he proposed a competitive model between internal and external feelings, "because both of them are also reinforced by their own durations, which only results in directing the sensitive attention; in turn, they are indiscriminately absorbed one after another, the weakest by the strongest, and the dominant ones sometimes entirely suppressing those which are not reinforced to the same extent."[54] However, this schematic interpretation, whose value we recognise today, was not based on precise anatomical or physiological data. Cabanis' knowledge in these areas went hardly any further than Haller's experiments and the discussions these had provoked; in truth, philosophical problems themselves interested Cabanis more than did their physiological bases although he had endeavoured to establish the physiology of sensations at the Institut, a discipline which two centuries later was to have its own chair at the Collège de France, in the person of Piéron.

*Bichat: The Passage from Organic Sensibility to Animal Sensibility,*
*and the Threshold Concept*

With Bichat's work, quite the opposite situation existed as his examination of vital properties sought a reliable experimental basis and a similar methodological framework to Haller's, even if their conclusions were ultimately different. His definition of different vital properties was secondary to the distinction he had established between the "two lives": the animal life or life of relations, and the organic life which would come to be called the vegetative life. The former was characterised by its intermittent nature which encompassed conscious or voluntary activity; by contrast, organic life, which involved the organs situated most deeply inside the body, and thus the best protected, was constant and continued, for example, during sleep, and was maintained even when the "animal" life was suspended as Bichat had been able to confirm by observing pathological cases of apoplexy or following serious traumas.

It was through the experimental study of the processes involved in death, and by thinking of death as occurring by degrees, that he was able to demonstrate that a certain number of functions were not directly dependent on the brain. In his *Recherches physiologiques sur la vie et la mort,* he examined in particular "the influence that the death of the brain exerts on that of all the organs," and showed that circulation, absorption and even processes of secretion (contrary to Bordeu's opinion) carried on for a fair while. For him, this did not exclude there being an interaction between the two "lives," but suggested that it occurred in an indirect and more complex fashion. To the body of evidence obtained by the observation of the acephalic foetus developing perfectly in its mother's abdomen or of animals lacking a central nervous system (which he had borrowed from Cuvier), and by experimentation on the chronological order of functional interruption in dogs, Bichat, as an anatomist, added a remark which would come to be of great importance for later work on the sympathetic system: "The majority of the viscera which ensure these functions receive no or almost no cerebral nerves, but many fine fibres from the ganglions."[55]

This distinction between two different nervous systems, which had only just been outlined by Pourfour du Petit[56] in the first half of the

18th century, was going to appear systematically in Bichat's works and support his distinction of the two lives. In the *Anatomie générale,* dedicated to analytically breaking down the organism into 21 tissues which were invested with different properties, he separately studied the nervous system of animal life (the brain, the spinal cord and the nerves coming from it) and the nervous system of organic life (the sympathetic and parasympathetic). This in itself was an innovation: "All the anatomists up until now have looked at the nervous system in a uniform way; but if we take the trouble to consider the forms, distribution, texture, properties and uses of the diverse branches of which they are composed, it is easy to see that they must be attributed to two general systems, essentially distinct from each other and having two principal centres—one [being] the brain and its dependencies, the other [being] the ganglions."[57] By analyzing the diverse forms of sensibility in the two systems, he established that there were marked differences between pain of a "ganglionic" origin and that coming from the cranial or spinal nerves: "Citizen Hallé has rightfully observed that the pains which we feel in the parts where nerves from the ganglion are distributed have a particular character; they in no way resemble those we feel in the parts where cerebral nerves are distributed. Thus, the distressing ache we felt in the lumbar region when suffering from uterine disorders . . ., intestinal pains, or epigastralgia, etc., etc , in no way resemble pains in external parts; they are deep and 'grip the heart,' as they say. It is known that there are essentially nervous colics which are certainly independent of any local disorder of the serous, mucous and muscular systems of the intestines. These colics manifestly originate in the nerves of the semilunar ganglia, which are spread throughout the projections of the abdominal arteries. They are genuine neuralgias of the nervous system of organic life; nonetheless, these neuralgias have absolutely nothing in common with the *tic douloureux* (trigeminal neuralgia), sciatic and other neuralgias of the nervous system of animal life."[58] Bichat found it hard to provide explanations for the origin of such visceral pains, but he emphasised that they differed from other pains in their symptoms, their progression and their duration.

The basic distinction between the two lives and, as a natural consequence, between the two types of nervous systems allows us to understand easily how Bichat's definition of the vital properties agreed with that of the vitalists and how it diverged from Haller's. Above all, he criticised the systematic association made between consciousness and sensibility since, for him, sensibility was only the ability to feel: "The concept of sensibility almost always reminds us of nerves in our ordinary way of thinking and this, in turn, leads to thoughts about the brain; thus, one hardly differentiates between these three things; however, it is almost only in the animal life that they have to be joined together."[59] Moreover, to the extent that the vegetative nervous system was not dependent on a single centre and that the novelty of Bichat's concept lay in the fact that one could "envisage each ganglion as a specific centre, independent of the others by its action,"[60] the sensations arising from this system as a general rule did not get as far as that central zone where all impressions come together, the *sensorium commune:* "In the organic life, sensibility is the ability to receive an impression; in the animal life, it is the ability to receive an impression, and also to refer it to a *sensorium commune.* . . . Thus, there is an organic sensibility and an animal sensibility: from one stem all the phenomena of digestion, circulation, secretion, exhalation, absorption, nutrition, etc. . . . From the other stem the sensations and perceptions as well as pain and pleasure which modify them."[61] Furthermore, the perfection in existing animality depended on the "dose" of sensibility which every living being enjoyed. But there was no difference in the nature of the sensibilities, but only differences in degree: the passage from one to the other occurred depending upon the position of the tissues and the organs in the body. Thus, the presence of a morsel of food, perceptible to the consciousness due to its qualities of taste and consistency when it was in the mouth, depended on organic sensibility as it passed down the pharynx to the oesophagus and stomach. Pathological situations such as inflammation, for example, caused a transfer from one sort of sensibility to another. To reiterate one of Bichat's formulae, "one is probably only the maximum of the other."[62] All the metaphors used by Bichat—dose, quantity or the sum of sensibility—referred to the realm of physical strength in general whereas the matter in question actually concerned vital force; nonetheless, these suggested that, in the same way as occurs in the transformation from pleasure to pain, a quantiative increase in one form of sensibility could bring about a qualitative change. It was this variability in the vital forces to which, according to Bichat, one had to include habit and individual characteristics, which explained the disagreement that existed amongst authors concerning the sensibility of certain parts of the organism, the dura mater and periosteum in particular. The same reasoning was applicable to the other vital properties defined by Bichat: the perceptible contractility of organs which was almost the same thing as the Hallerian irritability of the muscle fibre, and the imperceptible contractility of organs which would be closely equivalent to tonicity. Once again, Bichat criticised the Hallerian position on the dual level of principles and method: "To consider, along with the majority of authors, irritability to be a property exclusively inherent to the muscles, to be one of their distinctive characteristics as opposed to any other organs, and to explain this property in one word which indicates this exclusive seat, I believe, is not to conceive it in the way nature distributed it in our parts. . . . In this regard, the muscles without doubt occupy the first rung on the ladder of animated substances; they have the maximum organic contractility: but every living organ reacts in the same, albeit less apparent, way as they do to the excitant which is placed there artificially, or to the fluid which courses there in the natural state, in order to convey the substance of secretions, nutrition, exhalation or absorption."[63] This is what explains the choice of the term "contractility" rather than irritability, or the more general term "motility" which Bichat used not only to describe the dilatory movements of the iris but also the contraction of the arteries as well as the less perceptible contraction of the capillaries. In making irritability the exclusive property of the muscle fibre, Haller was prompted to conclude that everything which contracts is muscular. One point of the debate, for example, was to determine the nature of the arteries and the veins: for Haller, they were not themselves capable of experiencing pain, but the nerves which accompanied them could elicit it[64] and, in the general depiction of parts, they were placed amongst those which were both insensible and nonirritable. Bichat took the completely opposite view and recognised therein not only the presence of nerves, but also of a contractibility related to the presence of muscle tissue. He carefully distinguished this property from an elasticity which was a property of the tissue.

These debates on the vital properties, and on the definition of sensibility in particular, mobilised the whole scholarly world for over half a century and had major consequences both for physiology in general and for the comprehension and treatment of pain in particular. They made it possible to establish the specificity of the field of knowledge dealing with living beings as opposed to the physico-chemical sciences. They also permitted the experimental method to become the *sine qua non* of such knowledge. But more directly, as far as the problem of pain was concerned, they led to the conviction among physiologists and physicians, quite aside from accepted vitalist beliefs, that to feel and to live are one and the same thing—according to Diderot's axiom which was subsequently taken up by Cabanis—and, consequently, that everything that lives and feels is capable of suffering. This resulted in more attention being paid to the problem of pain, and this interest inspired the publication of works dedicated specifically to this problem such as those written by Petit, Bilon and Mojon, along with various other related endeavours.[65] The consequences were particularly important in relation both to the question of the "sympathies" and to therapeutics.

## The Doctrine of the "Sympathies," or the Travels of Pain

The word "sympathy," which really means "to suffer or feel with," generally refers to the fact that an injured organ or part of the body provokes pain somewhere else which may even be quite remote. Furthermore, the large number of pathological conditions which come within the framework of the sympathies explains why, in the past, the site of pain was not automatically interpreted as being the site of the ailment. The concept of sympathy had previously been recognised in the Hippocratic works and included in his adage that in the body, "everything conspires, everything co-operates, everything consents"; it had also been studied in depth by Galen. Nevertheless, and almost certainly due to the use the Paracelsians and the iatrochemists had made of it, the concept had been criticised in the 17th century because it could be viewed as a secret communication between parts, an inexplicable affinity, a mysterious correspondence: thus, it could seem as if it was a failure of rational thought, and as if it were upholding the occult qualities against which natural sciences had fought everywhere. Despite this suspicion, the concept of sympathy was solidly upheld in medicine because it accounted for certain empirical findings such as the fact that when the stomach is affected, it is the head that suffers and that when one kidney is diseased, there is also a dysfunction of the other. A further reason for it being upheld in medicine was due to the fact that it worked in tandem with the theory of "metastasis" or transport of morbific matter from one place to another. The sympathy was to the nervous realm what metastasis was to the humoural realm. There had been, in fact, an attempt to circumscribe sympathies to nerve projections on the assumption that their multiple ramifications, plexi and perhaps even their anastomoses, conceived like those of the vessels, could explain these "pain journeys" within the organism. However, the vitalist philosophy of sensibility being spread throughout the body in different forms and to differing extents, rather than being the exclusive property of the nerves, had once again placed importance on the doctrine of the sympathies. The renewal of Hippocratism had equally contributed to the gaining of legitimacy for this doctrine. Bordeu, in both his *Recherches sur le tissu muqueux* and his *Recherches sur les maladies chroniques*,[66] was the first to provide a large number of observations which went some way towards proving that pain and illness could be communicated from one half of the body to its opposite counterpart, or from the top of the body to the bottom or vice versa. In the absence of any precise knowledge of how nerves functioned or of how sensory signals crossed in the spinal cord, Bordeu could only propose the presence of nerve ramifications which had not been discovered by the anatomists, and assign a role in the transmission of sensations to the "cellular tissue" or the "mucous tissue"; this was a sort of conjunctive tissue found in and enclosing every organ and constituting a sort of medium which, by simple contiguity, was quite capable of propagating a feeling from one part of the body to another very distant part even when there was no nerve providing a direct link from one to the other. To some extent, this cellular tissue played the role of ether in Newtonian physics and, as was the case with attraction in Newton's work, no one would define what these communications were, but no one would deny that these actions exerted from a distance existed.

However, it was indisputably Barthez, a vitalist physician from Montpellier, who, at the end of the 18th and beginning of the 19th century, systematised this doctrine of the sympathies. He did this through the two editions of his *Nouveaux Éléments de la science de l'homme,* since the book's main focus was "the forces of the vital principle in man, their communications or sympathies, their unity as a system, their distinct modifications within different temperaments and ages and their extinction at death."[67] In order to give a better explanation of his concept of the sympathies, he distinguished it from synergy, a concept which he had been one of the first to shape. Whereas synergy was an "interaction between the forces from the different organs which co-operate with a particular organ,"[68] "the specific sympathy between two organs is established when an ailment from one noticeably and frequently brings about a corresponding ailment in the other, without this succession being governed by the laws of mechanics, or even by the general and known means of functioning of the living body."[69]

Barthez acknowledged three categories of facts concerning the sympathies: either there was no noticeable relationship between the organs, or there was a connection between them, or else they resembled each other in their structures and functions. In the first category, where nerves or vessels connecting the organs could not be found, he would list the well-known sympathy between the reproductive and vocal organs, those between the head and the liver, and would recall that, of all the organs, the stomach was the one with the most widespread sympathies: "We can see no reason why the sympathies of the stomach, if they relied solely on nerves, would continuously be linked with any organs other than the head and the nerves."[70] In the third category were to be found all the occurrences of sympathy between the organs of one half of the body for their symmetrical counterparts. These included ophthalmia in one eye being conveyed to the other, or the communication between one arm and the other. Barthez pointed out the relevance of this in therapeutics. However, it was the second category that was the most interesting, not so much because of the role it accorded to the intermediary tissues such as cellular tissue, but because he began to clarify aspects of pain in which the nervous and circulatory systems were implicated. It was also of particular interest because Barthez, when confronted with pains and convulsions being projected away from the sites of the lesion, began to think that everything did not pass through the brain, the *sensorium commune*, but that the spinal cord must play a specific role.[71] With regard to the problem of the sympathies, Barthez paid particular attention to the direction in which pain was propagated, whether it was radiating, and whether this involved ascending pathways or not. The explanation which he put forward was that there was a distribution of "sensitive forces" which relied upon a vital principle; he said nothing of the nature of this principle although he did not confuse it with sensibility. Under normal circumstances, there was a continuous antagonism between these forces in different parts of the body which, having reached an equilibrium, constituted good health. By contrast, any disruption of this antagonistic balance, due to an excess or lack of these forces, caused illness and pain. Barthez can therefore be situated in relation to a conceptual model of action and reaction to internal or external stimuli, although these actions and reactions were not conveyed by springs or fibres but by vital forces.

The doctrine of sympathies seems to have been characteristic of medicine during the Enlightenment on two levels: it relied upon empirical knowledge, where observation played an important role, and provided a theoretical framework which allowed the uninitiated person who was suffering to understand why pains changed place; this was more than just an explanation, in that it allowed a category of everyday occurrences to be defined. However, it was also the "daughter of the Enlightenment" due to its role in interpreting sensibility, transforming the living body into a dynamic medium where many pathways of communication and tracks crossed and met. At the dawn of an age in which important discoveries about the nervous system would follow one after another during the whole of the 19th century, the study of sympathies seemed to be a necessary programme to be undertaken in the light of pathological observations and the analysis of therapeutic results in man, from which physicians hoped to discover a good deal about the organism's secrets. However, gauged by the standards of physiology, then of experimental medicine, the study of sympathies was soon relegated to the level of outdated knowledge, obsolete curiosities and indiscrete questions, even though pains continued to travel within the organism, without there always being an adequate understanding of its pathways.

## Pain Therapeutics in the Age of Enlightenment

There were two possible attitudes to illnesses in general, that of "expectant medicine" in which one waited for the healing power of Nature to do its work, and that of "active medicine" in which one rapidly and energetically intervened. By contrast, there was hardly any possible choice where pain was concerned: the physician's first duty was to ease the pain, *i.e.* to treat the symptom even before he identified the causes: we examined the examples of gout and rheumatism earlier.

When reading medical texts, one is aware of the frequent use, often in large doses, of opium and its many preparations, of which the most frequently used was the tincture form—either Sydenham's laudanum or the white poppy-head syrups used for coughs. Resorting to opium when in pain became the norm and, in contrast to what had happened during the previous century, there was no longer any true discussion about whether it should be used or not. Certainly, there were some criminal acts, such as that of "the *endormeurs* (literally: 'sleep inducers') of the Midi," who put travellers staying at an inn to sleep with opium in order to rob and murder them.[72] Such acts were still vivid in the memory, as were a few cases of poisoning or suicide which were reported in the records of forensic medicine. However, the position of physicians who used opium for a very wide range of ailments, consisted of saying that opium was no more dangerous than any other medicine. One should particularly take note of the large quantities which were imported, mainly from Turkey,[73] and the extent of trade at the beginning of the 19th century. According to the testimony of Jacques-Louis Moreau de la Sarthe, in certain circles at the time of the Revolution opium was not used only for entirely therapeutic ends but rather for coming to terms with such difficult times, particularly during the Terror. It was suspected that the remedies of the quacks and "mysterious go-betweens" were only as effective as they were because they contained opium. The true debate on opium was more concerned with its fundamental effects: was it a sedative, *i.e.* something to calm one down, or was it, on the contrary, a stimulant? In the absence of a chemical analysis of its active ingredients which would not be achieved until the very beginning of the following century, the argument which had been launched by the famous enunciation of the English physician, John Brown, *"Opium, hercle, non sedat!"* ("Opium, by Hercules, does not calm you down!"), remained open. It was difficult to understand the seemingly contradictory effects whereby there was a stimulation of circulatory activity and a relaxation of intellectual functions to the point of sleep, and all the more so in that the doses employed remained very approximate. Brown's infatuation for opium had developed because of its efficacy when he himself had suffered an acute attack of gout which, after being treated with opium, had not recurred for several years. However, within his medical system, along with a great number of illnesses which were generally attributed to the category of illnesses caused by a plethora, he considered gout to be numbered among "debilitating diatheses," *i.e.* illnesses which occur due to a predisposition or existing "weakness" (an indirect weakness as in the case of gout), necessitating, according to him, the use of "diffusible stimulants" such as opium, ether or camphor:

"Opium, though much used in the cure of certain symptoms of diseases, was never understood by those physicians who, in books and lectures, assumed to themselves the province of directing the profession of physic. Every property they assigned to it was the reverse of the truth. Instead of allowing it to be the strongest stimulant in nature, they made it a sedative; and though they found great difficulty in finding a single sedative more, to help to make out their catalogue of a class of such bodies in Nature, they were confident that it was one. . . . Another property they ascribed to it was that of bringing on sleep; whereas it was the most powerful body of all others in producing and keeping up by watching state. . . . They also assigned it the virtue of allaying pain, but there is a kind of pain that it increases, and beside that aggravates every other. The pains that opium is calculated to remove are all these that depend upon general asthenic affection, as those of the gout, of chronic rheumatism, that of the gangrenous, as well as the putrid sore throat, all spasmodic and convulsive pains, all pains from pure debility as in the legs, ankles and soles, or in any part of the skin, nineteen head-aches out of twenty, which are in that proportion asthenic; the pain of any deep-seated sore or gunshot wound after every degree of asthenic diatheses is removed from the habit. It is an equal remedy against asthenic inflammation, whether local or general, as preventing their tendency to mortification and sphacelus."[74]

This ample list of "weakness" illnesses for which Brown prescribed opium because of its virtues as a stimulant demonstrates how widely it was employed, although his underlying interpretation was, in effect, a dual inversion of the generally accepted medical ideas: the majority of illnesses which Brown viewed as debilitating, *i.e.* those which, in his system, were dependant on there being an inadequate relationship between incitability and the incitation produced by the stimulus,[75] were, on the contrary, generally interpreted as illnesses resulting from an overabundance of the vital forces; thus, the opium which he used as a stimulant was normally used as a sedative! Consequently, opposing medical concepts cohabited within the therapeutics field and this particular medical practice, which viewed illnesses as being due to a lack of stimulation, contributed to the increased use of opium.

However, before he even began to consider what remedies he should use, the physician of this period had to weigh up which type of "curative indication" he should follow. Or to put it another way, faced with a pain which he had to relieve, was he to use a remedy similar to the ailment, or ought he to treat it with its opposite? Was it desirable to provoke an evacuation (bleeding, purgative, etc.) or perhaps, since this was a period during which pathology often resorted to explanations involving nerves, to consider the benefits of stimulating or relaxing?

The treatment depended above all on the answers to these questions. The polypharmacy which dominated, despite the efforts made to rationalise and simplify it, provided a multitude of remedies with similar or identical properties all derived from the accumulated knowledge of centuries. The choice of a sedative curative indication dictated that there should be a simultaneous or successive utilisation of all the remedies from one group—in this case, the "anodyne" (harmless) remedies which were used in a very wide variety of forms: pills, enemas, topical applications of liniments, ointments or creams, syrups and various potions. The patient's day was organised around the taking of such medicines, whose effectiveness in alleviating pain were not so useless as some argued.

The question of knowing whether pain should be treated with "the same" or "the opposite" was part of this system of "curative indications" which was initially linked to a medicine of humours and qualities. Originally, the "hot" pain caused by inflammation was to be treated with the opposing "cold" quality of narcotics. The abscess which caused pain was to be dealt with by its incision—this constituted an additional pain inflicted by the surgeon but was necessary for healing to take place; the same could be said about amputation in the case of gangrene, or the extirpation of cysts or the re-setting of dislocations. As we have already seen, the problem surrounding "the usefulness of pain" is very complex and may only be understood at that time through these principles. The different medical doctrines were generally at odds but, in this instance, agreed on the existence of certain vital, sensitive, irritable or incitable properties—all terms indicative of a belief in a vital energy which could be stimulated, consumed or worn-out; they therefore quite naturally favoured the use of therapeutics which were intended to "revive the vital forces."

The vitalists, whose philosophy became dominant towards the end of the 18th century in medical circles throughout most of Europe, albeit with variations from country to country, believed that nature should be observed in order to discover which way its efforts were directed to effect a cure. They also considered that nature needed to be actively assisted, through therapeutic measures involving perturbation, shock or stimulation. "Medicinal electricity," which was used from the middle of the century on in cases of paralysis and nervous illnesses, certainly developed largely as a result of a belief in the logic of "commotion" in both a literal and a figurative sense. Thereafter came the works of Galvani who discovered "animal electricity" as a result of a fortuitous construction of a "metallic ark" which united the muscle and the nerve of a frog,[76] and Volta whose achievements with electric batteries had two important consequences: firstly, to put a type of electricity which was more accessible and had a less violent effect than did the discharge from a Leyden jar at the physicians' disposal; secondly, it brought home the idea of there being an identity between electrical fluid and nervous fluid, which permitted an explanation of the beneficial effects of electricity in treating certain pains. As Bilon emphasised, electricity and galvanism caused nervous pains, but it was for precisely this reason that they were used effectively in all the diseases of this system: "One can see a sort of affinity between the remedy and the ailment since both of them act on the same organs."[77]

The therapeutics of perturbation, as extolled by Bordeu, may entail transforming a pain or a chronic illness into an acute pain in such a way as to bring about a "crisis" which freed the patient definitively of it; thus it was a question of intensifying the sensibility and stirring up a diminishing vital energy which in itself was not strong enough to accomplish the fight against the illness successfully. Thus, in the arsenal of remedies against pain, we also find methods which were aimed directly at inflicting pain in order to cure it: this category would include friction, flagellation and urtication; cautery was also widely used, not with the purpose of producing asepsis in the wound, but of provoking a beneficial discharge and awakening sensibility. In the very frequent use of setons and vesicatories, two therapeutic models were superimposed; these were related to a humoural aetiology and a nervous aetiology respectively: one was based on producing a diversion or displacement, be it of the sick humour or of an excessive focus of sensibility which one wanted to remove from the precarious area in which it was concentrated and to direct it to an organ of lesser importance to life, either more resistant or more accessible for medical intervention; the other one fell within a counter-irritant strategy of using stimulation to counterbalance an irritated or extremely sensitive part of the body by artificially inducing another part to be even more so. The article "Vesicatory" in the *Encyclopédie* makes references to commentaries found in the works of Hippocrates and Galen, concerning the use of painful remedies to cure pain:

"He [Hippocrates] therefore believed that pain predisposed the part to invoke and attract the disease, with the consequence that a pain produced artificially which was sharper than the original one, would diminish or annihilate this one, or would at least provide a beneficial distraction, a dislodging of the illness. . . . Thus it was through a very simple analogy that Hippocrates was led to use painful remedies and external *échauffants* ['blistering' remedies] to awaken or to stimulate nature when it became dormant, or when it was no longer sufficient on its own . . .; pain decomposes in favour of artistry into an infinite variety of intermediary sensations which can serve as epispastic agents ranging from positive or absolute pain to a feeling very close to pleasure."[78]

The most common vesicant used was a plaster made from cantharides powder, euphorbia powder, Burgundian pitch, leaven, wax and sometimes including mustard seeds. However, other methods which also fulfilled the functions of a vesicatory were equally extolled: for example, rubefacients (often with a base of nettles), caustics and above all, the technique of moxibustion[79] which had been taken from the Oriental world of China and Japan and which consisted of burning a sort of stick made from fabric and vegetable fibres on the skin in order to create a diversionary point. Acupuncture was known of, above all, through the tales of travellers and was included in the same category of stimulants: "Acupuncture consists of making over the entire body many little wounds by means of pointed instruments with which one pierces the surface of the body, by pushing them quite deeply into the flesh. . . . The effect of these punctures is to form several inflammation centres, to arouse the nerves in the mucous or cellular tissue which are benumbed and use this irritation of the skin to bring about nervous oscillations within this organ, which sometimes encourages critical deposits therein."[80]

These different therapeutic methods, involving the use of stimulation to alleviate pain, without doubt, established a closeness to pain and conferred an ambivalence on the experience which has disappeared in contemporary Western society. In de Sade's universe, its violence and cruelty claimed to be based on the laws of Nature itself and its characters acquired their energy and pleasure from contact with pain, though it was usually by inflicting it rather than submitting to it themselves. Without doubt, this sadistic world was conceived at a time when the love of life and the exultation of sensibility often adopted an explosive dimension to an extent summarised by the concept that it is pain "which gives new strength to the principle of life."[81]

To understand the behaviour and the testimonies of this period, one must accept a cultural context backed up by scientific corroboration in which the co-existence of divergent attitudes was not considered to be contradictory: on the one hand, there was the fear of pain which was deeply ingrained within every living being who consequently sought for a means of delivering himself from it by finding remedies; this attitude, which was legitimate among the suffering individuals, was also shared by the physician for whom the patient's pain was something intolerable and quite unacceptable in the eyes of reason. On the other hand, however, the physician—and not just the surgeon—was sometimes compelled to inflict pain in order to cure. This paradox was without doubt exacerbated by the medical ideas of the 18th century concerning the necessity of exciting the sensibility and of awakening the vital energy and could well be considered as a co-constituent to the practice of medicine. In this consubstantiality of medicine and pain, the only possible answer lies on the ethical plane, in the ultimate purpose inspiring every medical act.

# 6

## The 19th Century
### The Great Discoveries

In the 19th century, there were an increased number of break-throughs in the understanding of pain mechanism as well as a flowering of clinical disciplines and therapeutic innovations. These were such that the century was truly one of great discoveries in which the terrae incognitae were revealed in a decisive way allowing men to better understand and sometimes better relieve pain. In 1885, the famous *Dictionnaire Dechambre*, commenting on the history of advances made in the field of pain, took pleasure in reproducing a sort of law of the three estates, in the fashion of Auguste Comte,[1] which, if it was not acceptable as such to the historian, was nevertheless of great use to him: a "metaphysical" phase which, according to the Dechambre, was char-acterised by the confusion between the "physical" and the "moral," would be followed by a "physiological phase," dominated by experi-mental research on the structure and the functions of the principal organs which have a role in sensibility, while the final phase would be characterised as follows: "Finally, after forty years at the patient's bed-side or in the autopsy room, doctors for their part, both clinically and anatomically, pursued the study of the symptoms of pain. Here again, the immense and recent advances in neuropathology and experimental medicine have given and daily continue to give an abundant harvest of facts about the sensibility to pain, and on its abolition, its anomalies, its inhibition and finally its pathological forms. This very contemporary period is thus the clinical phase of a long story."[2]

This retrospective of facts acquired in the 19th century cannot satisfy the historian: in effect, what is seen in the *Dechambre* as a confusion between physical pain and moral pain is quite the contrary to the historian who can emphasise the interest of the relationship between these two aspects of the pain experience and conclude that the definition which is given to it is oversimplified. "Every painful sensation," remarked the *Dechambre*, "assumes the existence and involvement of the following organs: 1) an organ which is modifiable by pain stimuli, and an initial impression as a consequence of the excitation; 2) a system of conducting pathways for the impression to nervous centres, that is, special or non-special nerves for centripetal transmission; 3) a nervous centre of perception. Through this awareness, *which may or may not be associated with psychic activity,* the individual subjected to a painful sensation judges the nature, the variety or the site of the excitation, knows the physical qualities of the stimulant, and finally, experiences the actual feeling or the pain."[3] This concept either evades, or at least relegates to second place, the patient's own particular history—his or her feelings prior to the pain and affectivity, *i.e.* everything which is usually involved in the outlook and morale. By contrast, the medical texts from the first quarter of the 19th century seemed to pay a great deal of attention to the links between "physical" and "mental," as well as to the analysis and decoding of symptoms, as was shown for example in another famous dictionary from that period, the *Dictionnaire des sciences medicales,* written by Adelon, Alard and Alibert.[4] Furthermore, the interest taken in the concrete forms of pain and their semiological value arose precisely because of the development of the clinic. This without doubt preceded the age of experimental medicine, even if important work in experimental physiology had begun to develop in the first half of the 19th century in many European countries, for example in France with Magendie and Flourens and then Claude Bernard. Thus, not only is the proposed chronology hard to accept, but it reveals, no doubt accidentally, one of the faults or at least one of the major problems in the approach to pain during the 19th century, *i.e.* the lack of synchrony and limited concertation between clinical work, therapeutic research and experimental physiology. The negative and without doubt paradoxical side to the progress in the experimental physiology of the nervous system lay in the illusory belief that the passage from physiology to pathology could be accomplished without difficulty, simply by applying the results from one to the other. The position of someone such as Magendie was completely representative of the aims of experimental physiology: "What I wish is to establish with you some fundamental propositions based on a healthy physiology. If thereafter we have time to digress into the domain of pathology, we will be much better prepared to do so than is often the case in illnesses, and you will recognise the experimenter's hand in the suffering man, the animal in whom you produce similar sufferings at will. A person is struck down by a cerebral haemorrhage. He or she loses a certain faculty. Which part of the brain has been affected? Without hesitation you will reply and boldly describe the site of the lesion if, through your laboratory experiments, you have recognised the place which must be damaged in order to develop similar phenomena."[5] However, this concept of ailments in which the cause could be reproduced in a laboratory and thanks to which great advances in medicine were made, turned out to be more productive for the understanding of movement disorders (paralysis, etc.) than for problems of sensibility and the understanding of pain. For one thing, this limitation depended on the nature of a methodology which was based on sections at different levels of the nervous system, with the compression or irreversible destruction of a strictly defined zone thus allowing a study of the problems resulting from it. Such a method could not easily lead to the study of interacting and regulatory processes in which the nervous system, notably its sympathetic division, plays such a large role. On the whole, this experimental method which made it possible to understand physiological and pathological processes, and was really initiated by Claude Bernard, developed noticeably later than did the "clinical" phase which was described above. On the other hand, in order for the clinical experience to be of real value to the research on pain, it had to go beyond a framework of merely collecting and describing what can be observed, and take account of the questions which the experimental physiologist was asking. Alone, or with the help of the histologist, the 19th century clinician was able to identify a growing number of diseases and to distinguish one from the other with certainty. The activity which the *Revue neurologique* regularly related was very characteristic of this state of mind and type of research: it was the period when, amongst other things, it was understood that the awful pains of tabes dorsalis belonged to the terminal stages of syphilis, syringomyelia was described and neurology acquired a series of precise "signs" (for example Babinski's reflex) which permitted better diagnoses to be made.

However, this activity did not really address the question of the physiology of pain, and it was by a quite exceptional means that it became involved in one of the most important debates of the century concerning the explanation for pain, which set against each other, in an exclusive fashion, those who championed the central nervous system and those who believed that the explanation lay in the peripheral nervous system. The reasons invoked in this debate had more to do with the arguments and observations of physiologists than with a dialogue between them and the clinicians. The position of the specialised scientific and medical press reflected this split up to a point: the Académie de médecine[6] was the fief of hospital doctors and surgeons concerned with problems quite aside from those of experimental physiology, and who had preoccupations which sometimes seem a little ridiculous to us; those who belonged to the other field made use of specialist publications such as the *Journal de physiologie expérimentale* or the *Journal d'anatomie et de physiologie,* but also published in the *Comptes rendus de l'Académie des sciences,* more progressive in that respect than its counterpart, and the *Comptes rendus de la Société de biologie.* Even though one sometimes comes across remarkable memoirs from certain periods in the *Archives générales de médecine* (indeed, it was in this publication that Duchenne de Boulogne published a large part of his work), the general orientation of the review, which was aimed at practitioners, was not in favour of a link between clinical and experimental physiology at the research level. One could well describe the situation by noting that there were numerous contributions and notes on disorders involving pain, on the progress of the treatments, as well as important studies on the function of one or another nerve centres, but there was no attempt undertaken whatsoever to bring all these facts together.

As far as pain was concerned, thinking remained within the global framework of a "specificity theory" throughout the whole 19th century, from Johannes Müller to von Frey. One of the objectives of the present work is to understand why this theory in its many different forms remained dominant, even though it was at odds with certain clinical observations and with the appearance of new disciplines such as endocrinology which all strongly indicated that the phenomena of pain were a great deal more complex. Certainly, the specificity theory has known many forms, depending on whether the specificity refers to pain receptors, transmission pathways or centres. However, behind this concept, which must be distinguished from that of a specialisation of functions, there could well be an accepted *a priori* principle of orderliness holding that every organ is made for one function and one function only. Curiously, in the case of such principles which function unconsciously, the inconsistencies in the finalism in biology are assimilated with the economic principle of the division of labour. The development of the cellular theory which, for example at the level of tactile sensations, allowed a better differentiation of the different endings within the skin, reinforced this tendency in research to look for specificity even when it was not automatically implicated in the particular area being studied. At the same time, the work on cerebral localisations by such eminent experts as Broca, Ferrier and Charcot also reinforced this concept half a century after the failure of J. F. Gall. In addition to the previously mentioned problem dividing medicine and experimental physiology was a difficulty related to the actual pace and compartmentalised nature of the lines of research: it is undoubtedly impossible for a group of research workers, at any given time, to work simultaneously on different subjects, for example, on the sympathetic system, the spinal cord and the brain, etc. Similarly, however, it is without doubt beneficial that at a given moment in time there should be a convergence of effort on the part of the scientific community on a particular subject. Nonetheless, when it comes to knowledge concerning the nervous system, due to the complexity of its different levels and transmission pathways as well as the hierarchical nature of its processes, research concentrated on one particular sphere can lead to a situation where one cannot see the wood for the trees. If it is true that research is not possible without a clear definition of the problem which is to be tackled experimentally, it is equally necessary for there to be an overall reflection encompassing the various different areas of specialised research: however, institutions such as the Académie de médecine, which had precisely such a role, did not seem to play it fully. Solidly anchored philosophical positions, violently opposing materialism and spiritualism in France, also tended to thwart a better understanding of the pain mechanisms. The types of confrontation which then existed were concerned as much with the use of experimental methods as with research into localisations and specificities which were always suspected of reducing the mental to a physico-chemical level. These conflicts could very well have prevented a better grasp of the pain phenomenon, much more deeply and seriously than the conventional religious viewpoint on the value of pain, by contributing to a hardening of the positions of each side.

## The Period of Powerless Rebellions

*The Experience of the Untenable: Larrey and the Surgery of War at the Beginning of the 19th Century*

The complexity of the questions involved in the comprehension of pain and what was required to alleviate it led to the setting aside of sometimes rash generalisations, be they on the immobility and the conservatism of the medical body which was jealous of its power, or on how much suffering sick people could endure. Like all established bodies, the medical corps certainly was conservative, and it is necessary to introduce distinctions between different periods in the 19th century, between various groups depending on their intellectual training, their branch, etc., and between the elite and others. If it is clear that the great hospital surgeons and doctors—the ones who published and consequently with whose thoughts we are more able to become acquainted—criticised those of their colleagues who continued to believe that pain was useful for the success of an operation, it is then likely that this belief was more than just a vestige and that it was deeply embedded in their attitudes. However, the situation seems very different when one examines the reactions of the most enlightened physicians and surgeons.

The example of Baron Larrey,[7] who throughout the Napoleonic campaigns was principal surgeon to the Imperial Guard and Inspecteur général of the Army health service, tells us a great deal about the states of mind of the elite and demonstrates that surgery, particularly in times of war, plays a pivotal role in the development of progress. Thus surgery appears to a certain extent to have been the most advanced, rational, efficacious field of medicine, not only in terms of the technical prowess involved, but also from the point of view of the way it interpreted lesions and pathological facts. Larrey was a tireless organiser of the army's mobile ambulances, believing strongly that the faster one operated and the less one moved the wounded person, the better chance one had of saving him. Because he worried about equality, he always gave priority to the most gravely wounded when he was operating and gave no preference to rank. Throughout his *Mémoires de chirurgie militaire et campagnes,* one learns of the awful conditions under which it was necessary to amputate, and how speed and a skilled hand were often the only means of dealing with pain. In spite of the fact that Larrey very often amputated, he was not indifferent to the feelings of the wounded. Thus, during the Egyptian campaign, when he described the siege of St. Jean d'Acre during which General Caffarelli had had his elbow joint shattered by a bullet at point-blank range, he noted: "Such a disorder necessitated the amputation of the arm; the general himself demanded this; he also put up with it with extreme courage, and perhaps with too much concentration, because he did not utter a single word. Being very attached to this brave general, I operated on him with all speed possible in order to shorten his pains."[8] In his memoir on amputations, he compared his own results to those obtained by the surgeon Faure, notably in 1745 at the battle of Fontenoy. The latter being convinced that "there is more honour in saving a limb than in amputating it with dexterity and success"[9] preferred to delay and wait until there was no longer any hope of saving the limb. Although being in agreement with the principle of the conservation of the wounded limb whenever possible, Larrey explained the reasons which led him to act as quickly as possible, namely: the better chances of success, the benefits provided by the state of shock brought about by the wound in the first hours and which put the injured zone in a state of "stupor" before inflammatory reactions could set in, and the hope that the amputation would bring about a new state of shock which would lead to "a favourable change throughout the system."[10] In addition he was concerned that he should save the wounded from useless suffering and from prolonged meditation on their future mutilation which could only add to their terror: "After twenty to thirty days, if the prognosis is as troublesome, one is led to amputate: thus all the sufferings which the injured person endured were in vain; and besides the surgeon's art requires that nature, worn out by the previous recovery efforts, should expend further reserves. How then can the operation not be hazardous!"[11] The virtually daily practice of amputations, the habitual treatment of gaping wounds made by the Biscayan musket and the very dangerous conditions in which Larrey was obliged to operate, did not diminish his acute awareness of the sufferings of the wounded, who were sometimes operated on lying on the floor in a mixture of straw and snow, with the minimum of heat from a bivouac fire and with the expectation of the enemy's imminent arrival. For example, after the battle of Eylau and later in his description of the retreat from Russia, the pain of the wounded was unbearable: "Never has a day been as awful as this one; never has my soul been so moved; it was impossible for me to hold back my tears during those very moments when I was trying to bolster the courage of my wounded. It saddened me to see the death of some of these poor souls whose wounds necessitated the amputation of the thigh from its articulation in the pelvis, because the unfortunate circumstances in which we found ourselves, the excessive cold and the lack of a [proper] place had prevented me from carrying out operations which in themselves were very difficult and danger-ous."[12] From his experience as a war surgeon and his memory of the pains of gangrene caused by frostbite endured by the soldiers during their retreat from Russia, Larrey came to the conclusion that you had to do everything you could to spare patients and the wounded from pain: he was one of those rare few who were on the side of Émile de Gérardin in 1828 at the Académie de médecine[13] when the latter described the first attempts at anaesthesia which had been carried out in Great Britain. These reports were only met with scepticism and con-tempt by his other colleagues. One of the most difficult enigmas to resolve when dealing with the first half of the 19th century arises from the coexistence of the idea amongst doctors and surgeons that pain was unbearable and useless, and the no less assured conviction that there were no means at their disposal for lessening the pain of operations, whereas the historian has the distinct impression that the means were at hand and that occasions were missed. But is it necessary to believe in missed opportunities or could there not be other explanations?

### *The Moxa: "A Curative Means Whose Very Name Spelt Terror for the Patient"*

There are converging lines of evidence to suggest that doctors and surgeons who gradually relinquished the idea that pain might be useful for healing, tried to find some means or other for restricting, shortening or lessening the suffering. Such was the case with Jean-Baptiste Sar-landière, a hospital doctor who in 1825 had to pay to have his *Mémoires sur l'électro-puncture* published before attracting the attention of Percy and Laurent, Jules Cloquet and Magendie. In his work, he devoted a section on the use in France of Japanese moxa which was used in con-ditions which made it "the terror of the sick." This procedure, which came from the Orient—particularly China and Japan—and which had been known of for a long time in Europe through the second-hand accounts of travellers, Jesuit priests and learned Dutchmen such as Klaproth, consisted of placing on the skin of the patient, as close to the painful spot as possible, a cone or stick made from vegetable matter, which one then burnt until a sore developed. The theory behind this procedure, which continued to be used throughout the 19th century, varied, but was generally based on the Hippocratic adage according to which, when two pains coexist, the greater "obscures" the other, *i.e.* makes it disappear; however, medical theories on irritation, stimulation and above all the work on counter-stimulation carried out by Rasori prescribed a therapeutic regimen aimed at awakening the vital energy of the patient, and which had replaced the old humoural explanation of a diversion that allowed the discharge of tainted humours. However, the testimonies of those who had brought the moxa to Europe did not specify the nature of the vegetable matter, nor how to use it, and in particular failed to indicate that it was artemisia (and not cotton) which was required since this assured a slow ignition and reduced the pain to just a brief instant. All that was known was that many had to be used and in a repeated fashion in order to obtain results, and it was one of the most painful curative processes because, as a result of the use of the cotton cylinders varying in length from 3 to 7 centimetres, it left frightful sores which were painful for a long time.

He invited European doctors to show humility when faced with the knowledge of the Chinese and Japanese and to follow their example word for word: "Our cotton moxa acts by instilling, in a short time, a very uncomfortable feeling of heat in the cutaneous segment to which it is applied; this heat very quickly acquires a very high intensity, and this intensity lasts at least twelve to fifteen minutes";[14] by contrast, the Japanese moxa burnt very slowly, producing a more progressive irradiation of heat, and acted on the nervous extremities: "This sensation is in no way a pain, but, observed attentively and on a very sensitive part, it seems like a type of vibration, or even a feeling which could be considered between uneasiness and pleasure."[15] The advantages brought about by the new way of using the moxa retrospectively showed up the ordeals which patients had had to endure: held by several operators, whilst the doctor blew on the fire, inflicted the pain and showed himself to be ruthless in the face of the growing lack of docility on the part of the sick person; the doctor would then come out of the session feeling degraded and tried by the repugnance which he had inspired in the patient. This experience was unbearable for both sides since it likened the doctor to an executioner, and it must have played a role in the search for a less painful therapeutic regimen. It seems that the moxa in its old form or that defined by Sarlandière was successfully prescribed for a vast array of chronic pains and illnesses: neuralgia, coxalgia, rachialgia, as well as for all "hardenings" and (non-inflammatory) tumefactions of the muscles, joints, glands, vessels, etc. The way in which Sarlandière insisted that his method of electropuncture was not painful or, in any case, less so than ordinary galvanism, no doubt reflects a slow evolution in thinking which led to the evaluation of a therapeutic regimen in terms not only of its efficacy but also of the sufferings it could cause the patient. Henceforth, if there seems to have been little reluctance to use opium in cases of acute chronic pain, it was a very different story for surgical pain, as if two separate logics held sway depending on the situation.

### *"Avoiding Pain by Artificial Means is a Myth"* (*Velpeau, 1840*)

This apparently incomprehensible disjunctive reasoning had often been put forward as proof of the medical world's indifference with regard to the pain suffered by the sick and its tendency to under-estimate or despise the patients' complaints: at best, this was viewed as a wish to see pain accepted fatalistically as something to which one had to resign oneself; at worst, the suspicion existed that doctors saw pain as the necessary companion to healing and the indispensable auxiliary in the theatre where operations were often carried out in front of a public to be captivated by the speed and the elegance of the surgeon's skills. This situation was denounced by Velpeau in his *Leçons orales de clinique chirurgicale* which he gave at the Paris Hôpital de la Charité. His observations in 1840 are often quoted: "In all, avoiding pain by artificial means is a myth which can no longer be pursued today. Sharp instruments and pain are words which are inseparable in the mind of the patient, and whose association one must necessarily admit when it comes to operations. The surgeons must make every effort to limit the pain of operations while, at the same time, ensuring the success of the final results."[16]

Nonetheless, it was the same Velpeau who seven years later would be the most ardent defender of anaesthesia and who would be vigorously opposed to Magendie. Is it necessary therefore to talk about his blind spot in 1840 and his about-turn in 1847? It is true that the eagerness with which Velpeau started to administer ether was equalled only by the pessimism and the categorical tone he had earlier used to condemn all hopes of suppressing pain. However, he was not the only one who, around 1840, considered that an operation without pain was not possible, and similar observations can be found in the works of Gabriel Andral and others.[17] However, this declaration by Velpeau should be viewed in the light of a specific frame of reference: the immediate context of the phrase somewhat mitigates this opinion since Velpeau expressed his scepticism regarding an "extirpation of the breast" carried out under "animal magnetism," and also with regard to the use of a stupefying gas by the Edinburgh surgeon Hirckman,[18] which seemed to him like a return to the narcotic sponges of the Ancients. "One is very occupied with how to carry out operations painlessly," he said, but the procedures which he took seriously were compressions of parts above the point of incision, which seemed preferable to the ligating of the nerve trunks, a method extolled by the English surgeons James Moore and Benjamin Bell.[19] Secondly, all of his *Leçons* are driven by a clear concern to avoid causing the sick person pain, for example by choosing one procedure because it is less painful than another, as in the cure for hydrocele where he denies that pain has any therapeutic role during the injection and thereby denounces the error of some authors.

His *Leçons* are hallmarked by an approach which serves as a truly ethical behaviour code for a surgeon dealing with pain: respect for the adult subject who is in a healthy state of mind and who can sometimes refuse to be operated on against the advice of his family; taking each operation in its own right rather than like a mass-execution at an appointed day and hour virtually comparable to an auto-da-fé; giving explanations of the objectives and inconveniences of the intervention while avoiding the slow preparations of the past which had been like long meditations on pain. It was accepted that it might be necessary to adapt the rules of conduct slightly to make allowance for there being both faint-hearted and courageous people. Thus it was one's duty to trick the former on "the length and the severity of the pains which they would experience, as well as on the dangers they would be exposed to."[20] Moreover, "the surgeon must do everything, albeit without overstepping the limits of truth, to induce the patient to ask for the operation himself."[21] Velpeau proposed an ethic for the physician confronted with the problem of pain, in which both the rights of the sick person—who was considered as the subject—and the duties of the doctor were postulated at the same time: "If it is cruel to expose those whom one could treat in a gentler way to the action of the iron, it would be even less compatible with the well accepted interests of humanity to compromise the future health of the subject under the vain pretext of firstly sparing him some pain."[22]

This "economy of pain" in which the doctor had to weigh up the interests of the patient seems to us to be anything but an excuse for indifference to, or acceptance of, the patient's pain; it reveals a humanist attitude, which also obliged the surgeon to guard his students against the consequences of technical virtuosity: "Speed and extreme elegance rarely go together with prudence and safety. You have gained a minute, but your incision is too long or too short, too straight or too slanted . . . finally the wound lacks the qualities which it might have had."[23] In the past, the speed of carrying out the operation was considered to be the principal means of limiting the duration, if not the intensity, of pain: all efforts were turned towards reducing the duration of the pain, or, in other words, towards a better anatomical understanding of the tissues, types of instruments and courses to follow which would allow one to achieve one's goal as quickly as possible, producing the least amount of damage possible. Without any doubt, these rules represented a step forward by comparison with the previous century and went hand in-hand with the audacity of carrying out more serious operations. This portrait of the great surgeon of la Charité, with his preoccupations, obliges the historian to find numerous reasons which could have been behind the existing scepticism and misgivings.

### *Stupefying Gases at the Beginning of the 19th Century:*
### *Dying from Pleasure*

At the beginning of the 19th century, under the impetus of the La-voisian revolution in France and the work of Joseph Priestley in England, great progress in the quantitative chemical analysis of gas were accomplished. At the same time, research into respiration aimed at testing the effects of these gases on human and animal organisms was beginning to take place, particularly in England, with a therapeutic end in mind. One of the leading establishments for the application of chemistry to medicine was Thomas Beddoes' Institute of Pneumatic Medicine in Bristol, where the chemist Humphrey Davy also worked. Thomas Beddoes, who followed Brown's medical ideas concerning stimulation, practised a learned medicine albeit one which the officially accepted medical world dismissed despite its inroads. Humphrey Davy, with Beddoes' advice, proceeded tentatively and practised auto-experimentation with substances at the expense of his health and sometimes at risk to his life. He was also helped by Watt with the construction of breathing machines. What they were trying to find out was whether the inhalation of one gas or another would allow a patient to recover his strength quickly: thus they asked themselves whether the inhalation of pure oxygen had stimulant properties and whether it might not bring about an improvement in consumption.

Very often, the work of Beddoes and Davy produced letters from patients or colleagues which stated the benefits gained from the inhalation of these gases, and in a few cases, the dangers or failures. Thus the search was not specifically for a gas which suppressed pain, but each experiment brought a package of surprises and anxieties. The experiment was not really carried out just "to see" or purely by chance, since the chemist methodically experimented on all the gases from one family, trying to determine the effects on the living being of inhaling different quantities. However, nobody really knew how the experiment would turn out or whether the gas was "respirable" or lethal; it was not only "nitrous oxide" (nitrogen monoxide) and its "amazing effects" which were tested, but also oxygen, hydrogen, carbon dioxide, etc., inhaled in increasing doses. "In the consequence of the discovery of the respirability and the extraordinary effects of nitrous oxide, or the dephlogisticated nitrous gas of Dr. Priestley, made in April 1799, . . . I was induced to carry on the following investigation concerning its composition, properties, combinations and mode of operation on living beings."[24] The account given by Davy of the effects of inhaling each substance, which involved cautiously increasing the number, duration and quantity, shows that the questions posed by these experiments were of a totally different nature from those relating to the particular problem of pain.

On several occasions, Humphrey Davy emphasised how difficult it was for him to find the appropriate words to describe what he defined as an obscure pleasurable sensation which was unknown of until that time. The slight degree of vertigo which followed the first inhalation was soon followed by an unusual feeling of fullness in the head, with the loss of distinct sensations and will—a feeling similar to that which comes about during the first phase of intoxication but is not accompanied by pleasure. Repeating the experiment with a much greater quantity, Davy then obtained "a sensation analogous to gentle pressure on all the muscles, attended by an highly pleasurable thrilling, particularly in the chest and the extremities. The objects around me became dazzling and my hearing more acute. Towards the last inspirations, the thrilling increased, the sense of muscular power became greater and, at last, an irresistible propensity of action was felt; I recollect but indistinctly what followed; I know that my motions were various and violent."[25] Davy admits that had he not made a note of his reactions straight after his experiment, he would not have believed what had happened since the next morning all that he retained was a very confused memory. He was equally worried about the delayed effects and was surprised not to have any on the following day.

The details provided by Davy on the duration of the inhalation, which always lasted less than five minutes since if it had lasted any longer the tube he was breathing through would have fallen out of his mouth,[26] could explain why he did not in fact talk about the loss of sensations and consciousness, other than for a brief instant, and even stated that the absorption of gas did not hasten his bedtime. The hilarious reactions, which lasted from five to six minutes and manifested themselves by a vibrant tingling and laughing, were generally accompanied by a greater or lesser degree of intense pleasure, or by "sublime emotions which were linked to particularly clear ideas," which led him to inhaling the gas for fun and to amuse himself. However, he did note the power of nitrous oxide to calm pain and cited two examples concerning headache and one concerning toothache: he wrote "The power of the immediate operation of the gas in removing intense physical pain, I had a very good opportunity of ascertaining. In cutting one of the unlucky teeth called *dentes sapientiae,* I experienced an extensive inflammation of the gum, accompanied with great pain, which equally destroyed the power of repose, and of consistent action. On the day when the inflammation was most troublesome, I breathed three large doses of nitrous oxide. The pain always diminished after the first four or five inspirations; the thrilling came on as usual, and uneasiness was for a few minutes, swallowed up in pleasure."

One of the obstacles to the immediate use of nitrous oxide in surgical operations could have been linked to the fact that originally the new gas was seen as a substance with the same nature as the opiates, *i.e.* aimed at calming pain after intervention. This passage, which seems so simple to us, of an "after" to a "before" and a "during the operation" was not obvious then, and one cannot invoke as an explanation for this difficulty the problems of the purification, the dosage and the duration of inhalation. These problems were resolved quickly enough once the obstacles linked to the ideas held by society (and not only the medical world) about the nature and impact of surgical operations as well as the relationship between the body on which an intervention was to be made and the surgeon were swept aside. One could provide a clue to this problem, which was not only technical but also related to ways of thinking, by recalling the views held some twenty years previously by Antoine Sassard, a surgeon at the Paris Charité hospital: ". . . Let us see whether there exists some means for extinguishing or even stifling this natural feeling [pain]. What! Doesn't *analogy* show us how we should behave? Since it is clear that violent pains such as those caused by colics [and] cancers and other disorders are calmed and even destroyed by the use of narcotics, who is to stop us from administering, before the majority of our operations, a narcotic with a dosage appropriate to the age and temperament of the individuals, which would induce sleep after a certain time, and by repressing out the sensibility of the parts, could in certain cases prevent pain, and in other cases, diminish it and make it bearable."[27] Dreaming of this method, Sassard still thought in terms of analogies, as if the suppression of pain during the operation was not an aim in itself. This was the step which Humphrey Davy was about to take.

Davy's research, immediately known in France thanks to excerpts by Berthollet from Davy's articles in *La Bibliothèque britannique,* clearly showed in which theoretical framework and which set of hypotheses he placed himself: it concerned determining whether nitrous oxide was an ordinary stimulant and, following Brown's system, whether an overdose might not enfeeble the patient. Thus, the absorption of wine, consumed in quantity just after the inhalation of gas, brought about what in effect was a state of insensibility, accompanied by a loss of awareness of external objects; this lasted for a little more than two hours and wakening up was characterised by headache and nausea.[28] By developing a box which, by continually renewing the air, allowed one a period of an hour and a quarter to inhale the gas, Watt modified the conditions of the experiment. By so doing, the impressions obtained (pleasure, hilarity, the desire to react, etc.) were prolonged, making it possible to confirm that there was no perception of the external world with such a feeling of well being and intellectual power that Davy flew into a rage with the person who removed the box. He felt intense and sublime pleasure every time he breathed in the gas and indulged in it outside of his experimental programme. Over a period of about two hours, he experienced an intermediate state between that of being asleep and being awake which, for example, silenced the stinging pain from a cut.

Although nitrous oxide had been used to treat illnesses by the "absence of sensibility," Davy implicitly signalled its possible uses in surgery: "As nitrous oxide in its extensive operation appears capable of destroying physical pain, it may probably be used with advantage during surgical operations in which no great effusion of blood takes place."[29] The qualification arising from Davy came from the observation that the absorption of gas inhibits voluntary movement and relaxes muscles, which led to the fear that it would affect the contractility of the vessels and the tonicity of the tissues, thus making it impossible to stop a haemorrhage. Only a better knowledge of the vasomotor nerves and a comprehension of the functions of the sympathetic system would allow this fear to be allayed—but this would not come about until a much later date. One can see how, for the time being, many arguments came together to prevent Davy's research being viewed as anything other than strange suggestions or perhaps wishful thinking, but in no way systematically applicable. For reasons already stated, one must add the fact that "laughing gas" could be used for playful and not strictly medical ends, and in a climate where the spectre of poisoning hung over the use of this new substance. In the translation of his *Elements of Chemical Philosophy* into French (*Éléments de philosophie chimique*), Davy was already involved in rewriting the experiments in the following terms: "It has a sweet and sour taste and a weak but agreeable smell. One could inhale it although it would be appropriate for maintaining life. In 1799, I recognised that when you inhaled it, you felt the same effects as from alcoholic beverages; it is more often a fleeting poisoning, or a strong exaltation which it produces."[30] All the attempts which were to follow, whether they were concerned with isolating morphine, with ether, or a little later with chloroform, would be dominated by an apprehension of poisoning and, in the case of inhaled gases, by the fear of asphyxia. As a result of this, in 1875, Claude Bernard would again have to demonstrate that anaesthesia was not due to asphyxia. The new-born science of toxicology and forensic medicine would join together to give credit to misgivings of poisoning, though this was not entirely a conscious intention on their part.

### A First Step: The Isolation of Morphine

In the article on opium in the *Encyclopédie méthodique* of 1824, Jacques-Louis Moreau de la Sarthe, who was very much involved with the Ideologues, wrote: "It is entering France in considerable quantities, and Doctor Mérat, who made this remark, informs us, having consulted custom records, that some 2,000 pounds were imported in 1803 and 3,000 pounds in 1807."[31] The problem then was no longer one of knowing whether or not one should use it, nor one of whether it was a sedative or a stimulant. Rather it was a matter of understanding which "system" it acted on and how. Henceforth the emphasis lay on analysing its chemical components, and it was solely on the basis of a precise knowledge of its active ingredients that one could correctly appreciate its effects. The history of the isolation of morphine has been sufficiently well described to make it unnecessary to go into the details here:[32] let us remember simply the complexity of the process of "discovery," from the initial work of the chemist Derosne—published in *Annales de chimie* in 1803,[33] and which classified different substances under the same heading of "essential salt"—up until the isolation in 1806[34] of the "soporific principle" in opium by the pharmacist Sertuerner, a native inhabitant of Eimbeck (Hanover), who recognised this as being the active ingredient in opium and baptised it "morphine" in 1817,[35] as well as a series of chemical analyses carried out by Séguin, then Robiquet; these disclosed 11 different substances in opium. However, which one was really active and in what form? For a certain time, there was confusion regarding the powers of "narcotine," a crystallisable constituent of opium discovered by Derosne—it was not known if it had stupefying powers or whether it acted as a poison or medicine. Experiments carried out on dogs had brought on convulsions, violent vomiting and sometimes death. Derosne himself indicated the novelty and limits of his work by stating: "I propose to show that there is a particular substance in opium which it has been equally wrong to regard as an acid or as a mineral salt, and which has previously unrecognised properties. I will also indicate a few other substances which we have not identified in it. I will not give the exact proportions of the respective quantities of each as that is too difficult a task with such a complex substance; moreover, it would be absolutely useless since sales outlets rarely provide two completely similar types of opium."[36] It was particularly its alkaloid nature and vegetable origin which astonished the scientific community.

In 1804, Séguin, in an oral address to the Institut, referred to an unknown crystalline substance, but he lacked precision in his analyses and the results of his research were not published: Vauquelin, in the *Annales de chimie* of 1818, made a vain appeal for Séguin to be given credit for having first made the discovery.[37] In any case, Magendie did not hold this view: the discovery and first uses of nitrous oxide had been a British triumph and in the same way, the first work on the isolation of morphine was a German enterprise. The research led by Nysten throughout 1808 was concerned not with this crystalline matter to which he paid no attention, but with Derosne's salt or narcotine. At first, Sertuerner's preliminary work showing that there was a "vegetable alkaline base" in opium was not backed up by anyone else. It would be necessary to work with sufficient quantities in order to achieve consistent results which could be verified by other chemists. Before being able to use it as a medicine, it would be necessary to learn how to prepare and purify it, determine what doses to employ and verify its effects on the human body. Eventually, morphine was prepared as salts which resulted from the combination with acids—morphine acetate, sulphate and morphine hydrochlorate. However, it remained to be determined which of these forms was the most appropriate for the purposes to which it was to be put, and it was really by tentative trial and error that it was realised that the "energy" of the substance was totally dependent upon the state in which it was found. For a long time, Derosne persisted in thinking that his narcotine was nothing more than a modification of morphine, while Robiquet focused his energies on demonstrating that it contained neither morphine nor meconic acid. All the great chemists of the time, Orfila, Pelletier and Caventou, Vauquelin, Thénard, etc., as well as pharmacists, undertook the task of verifying the composition of opium and determining what quantities of morphine it contained.

In this period when "chemistry applied to medicine" was taking off,[38] therapeutic experimentation was often firstly autoexperimentation, whereby the chemist risked poisoning himself, and the passage from animal to man was not undertaken without legitimate fears. Doctors who used it spoke of trials carried out with the patient's consent, often when all other remedies had failed. Magendie noted that "this conflict of opinions" on the effects of morphine "doubtless stems from the fact that no account has ever been taken of the excipient and the influence which it exerts in the production of the effects. Morphine being, as we have said, almost insoluble in water, those who administered it in an aqueous vehicle cannot have observed a really perceptible action, whereas those who gave it with acids, saw all sorts of effects produced."[39] Dupuy and Lassaigne, vets at the École d'Alfort, undertook to test the effects of this vegetable poison on dogs and looked for antidotes. Morphine began its therapeutic career as "an unfortunate celebrity,"[40] and made its appearance simultaneously in the treatises on therapeutics and forensic medicine: there are few substances for which the uncertainties as to whether it was a remedy or a poison were greater. The precise chronology of events concerning morphine demonstrates the necessity of thinking again about the concept of "discovery" and discarding the view that even though the means existed, they were not exploited. Without even speaking about the technical, industrial, and commercial aspects of producing a medication, one can see in an exemplary way in the case of morphine what must take place between the moment when a new substance is isolated, the time when its nature and properties become known and when its effects on man can be predicted accurately, and that a whole process of experimentation, verification and clinical control is necessary.

In this venture, one decisive link came in the remarkable memoir published by V. Bally in 1828—"Observations sur les effets thérapeutiques de la morphine ou narcéine" ("Observations on the therapeutic effects of morphine or narceine"). By allying the rigour of chemical analysis to a vast medical practice in the hospital environment, dealing with between 700 and 800 patients, he was able to state the dose, the effects on the different "systems," therapeutic indications, and raise the problems of addiction. He chose to administer them in pill form, because, he said, "in this form, one is really sure of what one is administering, as of what one is doing. One may gradually increase the dose as required of these powerful substances whose use demands such unceasing attention and supervision. Through other routes, the medications are sometimes altered or too delayed; they undergo various chemical combinations and may also be neutralised."[41] Bally meticulously recorded the established facts when he drew up a list of the "systems" on which it has no effect (the respiratory system), those on which it acted (the brain, digestive and urinary systems) and those in which contradictory phenomena were found (the cardiac rhythm and circulation of the blood). He also noted indirect effects such as the anti-diarrhetic power of morphine. Thanks to this memoir and to the large range of observations which it describes, it is possible to get some idea of the pains which were treated with morphine (in doses which varied from $\frac{1}{16}$th of a grain for women to $\frac{1}{4}$ of a grain for men, twice daily, with a whole range of increases being permitted, as long as one proceeded slowly): neuralgia, pruritus in various parts of the body, and rheumatism. With a wholly clinical precision, Bally also made note of the symptoms which appeared following the taking of morphine so as to give warning of their alarming character: a feeling of vertigo which made it seem as though objects were spinning round and walls were closing in (Thomas De Quincey and Baudelaire, in "La Chambre double," had meticulously described this), an impression of electricity in the trunk and in the limbs, a complete awareness of what was going on around oneself though in a kind of torpor, hallucinations and a maintenance of intellectual faculties despite its sedative or calming power. In certain subjects he had noted the appearance of headaches and stomach pains, and there was already, in his memoir, a mention of a pairing which would go down in posterity not only during but after the 19th century—namely suffering and substance abuse which, by "wearing out" patients, necessitate a considerable increase in the dose of morphine. However, it is of some interest to emphasise the fact that before a specific action had been proven for it on the nervous system (the brain and its appendages), Bally had begun not only to prescribe it in the treatment of pain, but also for palpitations, coughs (against which he judged it to be ineffective), sugar diabetes and, with tragic results, on an apoplectic.

It is clear that in this phase of establishing a particularly active substance, the frontier between therapeutic and experimental acts was not clearly defined. When the commission for prizes awarded by the Académie, presided over by Double, established a list of six possible subjects in 1828, it proposed "a comparative study, in the various categories of living beings, of the mode of action of narcotic substances in order to determine by such means the indications and counter-indications of such an important class of medicines."[42] However, in the end, it was a question on the circulation of blood in the capillaries which was chosen. The problem of narcotics was very much on the agenda, but still not enough for priority to be given to mobilising all the necessary attention. However, in 1831, the Montyon Prize "in favour of those who have perfected the art of curing" was awarded to Sertuerner at the gathering of the Académie des sciences held on 14th March, "for having recognised the alkaline nature of morphine and thus having opened a pathway which produced great medical discoveries": the prize was two thousand francs, whereas the other laureates received from six to two thousand francs for the discovery of iodine and its uses.

## The Debates on Anaesthesia: 1847, a Revolutionary Year

Some of the features of the introduction of morphine into therapeutics were to be found again in the first uses of ether and chloroform to anaesthetize against surgical pain: tentative efforts were made to determine the doses and effects, mortal risks, fears of axphyxiation and poisoning. Despite this, from the very moment when the use of ether was first discussed at the Académie de médecine and the Académie des sciences, the method took off like a trail of powder, into the most distant provinces, while doctors and surgeons, who until then had remained silent, admitted having carried out trial surgical operations on "etherised" patients in the months prior to its official recognition. In contrast to the speed at which this new method spread from 1847 onwards—no doubt partly due to the pressure of public opinion—was the initial slowness and hesitation with which the first attempts had been carried out at the very end of the 18th century in England, and a little later in the United States.

The stages which preceded the use of ether as a general anaesthetic in surgery are relatively well known:[43] it is known that in the United States, "laughing gas" was sometimes used by travelling circuses at the beginning of the 19th century and that, on one occasion, quite involuntarily on the part of the manipulator, nitrous oxide inhaled by Indians brought about not only good humour but also a deep sleep.[44] Sometime later, it was ether which became a stimulant that was valued in the "frolic parties" in Jefferson and other towns. The properties of ether were not unknown because it was an ingredient of the famous "Hoffmann's soothing liquor" which aimed to calm people down particularly in cases of hysteria, and seems to have been used on experimental animals, really more to enable the operator to work peacefully rather than as a means of suppressing pain. In 1817, in his *Nouveaux Éléments de thérapeutique et de matière médicale,* Dr. Jean-Louis Alibert who worked at the Hôpital Saint-Louis in Paris and who was to become famous for his knowledge of skin ailments, saw potentially interesting properties in this class of substances: "Ethers have a very marked medicinal property, but the results of these have not yet been studied enough. These liquids act firstly as stimulants of the brain and the nervous system: but they end up bringing about a state of stupor, drowsiness and sleepiness. . . . I dare say there is more to discover about the mode of action of these amazing liquids than one would have thought."[45]

However, the passage from sleepiness to a complete and lasting insensibility to pain did not come about on its own: ethers were used for example to treat acute croup in children or in local applications for swellings and pains following bruising, or even for use against rheumatism: thus, neither the indications for using ether nor its mode of operation was specific against pain. Was it a question of having to absorb large doses? Immediately, the spectre of poisoning reared its head, as can be learnt by reading the *Traité de toxicologie générale* by Orfila.[46] It could be that, at least in France, forensic medicine, a discipline which was in the process of increasing its political clout, contributed to the growth of a climate of prudence and defiance regarding therapeutic innovations. This had come about as a result of recent cases of poisoning with opium and would be reinforced by subsequent accidents due to chloroform. After the passing of the First Empire, was the activity of the chemists, who would bear deeper examination, not more oriented towards industrial applications than towards the therapeutic ends at a time when the link between the two did not seem obvious? Finally, what were relations like between the chemists of the Académie des sciences and the pharmacists? And within the hierarchy of the medical profession, was the status of dentists not too subordinate for the joint needs of patients and operators in a particularly sensitive domain to be taken into account? Especially after 1820, what was the effect of the themes of the restoration and regeneration of society on the attention paid to pain? A complex set of factors, added to the tentative attempts to determine the exact properties of ether, information on the quantities necessary to bring about sleep and the risks entailed must be borne in mind when trying to appreciate the reasons why France was left in the background during the process which led to the discovery of anaesthesia produced by ether.

In fact, the first attempts, the first daring steps and first successes took place in the United States. On 30th March 1842, the American doctor and surgeon William Crawford Long removed a tumour from the neck of a patient who was anaesthetized with ether, without the latter feeling the least bit of pain. The patient was used to inhaling ether, and the doctor had carried out several previous experiments on his assistants. But he did nothing to divulge this information. Following one of those sessions of chemical demonstrations which formed part of travelling shows and where the public agreed to inhale laughing gas, the dentist Horace Wells had the idea of carrying out the extraction of a tooth when the patient was under the influence of the gas. He first carried out such an experiment on himself in 1844. He told his colleague William T. G. Morton and also Charles T. Jackson, the chemist and geologist, about this but an unfortunate experiment carried out in public at the end of the surgery lecture given by Dr. Warren at Cambridge College in Boston left him covered in ridicule. Throughout 1845 he continued to extract teeth under anaesthesia in the small town in Connecticut where he practised; meanwhile on 7th December 1846, Morton published the results of "his" discovery in Hartford in the *Boston Medical and Surgical Journal.*

In spite of the journey Wells made to Paris in 1847, and his attempt at getting himself recognised as the discoverer, the Académie de médecine and the Académie des sciences at first wanted to recognise only the work of Jackson and of Morton who in the meantime had replaced ether with nitrous oxide. Wells in his letter to the Académie de médecine, which was read out during the meeting on 23rd February 1847, proposed the interchangeable use of nitrous oxide or the vapours of sulphuric ether for performing operations without pain. This undoubtedly went against him: Orfila did not omit emphasising the negative effects resulting from nitrous oxide, for which Wells indicated a preference: "This gas however is not without drawbacks: everyone who has inhaled it, such as Mr. Vauquelin and Mr. Thénard, have suffered a great deal. I repeated their experiments, and I felt such intense pains in my chest and such a feeling of suffocation that I remain convinced that had I continued the experiment, I would not have survived."[47] Wells himself committed suicide using chloroform before he learned of the letter which the Société médicale de Paris had finally written to him, in January 1848, to grant him the recognition of having made the discovery and the first use, despite the pressure of Warren who had attended the first meetings of the Société. Jackson, for his part, laid his claim before the Académie des sciences de Paris, on 21st December 1846, of being the first to make this discovery; his case was presented by the intermediary of his friend, the geologist Élie de Beaumont. A meticulous reconstruction of the facts has already been carried out by historians and made it possible to assign his respective role to each of the four protagonists of this discovery: perhaps from 1842, Wells had had the idea of using nitrous oxide for dental extractions but he failed in the use of the procedure in front of Warren; Morton, who was a dentist in Boston, visited Wells to inquire about the procedure and was present during the unsuccessful experiment of 1844—however, on the advice of Jackson who told him about the properties of ether, he used it in 1846 and knew how to publicise its potential. The surgeon John Collins Warren was able to understand the importance of this discovery and had the audacity to want to use it publicly in the extraction of a tumour in the cheek, in October 1846, at Massachusetts General Hospital, Boston, even though Wells had failed two years previously. This time, the public success of a surgical intervention in which the patient did not feel a thing heralded a new era of operations as Warren would immediately proclaim.

As soon as this news reached France, a debate was instituted simultaneously in the two Académies, the Académie des sciences and the Académie de médecine which rapidly set up a "commission for ether." Moreover, certain protagonists such as Velpeau or Malgaigne were to be found in the two institutions. Certainly there were passionate discussions which went on from 1847 to 1848, in which reservations were expressed and questions asked about the conditions of use and secondary effects of ether. However, with the exception of the clash between Velpeau and Magendie at the Académie des sciences, there was very little confrontation. Ether, in contrast to chloroform which was beginning to be used at almost the same time, was not a new substance; as Mérat noted on 19th January 1847 during a meeting at the Académie de médecine, "the stupefying action of ether is well known in therapeutics. Several practitioners have seen patients fall into a kind of drunken state after having taken quite a strong dose. It is this property which causes it to be prescribed in nervous disorders and neuralgia, almost always successfully. The new use which American surgeons have made of it for operating could have been deduced from this well-known property of producing a sort of numbness; but the route of administration of ether with the help of inhalation is new, not only in its use itself, but also in its aim."[48] Mérat then recalled a note from Dr. Ducros, dated 16th March 1846, which was included in the *Comptes rendus hebdomadaires de l'Académie des sciences.* This commentary concerned the use of ether, administered by applying it inside the mouth using a paintbrush in order to put a hypochondriac to sleep in a much more effective manner than with opium. These remarks echoed those of Velpeau at the Académie des sciences: "We have known about this action of ether for fifteen, twenty, thirty years and more. The medical dictionaries, the treatises on forensic medicine, that by M. Orfila and his *Toxicologie,* in particular, clearly testify to the fact. What is new, is the concept of rendering patients completely insensible to pain before an operation, by means of inhaling ether. But nobody to my knowledge, had put forward this proposition before Mr. Jackson and nobody before the dentist Morton had used this method on a patient."[49]

In fact, the problem really concerned the idea of using ether for surgical operations and of agreeing to open an apparently lifeless body in which reactions were no longer there to guide the surgeon in his work. For the first to try it, this silence on the part of the patient which left the surgeon face to face with himself in a hitherto unknown silence was no less agonizing than was the necessity of persevering regardless of the pain and cries of the person being operated on.

In the weeks following the announcement of the new procedure, the only thing discussed in the Académies was the question of etherisation: at first it was a question of accumulating a sufficient number of facts to reduce the whole mystery surrounding the new procedure and to assure themselves that notwithstanding some "stubborn" subjects, etherisation did indeed bring about insensibility; during this first phase, the description of the symptoms (analogous to drunkenness, the agitated period before sleep, the gestures and speech of etherised individuals were noted down precisely as were the dreams about which they were questioned on awakening) prevailed over questions concerning the means of administering ether or on what doses should be used. It seemed that the technical problems of using inhalation had been resolved quite quickly thanks to Charrière's appliances and the various improvements made to them so as to allow the patient to keep breathing fresh air: an ethermeter was made by Dr. Massiart, and in order to make inhalation easier, adjustments were made such as the modification of the ferrules to allow a better adaptation to the morphology of the face. Priority was also given to the discussion of four aspects of etherisation: the maximum duration of inhalation and the most appropriate moment to begin the operation; the risks of death from the inhalation and the secondary consequences; the scope of the method and in particular a debate on whether it could be used in childbirth; finally, the ethical aspects which were brought up by Magendie in particular. However, the generalised use of the procedure also gave rise to physiological studies of great interest. These studies were concerned simultaneously with the question of what paths ether took in order to render the nervous centres insensible, with the disassociation between sensibility and movement, and with the specific action on muscle fibres dependent on voluntary controls.

The vast number of reports presented at the Académies during their first assemblies served above all as testimonials to the true facts. As early as the month of December 1846, Velpeau, who had personally received a letter from Warren, received propositions from Dr. Willis Fischer of Boston to carry out trials at the Charité, while similar trials were undertaken at the Hôpital Beaujon by Dr. Roux and Dr. Laugier, and at Saint-Louis by Malgaigne who was one of the most ardent defenders of the new method. Even though during the meeting of 18th January 1847 Velpeau was still wary, as of the following week he noted: "The experiments, which are becoming increasingly routine . . . along with their results have, with tremendous speed, increased so much since last Monday that I have no fear of being contradicted in the future, when I state that they reveal a very important fact, a crucial fact. . . . As facts like this reproduce themselves almost everywhere and in any hands, it is no longer permissible to regard them as exceptional, and no doubt that after inevitable tentative attempts, surgery will gain benefits, wonderful benefits from the inhalations of ether during surgical operations."[50] Amongst these procedures, one frequently encountered amputations, extirpations of tumours, Caesarian sections, operations on the eye, reductions of hernias, incisions into abscesses, etc. A kind of enthusiasm took hold in medical circles: as early as 15th December, Jobert, from Lamballe, proceeded with etherisations; throughout the months of January and February surgeons including Bouvier, Blandin, Chevallier, Roux, Cloquet, Gerdy and many others from different hospitals in Paris increased the number of operations which they carried out under ether; thanks to Landouzy, news arrived from Reims; news also arrived from Chambéry and even from Algiers (via a letter written by Dr. Malle, dated 2nd March 1847); meanwhile, Renault and the vets from Maisons-Alfort were carrying out tests on animals (horses, dogs and rabbits), trying to determine the number of etherisations possible in a day, the time needed to recover, and the maximum duration of inhalation. In this instance, the two Académies seem to have efficiently played their roles in spreading the news of scientific discoveries throughout the whole country. If in the first few weeks, certain surgeons described cases of subjects failing to respond to ether-isation, as well as great variations on the duration required to render the patient insensible (from two to twenty minutes), Amussat soon drew attention to the purity of the product used: "One must choose an identical ether which always has the same concentration and which, being used by all operators, would do away with one of the causes of the observed difference in the phenomena and consequently lessen the chances of unfortunate accidents of which it is prudent to be appre-hensive."[51] The first attempts were faced with a dilemma between the wish to prolong the time of inhalation sufficiently to achieve complete insensibility, and the fear of going beyond a certain threshold to an irreversible situation: this explains the debates on what type of opera-tions—long or short—were compatible with the use of ether. A voice was raised to proclaim that for operations of short duration etherisation was unnecessary, but straight away others retorted that minor opera-tions were sometimes very painful and that it did not make sense not to use ether. Certain academics asked themselves whether it was possi-ble to carry out operations which needed "the cooperation of the patient's will," as was the case in the removal of the tonsils where the mouth had to remain open for the duration of the procedure; soon, dentists added their evidence which proved that there was no serious obstacle in this respect. Blandin put an argument forward against etherisation which was more difficult to oppose: on waking, etherised patients did not know what had happened to them, but "it would be wrong to conclude that they have not suffered: everything points to concluding that, on the contrary, they suffer but have no memory of it."[52] These few dissenting voices were quickly silenced, as was shown, for example, by the discussion on the use of the inhalation of ether in childbirth.

James Young Simpson, the professor of obstetrics in Edinburgh, had used ether for the first time on 19th January 1847 in a case of "a defect in the structure of the pelvis," which required the "version" of the child: having been chosen because this manoeuvre meant that uterine contractions were secondary, this case and its happy outcome were presented to the Edinburgh Society of Obstetrics in January and February 1847, and having repeated the procedure several times, he made his results known in the *Monthly Journal of Medical Science* in March 1847 and then at the Edinburgh Medico-Surgical Society. However, having carried out experiments on himself and his assistants, he substituted chloroform for ether, and used it for a birth for the first time on 8th November 1847. To those who opposed him by quoting Genesis ("I will greatly multiply thy sorrow and thy conception; in sorrow thou shalt bring forth children"), he countered with the passage from Genesis in which God, wishing to provide a companion for Adam, makes him sleepy in order to remove one of his ribs and to close up the skin again, making God the first anaesthetist.

Trials followed rapidly in different Edinburgh hospitals and in less than a year Simpson had carried out more than 150 births under anaesthesia. The Press, and the correspondence within the scientific community also ensured that this news was spread rapidly throughout France: on 8th February 1847, the doctor and surgeon Paul Dubois who was in charge of the Department of Obstetrics at the Hôpital Beaujon, followed the example of his Scottish colleagues.[53] Although Simpson had emphasised that anaesthesia was dangerous for neither the mother nor the child, and that often it even led to a more rapid recovery after the birth, he still asked himself what effects ether had on the new-born as well as on the mother, and whether relaxing the uterine muscles carried the risk of preventing the contractions from achieving their purpose. Asking such questions was justified since several surgeons had noted that this state of relaxation was beneficial in the reduction of fractures and dislocations, and since the clear distinction between smooth and striated muscles had not yet been acquired: thus, since experiments addressing this question had not been carried out, no one could know whether uterine contractions were prevented by etherisation: certainly the innervation of the uterus seemed to depend on the sympathetic system and on contractions which seemed to be involuntary, but was the uterus exclusively dependent on the nerves of "the organic life" as Longet thought, or did it also receive nerves from the spinal cord, as Brachet had said; if the latter were the case, was there not a risk that labour would be modified by etherisation? Five cases of etherisation of primiparous patients were reported by Dubois—all of these having been chosen for study because of the slowness of labour (36 hours) and the suffering being experienced by the women. He noted that in all five cases, there was no modification in the abdominal and uterine contractions, but, on the other hand, he commented on a relaxation of the perineal muscles which aided expulsion. Everything pointed to the conclusion that etherisation suppressed the physiological pains of childbirth and allowed obstetric operations to be performed in the same way as Simpson had already experimented in Scotland. However, Dubois' conclusion was quite cautious: "In winding up, I will say that it seems to me that the nature of things must be opposed to the application in a general way, of the inhalation of ether to the art of childbirth. Everything is unexpected, everything is urgent in childbirth. But can the obstetrician continually have on him an appliance for ether? And, if he does not always have it at his disposal, the time necessary to obtain it for emergency situations might be better employed helping the patient. Moreover, all the issues surrounding this serious subject are far from being clarified by available information."[54] Without doubt, this confused testimony where "the nature of things" can indeed be interpreted in many ways, reveals poorly rationalised resistance; but Dubois' colleagues at the Académie de médecine did not accept it, and Velpeau along with Malgaigne were not wrong to emphasise "a sort of contradiction between the facts exposed by M. Dubois and the conclusions he draws from them."[55]

Even if the results of all the surgical procedures were soon to produce an opinion which on the whole was very favourable to etherisation (except in the cases of cystotomy to which we will return), it remained to be determined what moment was the most favourable for beginning the operation: was it necessary to wait for the complete insensibility of what was regarded as "the third stage" of etherisation, at the risk of not having finished when the patient woke up and with the persistent fear that this sleep might deeply modify the physiological mechanisms; or should one start during the second period, which was characterised by a state of drowsiness when the patient could still see and hear what was going on around him, but at such a distance that he or she no longer felt pain? The delimitation of these periods was made simultaneously by the differentiation of symptoms, and thanks to the experiments of Longet and Flourens, by the way in which ether successively reached the nerve centres, spinal cord, pons and brainstem. Before having carried out these experiments (mostly on dogs, rabbits and pigeons), Longet and other physiologists could have thought that ether worked exclusively on the peripheral nervous system. In fact, the treatment of facial and intercostal neuralgias, either by the inhalation of ether, or by the application to a painful part of compresses soaked in ether, had been known of and practised for some time, for example by Thénard on himself between 1812 and 1814, by the surgeon Honoré at the Hôtel-Dieu and then by Serres who, on 8th February 1847, gave a talk to the Académie des sciences about his experiments involving the application of ether directly onto nerve tissue.[56] This form of local anaesthesia was used a long time before injections of novocaine were employed at the end of the century and fitted in with the more general practice of using liniments or topical applications which were thought to act by absorption. Serres stated that the action of such local anaesthetics lasted for minutes or hours following contact, and he supposed that ether had not only a sedative action but altered the nerve tissue in a lasting manner. However, these various observations were concerned only with the peripheral nervous system.

But Longet, at the meeting held on 9th February 1847 at the Académie de médecine, presented the results of his animal experiments which proved that the central nervous system was susceptible to ether; during a first phase, which was called "the period of etherisation of the cerebral lobes" (Longet included the four medullary tubercles, the optical laminae and the striated bodies as well as the cerebellum under this grouping), the animal completely lost the ability to react voluntarily when pinched or wounded, but was still able to cry out and be agitated; on the other hand, when the pons Varolii and medulla oblongata were etherised in turn, there was no longer any perception of the tactile sensations: this was what corresponded to the phase of total insensibility in man and would soon be called the surgical period. The experiments had been carried out using the prior experience obtained from experimental physiology concerning the specialisation of the functions of the central and peripheral nervous systems; however, ether in turn became a powerful means of investigating these functions. As Longet came to state, "in ether, the experimenter is in possession of a new means of analysis, which (without prior mutilation and undue bloodshed), if employed shrewdly, allows him to isolate the site of general sensibility from the site of intelligence and free will."[57] It remained to be determined how much time separated the start of the surgical period from the moment when ether blocked respiration by acting on "the vital node" (the respiratory centre), and brought about death: it was at least in these terms that many physiologists, following the work of Flourens, posed questions on the dangers of etherisation. From another perspective, the physiologists who carried out animal experiments had commented that etherisation brought about a transformation of arterial blood into venous blood, and by so doing, gave some credibility to the fear of asphyxiation. At least this was what Amussat thought he had seen; however others, including the physiologist Gerdy, soon retorted that in this case asphyxia came not from the product itself but from the conditions of its administration, and from the fact that the animals had not been allowed to breathe pure air: it was the lack of oxygen and not intoxication by chloric ether which was to blame. In certain respects, physiologists had more reasons to be on their guard against the dangers of prolonged etherisation than had ordinary physicians and surgeons, and this must be taken into account when evaluating Magendie's attitude. Already, attempts had been made to find stimulants or antidotes which were able to check the action of ether: the use of either strychnine or electrical stimulation was proposed by Ducros on 22nd February 1847, as a means of instantaneously putting a stop to the effects of ether. Thus etherisation was propagated like a trail of gunpowder and in spite of the bad side effects which were sometimes experienced during inhalation (suffocation, convulsions, fits of coughing due to irritation of the respiratory apparatus by mucus), and in spite of a few respiratory accidents, nothing seemed able to restrain the progression of the new method.

Gradually as these studies were developing and the evidence was accumulating, surgeons became confident and came to prolong the use of inhalations for up to three quarters of an hour, albeit with interruptions. However, as a result, there were some deaths which could be put down to ether. Magendie, who had been absent from the preliminary discussions, seized the opportunity to express his disagreement with the human experiments which had been carried out by his colleagues: how could they know in advance what the effects of etherisation would be in view of the case of a patient who became raging mad and who, in a delirium and suffering convulsions, rushed at the surgeon: "That is what I do not find moral, since we do not have the right to experiment on our fellows."[58] His attitude has often been cited as an example of the medical world's resistance to change,[59] and it is true that his confrontation with Velpeau and Malgaigne was fairly ferocious; however Magendie was quite isolated in these matters, as is shown by all the debates during 1847 and 1848. The first confrontation took place at the Académie des sciences on 1st February 1847: Velpeau had just finished a series of observations and concluded with the forecast that etherisation would become the "source of an infinite number of applications, of a completely unexpected fruitfulness, one of the richest mines, where all the branches of medicine would not be slow to help themselves."[60] Magendie's argument simultaneously took several different lines: without doubt there was a clear irritation at seeing the non-scientific press, "the political newspapers," seizing upon the facts and broadcasting them widely to the public even though the scientific world had not yet taken up a fixed position. The *Charivari* in a lighthearted way and other periodicals[61] repeated the news of etherisation. This was one of the very first examples of the power of the press in a scientific debate, a power which drew strength from the hopes of the public which it in turn catalysed. Magendie was on his guard against confusing the issues and the risk that the public's impatience, relayed via the press, would put pressure on and alter the composure of the scientific debate. Besides, as we have seen, he was sensitive to the ethical problems posed by human experiments and wished there to be a prolonged and preliminary phase involving experimentation on animals. He had been struck by the account of disorders caused by ether intoxication, but in this regard, could only advance limited evidence, either indirect where rumour was involved or very specific: "This is the first time I hear resounding within these walls an account of the marvellous effects of sulphuric ether (for this could hardly be said about other ethers), a kind of narrative which the press seizes and takes further, thus satisfying this insatiable and avid need of the public for the miraculous and the impossible. What I see most clearly in these accounts is that, with the doubtless laudable goal of operating without pain, they intoxicate their patients to the point of reducing them to what one could term the state of a cadaver which one cuts or carves with impunity and without any suffering. Hardly is the experiment done, often even unfinished, before it is given publicity."[62] Magendie, in turn, recalled that the therapeutic use of ether was long-standing, but was critical of the new use which was being made of it; to him it seemed unethical to operate on a "dead-drunk" patient, as one of his colleagues at the Académie de médecine appeared to have done, without making excuses for doing so. In a state of wakefulness, the patient was still able to judge the heavy-handedness of a surgeon, how little attention he paid or what mistakes he made: thus, in the conscious presence of the person being operated on, there was a source of constraint or at least of a contract of obligation on the part of the surgeon while, in contrast, the etherised patient was delivered defenceless and at his mercy.

Certainly the benefit produced by the elimination of suffering could not be balanced seriously against the inconveniences but, not without reason albeit somewhat perversely, Magendie posed the question of the rights of the sick person and the rights of the surgeon. The purity of the intentions did not seem a sufficient argument: "It is not enough to wish for good, one must still guard against evil." Posterity sometimes looks back on Magendie's interventions at the Académie as a not particularly glorious episode which, according to the accounts of the debates, created prolonged laughter, and is a good indication of his colleagues' states of mind: he gave descriptions of cases where the administration of ether to young girls or women caused erotic dreams, "dreams such as one should not have" as the patients subsequently said. He also described cases where etherisation seemed to remove all inhibitions and changed reserved and bashful patients into unrestrained lewd women; thus the spectre of "uterine fury" (nymphomania) flourished and Magendie, "despairing that it should be supposed that [he] had caused the laughter,"[63] linked together as being equally shameful the delirium of etherised people, the convulsionists of Saint-Médard, Mesmer's magnetic tub and the partisans of modern magnetism! In fact, it seems indispensable, to understand the attitude of someone like Magendie, as there was nothing in his philosophy nor for that matter his epistemological roots which was inclined to opposing this innovation, to view it in relation to the previous conflicts in which he had been involved: he had been accused of cruelty because of animal experiments which he defended by arguing that he had used ether in order to lessen the sufferings of the animals. Thus, he was perhaps not sorry to find himself in his adversaries' shoes from an ethical standpoint, so to speak, in order to defend a cause which seemed much more important to him. Besides, he himself had always been in favour of extreme rigour in establishing scientific data, and carefully guarded against hasty deductions; he could thus consider with good reason that the proofs put together in favour of etherisation up until then were insufficient. This at least was his conclusion at the end of the debate with Velpeau. But perhaps in certain cases like that of pain, urgency was so important in the eyes of the medical community, that waiting was no longer possible. Having been put into a weak position at the Académie, Magendie was to see himself attacked almost immediately by Flourens on a much more serious matter: taking advantage of a note in which he studied the effects of sulphuric (diethyl) ether and chloric ether (ethylene dichloride) on the nerve centres, the latter contested Magendie's claims to have discovered the functions of the spinal cord's dorsal roots, and gave all the credit for this to Bell. Thus the debate was ultimately to return to questions of who had been first, and would divide the heart of the Académie des sciences into two camps, one which was very much the minority and dominated by Magendie, and the other whose principal leaders were Velpeau, Malgaigne and Flourens.

Flourens was a physiologist working on the nervous system who had in particular established the role of the cerebellum in the balance and co-ordination of movements. He carried out experiments on rabbits and pigeons to study the successive effects of various ethers on different nerve centres (in the order reached, the brain proper or cerebrum, the cerebellum, the spinal cord and the medulla oblongata, the site of the respiratory centre or "vital node" which, when touched, instantly produces death). Thus, he succeeded in demonstrating experimentally that "chloric ether" had the same power of producing insensibility as did diethyl ether, and following this result tested the "new substance, known by the name of chloroform."[64] In fact, the properties of derivatives of chlorine were discovered later than those of ether though chloroform had been tested and used concurrently from 1847 onwards, and then was used in preference to ether as an anaesthetic.

When Samuel Guthrie devised a new method for preparing ethylene dichloride, which was published in Silliman's *American Journal of Science and Arts* in 1831, he obtained a new substance from the residue of the preparation which he redistilled. This substance had a strong aromatic smell and sweet taste and was additionally discovered simultaneously and independently in France by E. Soubeiran who published an article in the *Annales de chimie et de physique* in October 1831 and in Germany by Liebig who inserted a very brief note in the *Poggendorff Annalen* in November 1831 before publishing a second article in the same *Annales de chimie* in 1832. If Soubeiran and Liebig were in agreement in stating that they had discovered a new substance and agreed on its qualitative characteristics, neither of them, on the other hand, understood its composition: after having described the procedure for distilling and rectifying the matter resulting from the dissolution of limestone chloride with alcohol, Soubeiran wrote: "This new etherised liqueur is a substance which is different from all the others that have been observed by chemists up to this time. Its elements are chlorine, hydrogen and carbon."[65] However, in no way did he indicate the medicinal properties of the new product, which he proposed calling "bichloric ether." The following year, Justus Liebig, in a memoir "Sur les combinaisons produites par l'action du chlore sur l'alcool, l'éther, le gaz oléfiant et l'esprit acétique" ("On the combination produced by the action of chlorine on alcohol, ether, ethylene gas and 'acetic spirit'"),[66] identified two substances which were to play a large role in anaesthesia: chloral (made from a mixture of chlorine, carbon and oxygen) to which he gave its name, and an oily liquid, which was a result of the decomposition of chloral using alkali and water, which he considered to be a mixture of chlorine and carbon, for which reason he called it "carbon chloride," and in which he could find no trace of hydrogen. One of the difficulties for chemical analysis was to distinguish the new product from ethylene dichloride, which was still called "Dutch oil." It was in 1834 that Dumas, writing in the *Annales de chimie,* came back to the analysis of the new substance described by Guthrie, Soubeiran and Liebig, gave it a definitive chemical formula and proposed that it be called chloroform.[67] On this occasion also, the discovery seemed to come about as the result of a complex process to which several chemists contributed. Discussions took place concerning the identification of the new product and whether it conformed to the theory of chemical elements, much more than on what were its medicinal properties: the split between physicians and chemists, which corresponded to the increased specialisations of their two professions, was in the process of being established; it is not by chance that alone amongst those mentioned, Dr. Samuel Guthrie, an amateur chemist, was preoccupied with its medicinal effects. However, medical practice in the United States as in European countries was dominated by a general concern with stimulation, and Guthrie saw in his discovery a new stimulant which was pleasant to take and did not have a depressing effect afterwards: that was what had resulted from trials which he had carried out on his entourage.

The first trials on the use of chloroform created a great deal of enthusiasm for the very reason that surgeons had noticed that its action was more rapid and longer lasting than that of ether and thereby offered greater security to the operator. Simpson was an ardent proponent of its use and a few years later, chloroform received a magnificent official endorsement when it was given to Queen Victoria on 7th April 1853, for the birth of her fourth child, by John Snow who had become a sort of specialist anaesthetist. However, before receiving this royal seal of approval, chloroform, in spite of the improvement in appliances for inhalation and the precautions taken, had been the object of considerable critical comments, and sudden terrible deaths had been reported in Great Britain as well as in France, where a legal enquiry had been undertaken. Doctor Gorré, the head surgeon at the hospital of Boulogne-sur-Mer, had put before the Académie de médecine the case of a sudden death caused by the inhalation of chloroform which happened before his very eyes on 26th March 1848: it involved a young woman of about thirty years of age, Mlle Stock, who had required the removal of a piece of wood which had penetrated deeply into her thigh as the result of an accident. A judicial enquiry opened straight away to make a careful study of the circumstances of her death, the origin of the chloroform, the amount used and finally the responsibility of the physician who on that day had been assisted by the patient's usual doctor and midwife. The Académie soon heard testimonies from the surgeon Robert of Beaujon concerning another case which had happened in Lyons whilst British journals, and notably *The Lancet*, reported similar facts. It is quite impressive to observe that the 1848 revolution in no way perturbed the course of events and discussions: the Ministry of Justice and the Ministry of Public Inquiry demanded that the Académie should make a report at once: "Humanity as much as justice being interested in the resolution of the doubts which still exist as to the complete harmlessness of chloroform, I do not doubt the urgency with which this society will proceed to settle the question."[68] The wording of the request left hardly any doubt as to what was intended and no one was really ready to renounce a method which had so completely changed the relation of man to pain. Dr. Delebarre Jr., in a note to the Académie on 18th July 1848, summed up very well the general state of mind: "It is a very serious thing to dispute the existence of such an admirable and useful discovery as the one in question. Before condemning it, would it not be better to look into whether the accidents, which in any case are very rare, and which have been attributed to ether or chloroform, might not have resulted more from the way in which these agents were used?" The Académie's report read by Malgaigne on 31st October 1848, on behalf of a commission which was almost the same as that for ether, concluded against all the evidence that chloroform was not directly responsible for the deaths which had been submitted for examination. However, the commission to which the problem had been referred had nonetheless worked hard, proceeding to autopsies which had consistently shown a state of flaccidity in the heart whose cavities contained no blood and, at the same time, gas in the venous system. However, these findings were themselves open to different interpretations: once the hypothesis of a post-mortem production of gas was set aside, the origin of the gases in the veins, which were doubtlessly the cause of the accidents, remained to be determined. For example, the subject was raised of sudden death due to pulmonary emphysema, either of a pathological nature or produced experimentally in animals: the case of Mlle Stock, who in addition had a heart with an extraordinary volume, did not really allow a conclusion which totally set aside the thesis of asphyxiation or intoxication by chloroform. The hypothesis of death due to heart failure which was put forward by Dr. Gorré and others, was also rejected on the grounds that the action of chloroform on the organism was poorly understood. Malgaigne proposed informing the minister that in some way could a toxic action of chloroform be blamed and that "without any application of chloroform" a certain number of sudden, unforeseen and unexplained deaths were regularly reported in medical practice, be it following surgical shock or even outside all pathological situations.[69] The analysis of the report shows that, all things considered, both on the basis of the small number of cases submitted for autopsy and as a function of the knowledge of the time concerning the mode of action of chloroform, the academics were not necessarily bound to endorse the hypothesis of a fierce cardiac depression although some of them such as Ducros had already thought of electrical stimulation for bringing a patient back to life,[70] nor were they bound to conclude that chloroform was harmful: they did not have to adopt this conclusion but they undoubtedly could have, had they wished. Instead they chose the interpretation which was the most favourable to the use of chloroform, the one which would reassure the public and acquit the product and the doctor; they were, one might say, betting on the future: "Before attributing to chloroform a death with which it may have had no relationship, would it not be better to look into whether the subject was not afflicted by some other well-known cause of death? Because if an adequate cause of death were thus demonstrated, the administration of chloroform could be no more than a simple coincidence; and if, after all, one judged that it may have contributed to the final result it would still not be necessary to accuse the agent itself; it would only be appropriate to establish a contraindication to its use."[71] This text is admirable for it defines not only a framework for medical thought which still holds today, but also under what conditions a medical innovation may be considered to be acceptable: included therein are two possible aspects of the medical choice when faced with innovation—one on a scientific and one on an ethical plane. Is the present argument based on rigorous and cautious reasoning where good logic must in effect conclude in favour of the undecidable, indeed of the negative, as demanded by the generally accepted rules for scientific truth? Or else is one dealing with a specious reasoning, in which literal respect for these rules quite simply leads to sophism? Secondly, there was the argument put forward by Malgaigne which took into account considerations of a probabilistic nature—"When after having experimented thousands of times, surgeons have no scruples of going back to it again, there is therein, dare it be said, a clear signal which must not be ignored."[72] His argument reduced fatal accidents to a practical problem of contraindication. The significance is obvious: the generalised use of a new practice which did away with pain for the majority of patients was more important than the death of a few individuals, even if everything had to be done to get round the unforeseen and comply with contraindications. The logic of the choice posed by Malgaigne clearly shows the terms of this equation in which, as a condition for progress in medicine, the majority prevails over the individual. Indeed one can find another good example of this in the first treatments for rabies which posed similar alternatives. The sacrifice of a few individual lives is unacceptable from the point of view of the individuals themselves as well as from an ethical point of view; however, it is accepted provided the intention is good and the outcome was impossible to predict. Without doubt, the strict respect for the right of each individual might lead to nothing being tested other than that which has an assured result; however, the same argument could be turned around, since the hypothetical survival of some individuals, who in any case might possibly have died from something else, would thus be assured at the price of the suffering of the majority.

The nature of these arguments may remind us in certain respects of another famous debate in the 18th century, namely one concerning inoculation. This debate set the academic and philosopher d'Alembert

against the majority of his colleagues within the enlightened ency-
clopedist movement: was it necessary to inoculate *en masse* against
smallpox, with the minimum but potentially fatal risks which this could
entail for those inoculated who might perhaps have never contracted it
or who, one could even imagine, would have caught smallpox naturally,
perhaps without any risk of dying? The alternative was to be against
inoculation in order to avoid causing the premature albeit unpredictable
deaths of some individuals, despite the fact that the majority of those
inoculated would be protected against this illness and above all against
the risk of its fatal consequences. A comparison of the argumental
framework of the two debates tells us a great deal about the evolution
in ways of thinking. As can be seen, the terms of the debates were
complex not only in that they set the right to life of an individual against
the advantages of the whole of society, but also in that they took into
account the probability of catching an illness and of dying from it.
There was at least one difference in the magnitude of the two debates:
it was that nobody can choose whether or not to catch smallpox,
whereas the patient (and the physician) can theoretically choose between an operation under anaesthetic and an operation entailing pain;
furthermore, the risks of death following an operation were great, much
more so than today, and the fact that operations were carried out under
anaesthesia did not on the whole seem to increase these risks perceptibly, except for the small number of fatal accidents already mentioned.
Thus one could interpret Malgaigne's attitude as being a legitimate
conviction that, all things being otherwise more or less equal, in other
words with a statistically negligible increase in the risk of fatalities, the
real choice was between suffering and not suffering. What is very
revealing when it comes to the way in which mentalities were evolving
is that as a result of their appreciating that an alternative existed, when
they weighed up the advantages against the inconveniences, patients
could come to view pain as being more important than life itself. To
our way of thinking, all the lessons have not yet been afforded by this
debate as formulated in the mid-19th century: in comparison with the
preceding century when, although pain was attacked in every possible
way, it was still considered to be the most overwhelming proof of the
persistence of vitality, it clearly became an issue which had to be weighed
up in the same light as life and death. We can find here one of the
principal meanings of the views held by Malgaigne, Velpeau and all
those who agreed with their outlook concerning anaesthesia. It also has

similarities with the crises or series of questions which have arisen in contemporary medicine, for example with respect to problems such as artificial life support, euthanasia, etc. These could well have come from the usually unexpressed co-existence of two different ethical models, which clearly cannot coincide: one model which values life above all else and another which renounces life at all cost, particularly if it must be endured with pain. The latter triumphed in the debates on anaesthesia.

A second lesson which the historian can gain from this episode in the history of medicine is that the choice which doctors in the years 1847–8 had to make was certainly not one between certainty and probability nor even, in a simplistic way, between the individual and the majority, but rather was—and we can well say that it still is—in its very essence, probabilistic. Even today, a number of debates on burning issues originate from this lack of recognition or simple acknowledgement that medicine is also an art of the probable, which in no way means that it is not "scientific" and even less that it is carried out without regard for the rational basis of causes and effects; quite the contrary, it can only become more successful and progress by employing this rationale, and it is because it has emphasised this basis that it has perhaps made us forget that it was simultaneously probabilistic: rationality, far from excluding the probable, has to take account of it. But managing the probable is a political act—in the full sense of the term—with legal consequences, which the academics of 1848 fully recognised and whose ethical demands were that the terms be put clearly. Malgaigne and the first users of ether and chloroform anaesthesia took the political decision to take a chance on surgery without pain and, by so doing, took upon themselves the risk of the odd fatal accident. It is only in retrospect that this gamble appeared to have been necessary and legitimate.

The attitudes of the powers that be prior to and after the 1848 revolution are no less instructive in providing an accurate picture of the relationships between science and politics in the modern era: because, as we have seen, the ministers for Justice and Public Inquiry ordered the Académie not so much to present the alternatives and the degrees of probability attached to each choice (all the reasons quoted by Malgaigne were part of the internal debate within the Académie), but to decree that chloroform was innocuous and allow it to be used. Thus governments, by turning towards scientists for security, delegated their political power and, by the very nature of the questions which they put to the scientists, gave them a role which they did not necessarily want to play. Malgaigne wrote: "To sum up, we will say that chloroform, by the speed and consistency of its effects, is an admirable agent of anaesthesia. Surgery has become less cruel and more powerful since it has been put to use." One must avoid becoming recklessly confident and for that reason, "it is indispensable to familiarise oneself with the use of this marvellous agent, and never to entrust it to careless or inexperienced hands." Thus relationships were woven between scholars and the government in which the probabilistic and political nature of choices, which was of interest to everybody and should have been presented as such to the citizens, had been diluted by the creation of a convenient division of their roles. Facts which were avoided in the debate and deemed impossible or inappropriate in the name of needing to "reassure the public" resurrected an attitude about the relationship between the governing and the governed where the right to know is not a concept necessarily shared by all. Thus, by seeing the debate on anaesthesia not simply as one between scholars and the authorities, but as part of a triad in which the third component, the citizens, had been eclipsed on behalf of "public opinion," one perhaps can reach a better understanding of some of Magendie's unease, and of the exact nature of the relationship between the revolution of '48 and this concomitant scientific debate. Of course, neither the declaration of an exclusive professional competence, nor claims for the rights of individuals, nor the democratic principle of an individual submitting to the will of the majority, nor the poor explanations concerning career strategies or schemes for power may be ignored. However, none of these facts are pertinent to the political situation of 1848, nor fully account for the re-evaluation of established views and firm stances adopted at that time. These included refusal by the authorities to take the uncertainties of scientific research into the public domain because of all the criteria by which it would be judged and, as a natural consequence, the will to use science as an instrument of public order and legitimisation.

The legal judgments handed down concerning accidents produced by inhaling chloroform shed light on the evolution of thinking as well as on the misgivings which people had about every minor operation. Thus the Breton affair, concerned with a patient who died on 15th February 1853, probably from ventricular fibrillation,[73] initially led to a condemnation of the two doctors, Triquet and Masson, who had carried out the procedure. In coming to their decision, the 6th Chambre du tribunal de police correctionnel of Paris upheld the accusation of homicide through carelessness, on the grounds that the locale where the operation was held had been badly ventilated and overheated, and that the doctor had failed to check the patient's respiratory passages which were constricted as a result of his state of anxiety and restlessness; the court added: "In view of the fact that chloroform is a dangerous and active agent, able to cause death directly, it must be used only with the greatest circumspection, which leads to the conclusion that one must use it only for the most serious operations, where the magnitude of the pain is such as to overcome the physical stamina of the patient, and in those cases where the immobility of the patient is an essential condition for the success of the operation; it has been established through Triquet's evidence that the operation which he intended to carry out on Breton was a very minor operation; consequently, it is wrong that the accused exposed the aforementioned Breton to death for an extirpation which posed neither danger nor pain of any acute nature; the use of chloroform, in the present circumstances, was an act of serious imprudence";[74] the two doctors were ordered to pay a fine of 50 francs. However, the affair went to appeal, with the Seine Doctors' Association calling for a special deliberation and finally the accused were discharged. This judgement set a precedent and was of major importance because of the subsequent events concerning medical responsibility: it shows the solidarity of the profession in defending a colleague by contesting the charge of carelessness in front of a court which seemed to take account of opinions which were less than proven; but even more, it contested the right of magistrates to legislate in which cases it was proper to anaesthetize, and acknowledged the possibility of the patient dying without it being the responsibility of the doctor.

For a long time, this episode among others defined the physician's legal rights and duties. By arriving at this decision, if only after an appeal, courts of justice of that period contributed to lifting the last barriers in the way of using anaesthesia. At the siege of Sebastopol in 1854, for example, chloroform was used by the health service of the French army. In the second half of the century, efforts were centred on the ways of administering anaesthetic agents and on the creation of apparatus such as the one made by John Snow[75] which permitted a better dosage than provided by the simple chloroform compress, and on mastering the risks incurred by patients to a greater degree, by the competence of the operators, in particular, and by using such procedures as electrical stimulation to revive patients who had fainted. In spite of this, a study published in Great Britain put the number of deaths attributed to anaesthesia in England and Wales between 1846 and 1946 at more than 25,000, with approximately a quarter of these occurring around 1880.[76] It is thus obvious that pain had to be totally unacceptable if an individual would choose to put his life at risk in order to be free of it.

All attempts at trying to evaluate the change in attitudes with regard to pain must take into account not only the medical discoveries themselves and the debates which they created, but also the way in which the new methods available were put into practice. That is another study, which entails undertaking methodical research into the quantities of medicines prepared or fabricated and consumed. In the absence—or near absence—of this type of breakdown in hospital archives, private dispensaries and industrial archives in a field which was still undeveloped at the overlap between medicine, pharmacy, chemistry, and industry, we can provide only information which must be interpreted with caution: the *Thérapeutique jugée par les chiffres*, published in 1876 by Drs. Charles Lasègue and Jules Regnauld, gave the amounts consumed annually in the hospitals of Paris and its conurbation, which were based on the expenditure registers of the central pharmacy of the Assistance publique. Though incomplete and uncertain, the figures at least give an indication of the evolution of practices and the perceptions which the doctors had of these: "Since the discovery of the properties of ether, for which chloroform was substituted, anaesthetic substances have become one of the dominant preoccupations of modern therapeutics. Chloral only made its appearance in the hospital services in Paris in 1869. Consumption at that time was 5 kg but rose to 360 kg in 1875."[77] A comparison of the quantities of ether and chloroform which were dispensed between the same dates is even more telling: the quantity of ether increased from 195 kg to 614 kg while the corresponding figures for chloroform were 141 kg and 308 kg.

This information, which came exclusively from hospitals in Paris, seemed to have been consistent with a more general movement towards the liberal use of sedatives and all sorts of remedies for pain: according to the same source, bromides were not really used for the treatment of the ill until 1855, and opium kept a dominant position (about 150 kg annually), "although [the use of] morphine in the hypodermic form was authorised more and more in that the quantity increased from 272 g to the amazing figure of 10,335 kg."[78] Moreover, these did not take account of the use of other substances, such as codeine, hemlock, belladonna, etc., against pain. Historians find themselves faced with converging information on the reactions of the medical world to pain, as well as on the expectations of patients: if one considers that the physiological explanation for the effects of these substances was still awaited, there was really very little reluctance, a lot of enthusiasm and finally great audacity at that time.

In fact, the massive and largely empirical application of anaesthetics had preceded by some twenty years the progress made in the elucidation of their modes of action: this was the task to which Claude Bernard's *Leçons sur les anesthésiques,* published in 1875, was dedicated. However, this work amounted to no more than a start. It was not enough to know that anaesthetics acted on the nervous system, one also had to be able to say on precisely which organic elements they exerted their effects and produce evidence for the route taken by the agent via the circulation to reach nerve centres in a selective way. In order to demonstrate this, Claude Bernard came up with the idea of semi-submerging two frogs in chloroform, one with its head and top part of its body submerged, and the other with its lower quarters submerged: after a few minutes, the two frogs were completely anaesthetized, which showed that the circulation distributed the anaesthetic throughout the body. However, if the circulation was intercepted by a ligature at around mid-body level and one repeated the aforementioned experiment, the results were entirely different: the frog whose head and upper part were immersed in chloroform was completely anaesthetized; on the other hand, the frog who had only its hind quarters immersed in chloroform was not anaesthetized, even within the distribution of the nerve trunks of the immersed part. This difference in results could be explained, in the absence of all other paths of communication between the front and the back, only by the transmission of anaesthesia from the upper to the lower quarters by neural pathways (brain and spinal cord); this double set of experiments, which conformed with standard experimental methods, "proves that the influence of the anaesthetic administered to the peripheral extremities of the cutaneous nerves, cannot travel along sensory nerves, and that, to influence these nerves, it must necessarily reach them by their central extremities. To summarise, the action of chloroform and ether is on the central nervous system, and anaesthesia of nerve centres suppresses the sensibility of sensory nerves whose origins have been stimulated, whereas the inverse is not produced, the action of the chloroform on the peripheral extremity or the trunk of the nerves being powerless to produce a generalised anaesthesia."[79] Using the technique of a cranial window,[80] he equally showed that anaesthesia was not explained by congestion of blood in the brain (hyperaemia): hyperaemia was a passing state, linked to the animal being agitated when the anaesthetic was administered—a state which would subsequently disappear and which could be seen even without any anaesthetics. Thus it was on nerve cells that anaesthetics worked, following a well-defined order, firstly on the brain which brings about the loss of consciousness and voluntary movements, then on the spinal cord (with a loss of reflex movements). The desensitisation of the peripheral nervous system began at the extremities of the sensitive nerves but could take place only if the anaesthetic reached the central origins of the nerves—their points of attachment to the spinal cord. It was perhaps because of this belief that Claude Bernard thought "there is no local anaesthesia,"[81] since the action on the skin or superficial mucosae could never travel along the nerve pathways to produce general anaesthesia, and consequently must have been working by a very different mechanism. The succession of psycho-physiological events recorded while the anaesthetic was acting led Claude Bernard to wonder if different categories of nerves existed and thereby touched on the problem of specificity. He also speculated on the possible use of an anaesthetic agent as a "chemical scalpel": there would be "firstly the nerves of the special senses, then those for the less well localised external sensations (touch, pain), then those for unconscious reflex actions, then finally those for completely automatic reflex actions. . . . This classification of centripetal nerves according to the extents of the action of chloroform is worthy of attracting our attention, because, in recent experiments, we have determined an analogous classification of motor nerves through the action of curare."[82] However this problem is not the most interesting subject in the *Leçons;* that status is accorded to his study of the effects of mixed anaesthesia.

Chance, or perhaps the reading of an observation made by Nusbaüm, a German surgeon from Munich who, due to the length of an operation, was obliged to substitute morphine for chloroform,[83] played a role in the discovery of the combined effects of chloroform and morphine—one potentiating the effects of the other. Claude Bernard gave a subcutaneous injection of morphine to a dog who was beginning to regain consciousness after chloroform anaesthesia: the animal went back to sleep, which was the normal effect of the narcotisation, "but it was certainly unusual for the insensibility due to chloroform to manifest itself again after having disappeared, since the return of anaesthesia could not be explained by another dose of chloroform since none had been given."[84] Furthermore, it was not clear why the insensibility lasted for many hours. The order of administration of the substances was relevant. If, instead of giving chloroform first and then morphine, the reverse was done, one could successfully suppress or arouse the sensibility of the animal with very small doses of chloroform, depending on whether or not the animal was made to inhale it. Interest was huge within the field of experimental physiology (as the techniques immobilised experimental animals and allowed the unrestricted study of the phenomena of sensibility) and for surgical practice: "It was to be supposed that this combination of the effects of chloroform superimposed on those which can be brought to the service of surgery by morphine, above all when it is used as we have been led to do, firstly by giving chloroform either as a subcutaneous injection or otherwise, then administering the chloroform, which therefore acts in much smaller amounts. Thus one obtains anaesthesia without having to go through such an intensely agitated period, and above all, without running as many risks of the accidents which can be produced by high and repeated doses of chloroform."[85] In the years which followed, their use was suggested for painful childbirths and minor operations as well as in the treatment of hepatic and nephritic colics, where they were particularly useful. The method brought about a state of analgesia, without loss of consciousness or risk and when surgeons wanted to obtain complete anaesthesia (with relaxation of the muscles), the doses used were lower and the action was quicker than those given for simple anaesthesia. This procedure of premedication using morphine, which was administered about 40 minutes prior to the anaesthetic proper, is still used today.

The *Leçons sur les anesthésiques* represent a synthesis of the principal understanding of the mode of action of these substances, whose analgesic and anaesthetic effects must be carefully distinguished. Amongst the most fashionable products of the 1870s, chloral in its hydrated form, which has such a characteristic smell of melon, was considered to be an anaesthetic of the same type as chloroform. Taking into account the chemical composition of this substance led to conjecturing that it broke down within the organism itself into chloroform and formic acid. Claude Bernard showed that chloral was not an anaesthetic but an hypnotic whose effects were similar to those of morphine: animals fell deeply asleep but retained "a sensibility which was obtuse and slow to react."[86] The massive use of this hypnotic in no way helped to solve the problem of knowing what psychological and physiological traces were left of a pain which was endured without being conscious of it and without retaining any memory of it on waking. There were several possible explanations for this infatuation with chloral which lasted a considerable time; it was used in operations and in the treatment of tetanic crises of different origins: administration via the subcutaneous route or, even better, intravenously allowed a better control of the quantities absorbed than did inhalation, and these other routes avoided the inconvenience of irritating the respiratory mucosa as well as the first phase of agitation; besides, the choice of chloral was also favoured by the realisation that it had antiseptic properties ("anti-fermentescible and anti-putrid"). However, Claude Bernard was on his guard against the risk of cardiac accidents and insisted on the need to use pure substances at all times, to appreciate the doses not as a function of the simple ratio of the quantity of the substance to the weight of the animal or man, but by taking into account its concentration in the blood stream and by splitting up the injection(s). He also questioned severely the dogma of there being similarities in the effects of substances belonging to the same (chemical or natural) "family." Although he did not succeed in explaining the physicochemical reactions which occurred at the level of the nerve centres, he did achieve a better understanding of the physiological and therapeutic effects of each substance by linking experimental physiology and pharmacology together.

### Local Anaesthesia: The First Steps

Another consequence of the fatal accidents due to chloroform was the study of alternative means of suppressing pain. It would be a mistake to believe that local anaesthesia dates only from the time Carl Koller used cocaine in ophthalmology in 1884. The sedating and desensitising applications of opiates, and particularly of morphine, either on the epidermis of painful areas, or directly on the afflicted nerve were known to reduce suffering. In 1854, Arnott in England[87] and Velpeau in France used a refrigerant mixture (4 parts salt, 1 part ice) which they applied to the part in which they were going to make an incision or which they were about to amputate; but the effect of cold did not last very long. In the absence of a precise theory on the action of anaesthetics on the nervous system, doctors and surgeons proceeded using multiple, tentative trials. While the Russian, Nicolai Ivanovich Pirogoff (1810–1881) tried to introduce ether by the rectal route with a view to bringing about general anaesthesia,[88] hospital surgeons tested the effects of the new volatile substances by spraying them onto the painful parts: in Dublin, Hardy carried out trials which were hardly conclusive using chloroform (for therapeutic ends, especially in cases of cancer of the uterus, and not directly for surgical purposes); meanwhile in France, at the Hôtel-Dieu, Guérard applied ether drop by drop onto the part of the body which he was going to incise and with the help of Richardson's apparatus, directed a stream of air onto it which chilled the area considerably in three minutes: without producing insensibility, this technique greatly reduced the pain.[89]

The development of local, then loco-regional anaesthesia was linked both to the isolation of new substances and to the establishing of new techniques for penetration into the organism. The Spanish conquest of the New World had brought to Europe, via the account of doctors, Jesuit priests and travellers, knowledge of the exceptional virtues of the coca leaves which the Indians chewed all day long. However, their possible effects on Europeans were in doubt. The article published by Freud in 1884, *De la coca*, provides one of the best historical accounts of this problem.[90] The year 1859 was, in every respect, a memorable year: Paolo Mantegazza, who had lived in South America for many years, published his research on the hygienic and medical uses of coca in his *Sulle virtu igieniche e medicinali della coca*,[91] in addition, the return of the Austrian frigate *Novara* from an expedition that same year brought some coca leaves to Vienna from which one of Wohler's pupils, Niemann, isolated an alkaloid, cocaine.[92] By so doing, he was carrying on the work started by the German Gaedicke who, as early as 1855, had extracted an essence *(erythroxylon coca)* from coca leaves. At first, the substance was recognised for its stimulant effects (increasing the resistance to tiredness and malnutrition, suppressing the appetite), as well as for producing a feeling of well-being and of having keen intellectual faculties in all those who—following the frequent practice of auto-experimentation—had tried it themselves.

It was within this perspective of therapeutic stimulation, and not in connection with anaesthesia, that the first experimental work on man and animals was carried out by Al. Bennett in Great Britain and von Anrep, in particular.[93] In the 1880s, American journals, notably the *Detroit Therapeutic Gazette*, reported cases where detoxification from morphine and alcoholism could be achieved using cocaine without bringing about any new addiction. Indeed Freud himself wrote that one could even take doses of it over a moderately long period without producing any particular effect. It was in this context of research into a stimulant for dealing with depressive states and for weaning addicts off morphine, that Freud used it—notably as a psychotonic. However, following on from other authors, Freud in 1884 described a number of therapeutic indications: digestive problems which cocaine eased by stimulating intestinal movements and by lessening sensibility as mentioned by Mantegazza, states of cachexia, treatment of asthma and for local applications: "This property of cocaine and its salts to anaesthetize the skin and mucosa in the areas where the product is applied as a concentrated solution tempts [us] to resort occasionally to this means, above all in cases of affliction of the mucous membranes,"[94] particularly those of the larynx. This indication agreed with remarks made in 1862 by Schroff on the desensitisation of the tongue by the action of coca. As in the case of general anaesthesia, a good twenty years or so were needed from the time when cocaine was isolated until it was used as a local anaesthetic in surgery: a period of experiments on the physiological action of the alkaloid without preconceived ideas, the finalisation phase carried out by the firm Merck at Darmstadt and the particularly high cost of the product are among the many reasons which might explain this latency period; however, the main reason could well lie elsewhere. It was the convergence between a general interest in cocaine stimulated by Freud's publication in 1884 and the preoccupations of one of his younger colleagues, Carl Koller, who was working in ophthalmology for Stricker and was looking for ways of immobilising the eye in order to carry out surgical operations more easily. The local action of the cocaine was tested by Koller firstly on a frog's eye and then on himself, and was presented on 15th September 1884 by Bretteuer on his behalf, at a meeting of ophthalmologists in Heidelberg. It was subsequently reported in October of the same year by Koller himself and his friend Königstein before the Medical Society of Vienna.[95] Surgeons were far more interested in the suppression of the reflex movements of the cornea due to its desensitisation (the patient was no longer aware of the contact of the hand or surgical instruments) than in the suppression of pain which, in any case, could be achieved under general anaesthesia. The news spread throughout the whole world in less than a month, but in the end, its use specifically against pain was more the work of the American surgeon Halsted than of anyone else. Halsted,[96] a surgeon in New York, was best known for having introduced gloves into the operating theatre and for having preached bloodless surgery. He was interested in the problem of the struggle against pain and gave hypodermic injections of hot water to produce anaesthesia locally. From December 1884 onwards, he tried to anaesthetize the cubital nerve and then the orbital nerve of his assistant R. J. Hall; finally, he experimented on himself. This generalised practice of local anaesthesia against pain was particularly useful when having to operate on young children or on adults who constituted serious cardiac risks; it also taught him, at his own expense, that cocaine was not as innocuous as Freud had believed, and that it caused addiction phenomena. However, local and then regional anaesthesia opened up a new era in the struggle against pain which was marked at the beginning of the 20th century by the establishment of new derivatives of cocaine, namely procaine in 1904, developed by Fourneau, and novocaine the following year.

But before reviewing this golden age of local anaesthesia it must be remembered that, to some extent, it preceded general anaesthesia; however, it was only when, quite aside from problems linked to the costs of the new substances, new techniques made it possible to administer hypodermic and intravenous injections for therapeutic purposes that local anaesthesia was really able to come into everyday use. Although these injection techniques had an evident use in general anaesthesia, this could also be achieved by the respiratory route and indeed was done in this way right up until the middle of the following century. By contrast, the injections constituted a preliminary which was absolutely indispensable for local anaesthesia if one wanted to go beyond a simple, superficial action on the skin. Amongst the technical developments, it must be remembered that syringes and needles were improved (starting from a "clyster" or enema syringe, which had to be made smaller and easier to handle, then needles made from silver or gold which became finer and finer and, above all, hollow): John Wood in Great Britain,[97] then Pravaz in France, who had it in mind to find a cure for aneurysms,[98] greatly improved the materials used for injections. Numerous problems had to be overcome regarding asepsis which prompted misgivings even amongst those, such as Malgaigne, who had been the most ardent defenders of general anaesthesia. However, subcutaneous injections, which allowed the deep penetration into cellular tissue, then intravenous injections were used in the 1880s for problems outside those of anaesthesia, as a means of acting directly on painful regions, for example on the stumps of amputees where neuroma occurred, and for certain neuralgias.

The impetus given by the work of the physicians, surgeons and chemists who contributed to the discovery and the use of anaesthesia deeply changed the relationship of men to pain by putting an end to its inevitable nature. But this also had important consequences on the actual comprehension of the transmission mechanisms of the sensation and thus stimulated research into finding an explanation for pain at a theoretical level.

## Experimental Physiology and the Explanation of Pain: The Search for Specificity

Gall's and Spurzheim's research on the specialisation of functions in the different areas of the brain was only a qualified success[99] due as much to the materialist implications of this work which began to assign a place in the brain to intellectual and moral functions, as to their excessive suggestions of relationships between cerebral localisation, the shape of the cranium and cerebral convolutions and an individual's character or behaviour.[100] Nevertheless, decisive work did take place on the specialisation of the functions of the anterior and posterior roots of the spinal nerves. Before the work of Charles Bell and that of François Magendie, well-established pathological facts such as the maintenance of sensibility in an area in which movements were suppressed or vice versa could not be explained. Nor could the existence of separate sensory and motor nerves be confirmed although such an hypothesis had already been formulated by Galen, or whether a single nerve could carry out different functions. The solution to this problem led to theories on the specialisation of fibres and nerve centres, which were to play a decisive role in the interpretation of pain.

### Sensibility and Movement: The Specialisation of Functions

The story of the discovery of the motor function of the spinal nerves' anterior roots, and the sensory function of their posterior roots had been told numerous times, notably by Claude Bernard. He was able to settle the quarrel about whether Charles Bell or François Magendie was first to make this discovery, a dispute which had been reactivated by Flourens at the Académie des sciences in 1847.[101] In 1811, Charles Bell published a brochure for the use of his friends called *An Idea of a New Anatomy of the Brain* in which he took up Willis's ideas on the different functions of the cerebrum and the cerebellum and established the specialisation of the functions of the roots of nerves on the basis of their origin: the anterior root coming from the cerebrum was naturally in charge of assuring movement and sensibility, while the posterior root coming from the cerebellum accomplished functions of nutrition. The only experiment which might lend credibility to the latter "discovery" involved the convulsive reactions seen in an animal after its anterior root had been prodded: "On laying bare the roots of the spinal nerves, I found that I could cut across the posterior fasciculus of nerves which took its origin from the posterior portion of the spinal marrow, without convulsing the muscles of the back; but that, on touching the anterior fasciculus with the point of the knife, the muscles of the back were immediately convulsed."[102] Thus, it is clear that Charles Bell did have the first idea for an experimental procedure which would allow sensibility and movement to be dissociated in the roots of spinal nerves. It is also clear that he noted that the anterior roots had a motor function. However, he could not go any further for the reason that he carried out experiments on the convulsive reactions after stimulation of recently killed animals and thus had no way of testing their sensibility. In fact, Charles Bell proceeded more by anatomical deduction than by experimentation on living animals, which he found very repugnant; he came to conclusions on the basis of the regular or irregular arrangement of nerves as well as of their origins which could be simple or double. It was still within such an interpretational framework that he published a paper in Magendie's *Journal de physiologie expérimentale*.[103] In the second volume of his *Journal,* Magendie presented two studies which provided experimental demonstrations of the sensory function of the posterior roots: experimenting on puppies by surgically sectioning the posterior roots on one side of the lumbar and sacral cord in such a way as to preserve the paired root on the other side as a reference point, and having taken care to leave the cord intact, Magendie concluded: "I observed the animal: firstly I thought that the limb corresponding to the cut nerves was entirely paralysed; it was insensible to pinpricks

## ⊶ THE CHURCH AND PAIN IN FRANCE ⊷
### CONTINUITY

Debates about anaesthesia and, more generally, the increased number of pain-relieving methods available during the 19th century posed the question of the Church's position in relation to pain. This complex problem within an institution that was far from being homogeneous may be examined on at least three different levels: its theological foundations which extended beyond the framework of France, the existing French tradition of interpreting the Scriptures and finally the behaviour of its priests, bishops, and nuns at patients' bedsides and while tending the suffering. From a doctrinal point of view, the Catholic religion maintained a consubstantial relationship to pain rooted in the Passion and death of Christ borne to save mankind. This sacrifice, the fundamental essence of Catholic dogma, undoubtedly only had a meaning in relation to the Resurrection, but whether emphasis was placed on the sacrifice or on the belief in life after death, this cardinal event certainly altered the Christian relationship to pain.

The belief in the redeeming virtues of pain, the idea that suffering individuals were closer to Christ, that their anguish could be offered up in penitence for earthly sins, or even that God put only his elected few through terrible trials were all recurrent themes throughout Church literature. Mgr. Parisis, Bishop of Arras and then Langres—who was quite a liberal clergyman on the whole where Church and State relations were concerned and also with regard to education—stated in his text on pain, *Sur la douleur*, published in 1864: "Yes, moreover we know and believe that your elected officials view their crosses because they have been deemed worthy of emulating the divine model; and they become perfected because virtue may only be grasped through ordeals, and crosses serve as crucibles to purify them" (p. 19). The tradition which emerged from St. Paul's writings placed considerable emphasis on this vision of communion with God through suffering, as this passage quoted in the very recent *Lettre apostolique Salvifici Doloris* published in 1984 illustrates: "I complete through the trials of my own flesh what was lacking in Christ's ordeal for his Body which is the Church." This attitude, which considered pain as a means of discerning one's own faults and of breaking away from worldly preoccupations, closely linked pain to original sin. According to Ecclesiastes, the increase in pain was also an increase in "science." In the Catholic *Dictionnaire de théologie* by Vacant, Mangenot, and Amann, published between 1935 and 1972, there is no entry for "Pain" nor for "Suffering," and the problem of physical pain was included either under the entry for the French word "Mal" (evil, adversity), or the entry "Providence." Indeed, it appears that pain which afflicted the just was seen as a virtual denial of Providence, and the figure of Job—symbol of the trials endured by the Just—whose lamentations reflect humanity's rebellion against every adversity, had a deep influence on Western culture.

Though the views of certain Catholic theologians and intellectuals seem to have evolved considerably in this respect, as is evident in the efforts undertaken in the *Concilium* (1976) journal, or the recent work by François Varone, *A God reputed to relish suffering,* which radically denies any value in suffering in the eyes of God, the presence of evil or adversity in the world and especially of physical pain, particularly when it strikes children, has been the subject of innumerable attempts at justification. The Church has always had a dual system of explanation when it comes to physical pain: on one hand, there has been the Augustinian tradition which holds that "No one is unhappy unless he deserves to be" and interprets pain as a punishment for the wicked and a foretaste of the final retribution; and the other view which considers pain as a means of moral progression and of salvation. This dual position is evident once again in the Catholic *Dictionnaire apologétique de la foi* presenting the faith's justifications and its arguments countering the objections made from the social science viewpoint. A number of editions of this work have been published and it has been thoroughly revised by A. D'Alès, professor at the Institut catholique de Paris: "Illness and death, which prepare and consummate each individual's elimination in a living species, are certainly ills where individuals are concerned. In terms of the species, however, they are the fundamental principle of its conservation by ensuring the renewal of successive generations. . . . In this respect, it is possible to begin to understand how pain—the only physical affliction whose existence poses a real problem—has a place in the creation of an infinitely righteous God. . . . Pain, which goes against the inclinations of sensitive nature, cooperates unfailingly in the maintenance and re-establishment of the universal order because it reinforces the Divine Commandments and serves to exercise virtue" ("Providence" article, Book 4, 1923, p. 436). These texts, which again repeat Parisis' theme that "We are destined to endure pain," resound like the direct echo of an opinion constantly reiterated throughout the centuries.

In the course of the 19th century, through a type of paradoxical apology, pain even became a weapon aimed at the irreligious who thus would have been unable to comprehend its deeper meanings according to the Church. The Church's ambiguous position, as Roy Porter underlines, closely reflects Western culture's general stance in this respect ("Douleur et histoire en Occident," *La Douleur,* under the direction of Geneviève Lévy, Éd. des Archives contemporaines, Paris, 1992, pp. 95–116), and it is precisely due to its professed outlook on the pain question; this frequently amounts to tenuous explanations for pain or efforts to confer a meaning on it, and prompts various practical attitudes to deal with it such as acceptation, resignation, or even passivity when faced with an experience which is ultimately at the very core of human existence; however, at least in France, one only rarely comes across a definite position with respect to the use of anaesthetics or analgesics. The real issue beyond all the various homilies, pastoral sermons, encyclical letters, or Lenten messages therefore is to uncover how the passage from one approach to the other actually came about. For this, it would be necessary to question the patients who experienced the pain in order to have eye-witness accounts not only of what they personally felt but what the clergy said or did at their bedside beyond the expression of a strictly theological view of pain.

In the second half of the 19th century the attitude to pain became thoroughly tied up with a number of other social issues, including the organisation of the working classes and education, for instance. The resignation to pain was then part of a more general political and social outlook; such a juxtaposition between political submission to the established social order and resignation to pain is clearly evident in the *Dictionnaire des sciences philosophiques* by Franck: "Every sensitive being is condemned to pay a heavy tribute to pain as every living being suffers and wails. . . . We will show that ills and adversity stem from the very nature of mankind and the world in which he proceeds, and not because of accidental causes which are dependent upon his will. By this means, we wish to demonstrate clearly that the prospective happiness proffered to us as the result of social reforms imagined by certain entities will forever be pursued by man though all his efforts will be unavailing; it is therefore senseless to make of it life's sole or major goal, or the real fundamental basis of any moral or political science" ("Mal" entry, Book 11, p. 1017). This association is also a reminder of the perpetual shift back and forth between spiritual and temporal matters which the metaphorical language often used by the Church tended to reinforce, particularly when addressing a relatively uninformed public.

A certain number of Catholic physicians have carried out in-depth studies on the relationship between medicine and religion. Where the 19th century is concerned, it seems from several accounts consulted, including *L'Ami du clergé, La Revue catholique de Louvain, Le Médecin chrétien,* by Mgr. Ange Scotti, written in 1821 and re-edited in 1881, or again the classic work by Father J. C. Debreyne, a Trappist monk and physician, *Essai sur la théologie morale considérée dans ses rapports avec la physiologie et la médecine,* which has been re-edited innumerable times, that the physician's attitude to pain was not a major preoccupation. It appears that in France those who wrote about the relationship between medicine and religion—as was the case a little later with Dr. Henri Bon in his *Précis de médecine catholique* (1923)—were much more intent on examining such questions as abortion and the more traditional issue of the life of the foetus, the last sacraments for the dying or the baptism of newborns, rather than pain. The Church was preoccupied with condemning onanism, as is obvious in the long diatribes written by Father Debreyne, as well as with assailing magnetism on the basis of disloyal competition with its apology on miraculous cures and the risks to the wills of magnetised subjects; above all, it was concerned with questions about "sacred embryology" and reiterating its condemnation of abortion and establishing the proper attitude when confronted with a pregnant woman who had just died (to determine whether baptising the child in its mother's womb had any validity whatsoever). Along with its involvement in educational quarrels and the achievement of Italian unity, it was intent on defending its temporal powers and in maintaining a certain order within society; the question of pain was apparently viewed in the light of the latter preoccupation as is confirmed by the analysis Michel Lagrée undertook for Brittany in "Le langage de l'ordre: la souffrance dans le discours d'un évêque français au XIX$^e$ siècle" (*Concilium*, 1976, pp. 27–35).

This fundamental concern with ensuring that birth and death occurred within the bosom of the Church as the requisite condition for salvation underlies its most obvious attitudes. Thus, when tending to the dying, the behaviour expected of the physician was one of maintaining the patient's lucidity until his confession could be heard and until extreme unction was administered, or until his "final dispositions" were made; once these rites were accomplished, however, he had no reason to try to extend a life which was subsiding in pain. Along these lines, if pain jeopardised the patient's lucidity or his reason, if it threatened to provoke his suicide, then these were added reasons for the physician to provide him with relief. The question of anaesthetics and analgesics appears to have been part of a broader view concerning new medications and the ethical rules to be observed in circumstances warranting their use. Mgr. Scotti wondered if remedies whose effects were dubious should be administered, for instance? The answer was of a general nature because, according to a solidly anchored belief, if God dispensed ills, he also provided the appropriate antidotes and the means to relieve such disorders. The formula recommended by the Latin physician Celsus: "It is better to try a dubious remedy than to give nothing at all," was reintroduced with a few added restrictions: "A genuine necessity must exist for it to be employed. In other words, it may only be used if the art of medicine provides no surer means of curing the ailment and if the doubtful remedy can also be viewed as being reasonable. In addition, it would be true to say that it would be a sin to employ it if another remedy existed free of any hazards to the life of the patient. Among the multitude of doubtful remedies which exist, simple reason dictates that only the least questionable be used, the ones whose repercussions are the least serious in case of failure and which allow for a maximum number of precautions to be taken against unfortunate consequences. The physician's first duty is to avoid harming his patient, while the second is to relieve him. In short, he should only choose a medication if, according to his calculations, its utility is probable" (*Le Médecin chrétien ou Médecine et religion*, pp. 392–3).

These comments explain quite clearly the twist taken by certain discussions on anaesthetics in the Académies and the physician's relative freedom of action and arbitration as a result of the notion that he could carefully assess and calculate the remedial pros and cons; it was imperative to demonstrate at all costs that anaesthesia presented no greater added risk than any ordinary operation, that it could in fact increase the patient's chances of survival, and that it in no way harmed the child about to be born. It was the question of the supremacy of life which became essential in France much more than any need to respect the Bible's words "In sorrow thou shalt bring forth children." There has been a tradition of critical analysis of religious texts in France since at least the 17th century which was intensified by philosophers during the Age of Enlightenment, and which resulted in a certain detachment from more literal interpretations of the Scriptures. This position undoubtedly differs from the situation in Protestant countries which were more focused on a return to the Scriptures and the scrupulous observance of the Word as the only authentic revelation. Such a hypothesis on the contestable differences of approach between France and Great Britain might in addition be confirmed by their respective attitudes about that other "curse," *i.e.* work. The nobility in Catholic countries not only had never considered work as an obligation but, quite to the contrary, saw in work a decline in status; in Protestant circles, however, the literal interpretation of sacred texts was one reason which led to a staunch work ethic. Is it possible, therefore, as a result of this divergent context to conclude that a painless labour—which, as we saw, presented no real difficulties medically speaking, nor was it the object of any specific theological stance in France—became generalised without coming up against obstacles? In reality, the situation was considerably more complex and it may have been less due to the Church's official position in this respect than to the fundamental attitudes of its representatives and perhaps even more to the oppressed status of women in society, to their absence of legal autonomy or, in other words, their total dependence on political factors. Through a type of role sharing, the theological view about pain and the attitude of resignation to pain were appended to the social status attributed to women.

The weight of the Church's view on this problem was, as Jacques Léonard has shown so clearly, also partly due to and reinforced by the "nursing nuns" who, especially in rural areas, provided emergency first aid and served as intermediaries between the local populations and doctor, a role as go-between which was not without its conflicts. Wasn't compassion for the suffering, which had transformed charity into a duty, the driving force behind the establishment of hospitals in the Christian world along with poorhouses and other charitable institutions? These took on a specifically religious function, however, as well as a role of assistance, the two becoming inseparably linked as the first might not have been so easily exercised otherwise. The Gospel according to Matthew (8, 17) recalls that Jesus Christ "has assumed our infirmities and taken charge of our ailments," and the existence of a sacrament for the sick imposed a particular duty of giving special attention to those who were suffering and those who were dying. This obligation on the part of the Church gave rise to the omnipresence of religious symbols through hospital wards and to attitudes aimed at influencing the beliefs of the patients. Moreover, it was generally the nuns who held the keys of the hospital dispensary and who were in charge of superintending the wards as they usually had the most experience and sometimes a modicum of education. In addition to what tended to be a naturally parsimonious inclination in dispensing the pharmaceuticals, there was another practice described by Jacques Léonard through the example of Brittany which consisted in misappropriating remedies from the dispensary to either hand them out or sometimes sell them in the course of rounds to help indigent sufferers as well as to proselytise. We know of such practices through hospice records and because of all the research carried out by Claude Langlois for his *Le Catholicisme au féminin* (Éd. du Cerf, 1984) to show the vitality of the French religious communities run by mother superiors during the 19th century, their innumerable social roles and the positive image they generally enjoyed.

We know of them also because of Dr. Bourneville's fight during the Third Republic to obtain the secularisation of hospitals and the establishment of schools to train lay nurses. Bourneville's inflamed lectures in favour of laical hospitals sought to denounce the overbearing presence of nuns within hospitals, their desire for control (managing stores, supervising staff, transforming part of hospital facilities into convents or chapels), and particularly the conflicts between the doctors and sisters when the latter attempted to substitute themselves for the physicians or even disobeyed their prescriptions. By creating secular nursing schools, Bourneville sought to ensure that the sisters' replacements would possess fully-recognised credentials and competence superior to that of the nuns who often had no training whatsoever. In order to provide an irrefutable account of this determined religious presence in hospitals—there was even a time towards the close of the 19th century when, in state-run hospitals in Paris, the patient's chart at the foot of the bed included not only his temperature variations and diagnosis, but his religious beliefs as well—and the conflicts it provoked, we might mention the commentaries made by Dr. Armand Desprès in his book *Les Sœurs hospitalières. Lettres et Discours sur la laïcisation des hôpitaux,* published at the height of the clash in 1886. This testimony is not questionable as it was made by an adversary of secularisation, a defender of the usefulness of nursing sisters, and the leader of the hospital physicians against Bourneville's projects who saw in hospital nuns an ever present and available staff who could be directed to perform any menial chore. His words are eloquent: "Yes, the sisters speak of religion to the patients; it is the very nature of nuns of any denomination to try and convert others to their practices. Some patients accept this usage and there is nothing to add where they are concerned. But for those who are against it, they should not have to endure it and must be protected from such proselytising. . . . The religious nuns were, in fact, not the only ones to blame as institutional almoners [lodged in the hospitals] often berated them for not being zealous enough" (Desprès, *op. cit.,* 1886, p. 219).

In order to clearly determine the Church's position with regard to pain during the 19th century in France, a few glimpses of which we have provided above, it seems necessary to take into account a number of distinctive roles and different outlooks. These existed on several levels, and reflected the various considerations of the closely-linked network of multiple voices making up society, including those of the theologians, the Catholic physicians, the priests, the hospital nuns and visiting sisters. That is undoubtedly why, in reaction, the image of the independent free-thinking physician became so thoroughly established as well.

and the strongest pressure, it also seemed immobile to me; but soon, to my great surprise, I saw it move in a very obvious way, although sensibility there had been extinguished for good."[104] He repeated the operation but this time proceeded with sectioning the anterior roots and observed that the limb was immobile and slack, but that sensibility was intact. In the second study, "Experience sur les fonctions des racines des nerfs qui naissent de la moelle épinière" ("Experiment on the functions of nerve roots originating from the spinal cord"), Magendie was operating with nux vomica, a poison which has the effect of bringing on violent tetanic convulsions: the results (that the convulsions continued after the sectioning of the posterior roots) agreed perfectly with those obtained in the previous experiment. Magendie was always trying to obtain direct proof rather than making mere deductions, concerning the specificity of function of the posterior roots: pain such as that shown by the animal when its posterior roots were pinched, jarred or pricked was proof that sensibility was devolved to them. Pain in animals was the base for physiological experiments and for a long time remained the tool for obtaining knowledge of the physiology of the nervous system. The procedure aroused a great deal of hostility in Great Britain[105] as well as in France. However, Magendie's experimental honesty led him to moderate his results in his second study: "These facts therefore confirm those which I gave; however, they seem to establish that feeling is no more exclusively found in the posterior roots, than is movement in the anterior roots."[106] These reservations, which were used by François-Achille Longet and Pierre Flourens against Magendie, could be explained only by the discovery of recurrent sensibility, which Magendie described for the first time at the Académie des sciences in 1839, and again in 1847:[107] in a short study, he accounted for the apparent contradictions of which he was accused by pointing out the return of some posterior root fibres through the anterior root. Magendie emphasised that this anatomical fact was not the result of a coincidence or an illusion and provided weak sensibility from the anterior roots. It could also explain the way in which Bell had interpreted his own experiments. In order to discover the cause of this recurrent sensibility, Magendie thought of cutting the ventral (anterior) root and successively irritating the two ends: if the irritation from the central cut end of the anterior root (*i.e.* the one still attached to the spinal cord) brought about pain, it would be proof that the sensibility of this root came from the cord; if the irritation from the peripheral cut end brought about only pain, this would be proof that the slight sensibility of the anterior root came from the posterior root. It was this second hypothesis which was correct and which confirmed that there were no exceptions to the previously established law of specialisation. This discovery of recurrent sensibility was equally disputed by his pupil François-Achille Longet, who attended classes at the Collège de France and who claimed to have thought up the experiment.[108] In the various controversies which set Magendie against his contemporaries, Claude Bernard sided with his teacher, while Flourens regularly sided with opponents. The political and religious convictions of the protagonists were not absent in these disputes.

### Johannes Müller and the Specific Nerve Energies

On the other side of the Rhine, the study of sensory mechanism benefitted from the combined results of microscopic observation and experimental physiology. While Remak's observations provided proof of the existence of the two sorts of fibres, grey or nonmyelinated fibres from the great sympathetic system and white fibres,[109] Johannes Müller in his *Manuel de physiologie,* which was reprinted several times and in many languages, provided a synthesis of contemporary knowledge which led him to conclude that specific energies existed in nerves. A question which frequently came up in his *Manuel* was, all things considered, the self-same one which had been formulated almost a century earlier by Charles Bonnet in his *Essai analytique sur les facultés de l'âme:* namely how did one explain the clarity of sensations and how could the brain distinguish one from the other under normal (healthy) circumstances? In order to push this reasoning as far as possible, just as Bonnet's detractors had done, it was necessary to conceive not only particular fibres for each of the five senses, but to suppose that there was a particular fibre for each quality or particular nuance of each sensation, *ad infinitum*—a fibre for the scent of a rose, one for the scent of a carnation, etc. Such a concept would be possible in a context where each fibre was preformed for a predestined use, as Bonnet's work claimed. However, this hypothesis came up against all sorts of difficulties when attempts were made to give it an anatomical basis.

The questions took root in Müller's work through the strong conviction that paths of nerve fibres were rigorously ordered: from an anatomical point of view, he maintained that "the constitution of the nervous system can, in the final analysis, be reduced to a single problem: from where do the primitive fibres which are contained in a bundle arise and where are the endings of these fibres to be found? [As these] are independent and isolated from each other, from their origin(s) to their terminal(s)."[110] He made this anatomical separation the condition for distinguishing sensations. The solution to this problem was a prerequisite for further research on specific fibres for pain or particular receptors—the nociceptors—for painful sensations. If the fibres were anastomotic like nerve funiculi, then according to Müller, what would have resulted would have been a confusion and a lack of differentiation of the sensations instead of the brain perceiving a distinct sensation as in the case of transmission by a single fibre from a particular place. The laws of the "mechanisms of sensory nerves," described in the following chapters, expressed this whole idea of there being a specialisation with respect to the effects of a given stimulus applied to an area restricted to the branches of nerves found below the point of stimulation; from this anatomical distinction, he developed an argument in favour of the reality of painful sensations felt by amputees, since all the primitive fibres were still found at the level of the stump and in the nerve trunk which remained intact and since the sensation received by the brain was the same regardless of whether the stimulus had its effect on terminals, on the middle or on another point of the nerve pathways.[111] Without doubt, the fact that neuralgias sometimes spread beyond the anatomical distribution of the nerve could be seen as contradicting Müller's theory: however, he turned the problem round by drawing attention to the fact that in most cases, section of a nerve did not suppress pain, which indicated that the pain was not one which arose from the morbid state of the peripheral terminals, but was one produced at the level of the nerve trunks or their origins—in exactly the same way as, in certain cases, a patient can feel very violent pains in the extremities of the limbs, while these extremities are completely insensible to external stimuli. Other pathological findings, such as the irradiation of pain, which can be communicated from one afflicted finger to the others, or the phenomena of associated or concomitant sensations (as in cases of burning, nerve tumours, etc.) equally caused problems: Müller resolved these by imagining a model in which each posterior root ganglion would function like a sort of semi-conductor of the sensation—incapable of propagating weak irritations within itself but, on the other hand, reaching all the fibres which pass through the particular ganglion when the irritation is quite acute. This electrical type of scheme[112] was borrowed from the one which Reil had proposed to explain nervous transmission in sympathetic ganglia. It assumed that the power of conduction would be modified under certain conditions (an increase in stimulus intensity corresponding to a large accumulation of electricity at a given point). Müller hesitated between this explanation and another involving reflexes but with the slight difference that the reaction to a feeling produced via a sensory nerve would then occur in the sensory and not the motor nerves.

Different cases of confusion between sensations, independent of pathological circumstances, provided Müller with a chance to reflect on the origin of this loss of clarity. He put this down not to the mixed nature of the nerve fibres, but to the small numbers of fibres which were found in a given part of the body: thus going back to the measurements made by E. H. Weber of the distances between the points of a pair of dividers which could be distinguished on different parts of the body,[113] he concluded that precision and clarity of sensations depended on there being a large number of fibres in a given part, as well as on the nature of these.

But the question of the distinction between sensations turned out to be more complex and one which could not be answered completely by an anatomical explanation concerning the isolation and strict localisation of the fibres. In order to resolve the problem of specialisation, it was necessary to determine whether sensory nerves were the same whether these were found within the optic nerve, the olfactory nerve or the papillary body which seemed to be concerned with touch. It was essential to explain why sensations carried by these nerves were always specific and why, for example, we do not smell with the optic nerve, or why this nerve perceives light and space, rather than providing the sense of touch. Müller referred to the hypotheses of specific nerve excitability and of there being a special relationship between a nerve and a given type of stimulus, which he called the homogeneity of a stimulus to a sensor organ. He commented on the inadequate nature of the response, since certain stimuli such as electricity, although of an invariable nature, produced different sensations depending on which nerves they were applied to: "The specific excitability of the sensory nerves is thus not enough to make sense of events and we are obliged to attribute to each of these nerves, as Aristotle did, specific energies, which are their vital qualities, in the same way as contractility is the vital quality of muscles. . . . Thus the sensation of sound is the specific energy of the acoustic nerve, that of light and colours is the particular energy of the visual nerve, etc."[114] This idea of the specific energy of nerves was not revealed by experimental physiology, but was a throwback to an unknown property of the nerves which Johannes Müller recognised as being of a metaphysical nature which seemed to be linked to its vitalism. Müller did not know whether the cause of this energy was to be found in the nerves, in the spinal cord or in the brain. However, in the absence of a more precise histological differentiation of the nerve cells and in the absence of an understanding of cerebral localisations, this concept allowed him to uphold the postulated specialisation which seemed to him to be necessary for an understanding of the individual nature of sensations. This concept of the "sense mechanisms" was summed up in ten laws[115] concerning the transmission of sensations: if the effect of external perturbations necessarily has to act on the state of our nerves in order for us to receive an impression of the cause (Law I), and if the impressions vary according to the qualities or the states of different sensory nerves (Law V), "the specific sensations of each nerve can be evoked both by several internal and external influences" (Law IV); where internal sensations were concerned, Müller stated that "the central parts of the sensory nerves to the brain are capable of feeling, independently of the nerve cords or conductors, sensations which are determined appropriate to each sense" (Law VII); what reached the brain *(sensorium commune)* were not only modifications in the states of the nerves but, since the latter were modified by external conditions (chemical changes produced by heat and electricity, and movements), but also indirect information on "the qualities and changes in the external world and this occurred in a way appropriate to each nerve according to its qualities or sensory energies" (Law VIII). The following laws deal with how intellectual faculties influence the perception and interpretation of sensations: our imagination and our experience (and not the nerves themselves) inform us that modifications in the state of the nerves reflect modifications in the external world (Law IX): it is thus "the soul" which forms images and ideas, and which gives them precision and clarity (Law X). The description of specificity was particularly clear in the following propositions: "A single internal cause produces different sensations in the diverse senses, due to the individual nature of each of them" (Law II) and the same principle is applicable for external causes (Law III); as a result of the specific mode of sensation appropriate to each sensory nerve, "one sense can never be replaced by another sense" (Law VI). However, the question of the relationship between anatomical structures (at the peripheral as well as at the central level) and the principle of specificity remained unanswered.

When it comes down to the specific problem of pain, his views are not in the least ambiguous: we have examined the non-anatomical significance he conferred on this specific energy. In addition, he did not consider pain to be a specific sensation but a particular modality of the sense of touch which existed just as much in the internal parts of the organism—thus accounting for cenesthesia—as in the cutaneous tissues: if "the sensation of pain seemed to be determined by the violence of the stimulation of touch,"[116] its intensity and quality were modified by the attention placed on it, its duration, its extent as well as by prior experiences or habit: the sensation of heat could provoke a more unbearable pain depending on whether a finger or a whole hand was plunged into water and on whether the previous temperature was already hot or whether there was a violent contrast. However, these relative sensations which depended on subjectivity were not produced only by external objects: "Of all the senses, touch is the one where subjective sensations provoked by internal causes are the most frequent. The feelings of pleasure, pain, cold, heat, lightness, weight, tiredness, etc., can all be determined by internal states. Neuralgias, shivering, tingling or the states evident in the genital organs which occur spontaneously during sleep are striking examples."[117] As they were of internal origin, pains felt by hypochondriacs or hysterical people were not imaginary as, though pains could occur without an external cause, they are never imaginary.

Thus Johannes Müller proposed a theory for pain which tried to take into account the combined data from experimental physiology, histological observations, clinical investigation as well as numerous pathological findings and also integrating the psychological dimension of pain. While proposing a theory of specific nerve energies which was based neither on the anatomical specificity of nerve fibres and their terminals nor that of nerve centres, he conceived of pain involving not only a certain type of transmission of an impression from the periphery towards the centre, but also a series of complex relays which could bring about modifications of the sensation.

### The Discrimination of Tactile Sensations: Clinical Observations and the "Sense of Pain"

In the years which followed, anatomists and histologists discovered a series of organs within the skin which gave an anatomical basis to the notion of specialisation: after Pacini's discovery in 1840 of tactile corpuscles which would be found to correspond to sensations of pressure at a later date,[118] Meissner and Wagner discovered other receptor organs, in 1852, found in the superficial cutaneous tissues, which they identified as corpuscles for touch. During the session of 17 May 1852 at the Académie des sciences, Flourens read out an extract from the work of these two German scholars which distinguished, in what until that time had been called the papillae of touch, two different sorts of papillae, some vascular and the others neural; these contained "a small body formed of horizontally superimposed membranes having, between these membranous layers, numerous oblong grains with a dark outline. The membranous layers, as well as the grains are reminiscent of analogous formations in the corpuscles of Pacini. The touch corpuscle is surrounded by an excessively fine striated envelope. . . . These corpuscles must be the apparatus of touch since they alone receive nerves. They should be called tactile corpuscles."[119] The identification of different nerve structures in the dermis opened the era of research into the dissociation of sensations within the larger category of touch in general. The principle of "anatomical deduction," in which the function of an organ was deduced from its structure, could function fully: if the dermis contained neural corpuscles which had different forms, structures and terminals, was this not proof that they must also have different uses, allowing some of them to perceive hot and cold, others pressure, and still others pain? If, towards the middle of the 19th century, this stage had not yet been reached, it nonetheless remained true that Wagner's discovery corresponded perfectly to the principles of rational and economical hypotheses which held that each organ corresponded to one function and one alone.

At about the same time, and as a result of the observations made in relation to anaesthesia, a number of physicians underlined the need to dissociate sensibility to pain from the functions of other senses; it was Gerdy who, in 1847, on the basis of auto-observation, first remarked at the Académie des sciences that the dulling of general tactile sensibility lessened the pain during operations, whilst the other senses (smell, taste and actual touch itself) were not paralysed.[120] The following year, J. H. S. Beau, a physician at the Hôtel-Dieu, published a series of observations in the *Archives générales de médecine* which ultimately distinguished the sense of touch from that of pain. It was purely on the basis of clinical experience that Beau arrived at this dissociation which he observed as much in lead poisoning (a saturnine condition which caused intense neuralgias) as in hysteria or hypochondria; moreover, he was warned against the lack of thoroughness of some clinical examinations in which the physician, pricking, pinching or running a feather over the afflicted areas, was content to simply ask the patient if he could feel anything: the reply was normally positive as the patient retained the feeling of touch, but it would have been necessary to pursue the investigation further since the patient felt no pain despite having been pinched vigorously. Following pathological observations, he concluded that there existed "two types of anaesthesia; an anaesthesia of the tactile sense and an anaesthesia for the sense of pain." He added, by way of a note, that "one could more briefly call the latter analgesia, a word which can be found with this meaning in Castelli's dictionary. The word anaesthesia would be applicable to the loss of the tactile sense."[121] Could there have been more to this than simply observations which were becoming generalised as a result of the development of anaesthesia? The answer is affirmative for several reasons: the most obvious stems from the analysis of pathological circumstances occurring quite naturally—particularly in cases of hysterical anaesthesia which had been the growing subject of debate[122]—and no longer from provoked or artificial causes as was the case with anaesthesia using ether or chloroform; however, the most important reason lay elsewhere: in the belief that the suppression of sensibility could also encompass the mucous membranes and all the innervated tissues, and in the assertion that there was a chronological order in the abolition of sensations, with the feeling of pain disappearing faster than did that of touch, a fact which ultimately could be interpreted within the framework of an evolutionary theory, and, above all, in coming to a scientific hypothesis which derived physiological conclusions from pathological evidence: "Since pathological observation forces us to distinguish the anaesthesia of pain from the anaesthesia of touch, we must, I believe, also distinguish the sense of touch from that of pain in physiology. If, with the help of some kind of tool, a desk ruler or a cane, for instance, one hits oneself on the foot or a toe which is already sensitive due to a corn or a callous, two perfectly distinct sensations are felt if the blow is hard and strong enough to be painful: firstly, the sensation of contact or of a forceful blow, followed by the sensation of pain."[123] He explained this time-lag as resulting from two different modes of action, one direct and ascending in the case of touch, the other successively ascending and descending, or reflex, in the case of pain. It is clear that, as was the case with Müller's work, that of Marshall Hall concerning the role of the spinal cord in the constitution of the reflex arc had in fact promoted the cord from the role of simple point of convergence of the different nerve strands to the status of full-fledged nervous center capable of modifying the transmission of sensations. Beau went as far as he could in considering the consequences of his observations, since he concluded: "One can thus acknowledge the feeling, or if one was allowed to use this word, the sense of pain, just as one acknowledges the sense of touch, or as one acknowledges the special senses of sight, hearing, etc."[124]

In the wake of Beau's work, Octave Landry, a resident intern in Paris hospitals, working at Beaujon and a member of the Société médicale d'observation, in turn published in 1852 *Recherches physiologiques et pathologiques sur les sensations tactiles:* these studies were based on five sets of observations gathered at Beaujon, either directly or from cases in Dr. Sandras' wing, and enabled him to demonstrate a dissociation between tactile and painful sensations, thus confirming the previous work. For example, in one case referred to as "Ward St. Clair, No. 60 (Hospital Beaujon, in M. Sandras' wing), Baudin woman (Anna), 19 years old; entered 16th July 1851, for chlorosis complicated by hysteria, with obstinate nervous vomiting. The sensations of contact, though generally a little dull, are conserved throughout however. Painful sensations are abolished everywhere; a pinch, a pinprick are nowhere acknowledged."[125] He also described cases of hyperaesthesia to pain and, in another patient, the fact that cold water was perceived as freezing while water at room temperature seemed burning hot, though the other sensations (touch and pain) remained normal. Using a rigorous analytical method which involved recording the reaction time for each sensation, he tried to clarify the numerous sensations which were attributed to touch (contact, quivering or vibration, tickling, weight, consistency, humidity, etc.). Though he regarded temperature discrimination, pain and contact as the only cutaneous sensations, he felt moreover that it was necessary to acknowledge the existence of a special muscular sense which made one aware of muscular activity when other sensations had been abolished and which was situated not in the skin, but in the muscle tissue itself. However, rather than linking this muscular sense to cenesthesia or general sensibility, he made no distinction between the nature of internal sensations and cutaneous sensations. He concluded his work with an hypothesis on anatomical specificity which of necessity corresponded to the four basic sensations: "I have established previously that the four sensations of contact, pain, temperature and muscular activity are essentially different and distinct from each other. . . . After having shown by direct observation how well-founded this viewpoint is if one recognises the independence and individuality of each of these sensations, am I not justified in asserting that their organs have different nerve filaments? Because, I repeat, it would not be physiological to assume that the same nerve fibre is capable of reacting, often simultaneously, in several ways, and thus at the same time providing the sensations of touch, pain, temperature and muscular activity."[126]

These types of investigations, less linked to psychological analysis than were Gerdy's studies on the physiology of sensations and based mainly on clinical observations but lacking the assistance of a microscope, might ultimately have led to the analysis of cutaneous sensory receptors and pre-empted the work of von Frey and Goldscheider.

However, neither Landry's observations nor Beau's remarkable memoir achieved the importance they undoubtedly deserved, and this brings to light the progressive separation which occurred between experimental physiological research and clinical research. There were already differences in prestige and socio-professional status between the luminaries of the Academies—above all the Académie des sciences—and obscure hospital physicians. To these differences were being added, increasingly clearly, a divergence in their lines of questioning as well as the existence of a specialised and very diversified press, which accentuated the disciplinary rifts: on the one hand were the *Journal de physiologie expérimentale*, followed later by the *Archives de physiologie normale et pathologique*, not to mention a large number of neurological reviews; on the other were the *Archives générales de médecine,* the *Gazette médicale de Paris,* the *Bulletin de thérapeutique*, etc. This remark applies equally well to Beau as to Valleix, Trousseau, Dieulafoy and to all the great clinicians of the second half of the century. Two parallel lines of investigation were to develop: this did not signify that each side ignored the other, but showed the direction of the "epistemological vector": it was experimental physiology which posed its questions to the clinical world and asked for either a verification of its results or for sets of observations that might serve to initiate research; however, the clinical world was not liable to suggest new lines of research to experimental physiology. This is without doubt the price of the success obtained by the latter, and it also undoubtedly testifies to the lack of development in the field of experimental medicine which Claude Bernard so ardently wished.

### The Wallerian Method of Degeneration: A Turning Point in the Study of the Nervous System

Without the Wallerian method, for which its author was awarded the Montyon Prize for experimental physiology by the Académie des sciences in 1852, there is no doubt that the physiology of the nervous system would not have enjoyed the progress which it did from the middle of the century onwards. Augustus Waller, trained in France where he studied medicine and worked under Flourens, in Bonn, and in Great Britain. It was he who devised a method which, thanks to later established staining techniques, allowed an understanding of the ascending and descending pathways of nerve fibres based on the order in which they degenerated. He dedicated several memoirs to this question which he presented to the Académie des sciences in 1851 and 1852. This relatively simple process consisted of sectioning different parts of the nervous system in such a way as to interrupt its communications with its centres and then observing the degeneration process. It was already known through experiments on rabbits and frogs that such sectioning produced a disorganisation of the nerve fibre, but no one had taken sufficient notice of the difference in the course of events between the two sectioned ends: the one which was still attached centrally retained its normal state, while the peripheral end degenerated. Even when healing joined up the two parts after a while it was, according to Waller, necessary to wait for the "regeneration" of new nerve fibres to pass through different embryological stages before physiological properties would reappear.[127] Waller had the idea of sectioning the second cervical pair of spinal roots in a dog and, when it came to the posterior root, making sections on either side on the spinal ganglion. In the case of sectioning the anterior motor root, the peripheral end degenerated, which was shown by a grey coloration, and the same phenomenon was observed after the section of a mixed nerve; when a section was made between the spinal cord and the spinal ganglion, it was the medullary end which degenerated, while when the section had been carried out beyond the spinal ganglion, the end of the nerve attached to this ganglion remained unaltered and the peripheral end degenerated. Consequently, Waller was able to conclude that the spinal ganglion which is found on the posterior root plays the role of a nutritive and "trophic" centre for the nerve, whereas it is the cord which plays this role for the anterior root.[128] The microscopic examination carried out in every case at regular intervals revealed proof of the process of granular decomposition. This method led to the understanding of the origin of nerve fibres and the ascending (centripetal) or descending (centrifugal) pathways of neural conduction: it was to prove a powerful means for understanding the pathological and physiological phenomena of sensations.

### *Claude Bernard and His Experiments with Curarisation: Specificity or Specialisation?*

Claude Bernard's experiments on the effects of curare dated from 1844, when he discovered the maintenance of the heartbeat and the suppression of reflex movements in poisoned animals. He re-examined his evidence a number of times and then communicated his original findings on the "physiological action of poisons" to the recently founded Société de biologie in 1849 and undertook to explain the selective mode of action of this poison: it was lethal when it entered the blood, via a wound for example, though it was not capable of absorption through the intestinal mucosa and consequently had no effect, at least within certain doses. Curare was a poison of botanical origin, often used in South America where the Indians coated arrows with it. It had been known of since the previous century thanks to the tales of naturalist travellers (notably La Condamine, De Pauw and Humboldt) and, having aroused the interest of chemists, it was brought back to France in 1832 by Boussingault, Roulin and Goudot. Claude Bernard knew that it provoked death without convulsions or twitching but he had no precise idea of the mechanism by which it acted. The comprehension that death occurred through the paralysis of the nerves controlling respiration came only later, and Claude Bernard succeeded in keeping a curarised animal alive by practising artificial respiration until all the poison had been eliminated.

The conceptual progress made by Claude Bernard has already been magnificently retraced in a rigorous analysis of his published texts and experimental notebooks;[129] there is therefore little point in going over it in detail. However, the questions discussed by Claude Bernard in his

*Leçons sur les effets des substances toxiques et médicamenteuses,* which were given at the Collège de France in 1856, are of such importance for the comprehension of neuro-muscular phenomena, that we must take the time to go over them. Retrospectively, in terms of their interpretation, his experiments appear to be an additional link in the specificity theory which was soon to be questioned: it is precisely for this reason that this work is worth examining here. Curare had been used by Claude Bernard as a chemical means for analysing and dissociating physiological phenomena which could be substituted for the scalpel; its effects seemed the opposite of those of strychnine, with which he had also worked, in as much as while the latter over-stimulated the nervous system and exhilarated it, curare seemed to abolish the nervous system's properties. This study allowed him to put an end to a debate which was over a century old, on the question of irritability or sensibility: "This action of curare also allows it to be used for analysing the properties of the motor and sensory systems and for determining whether muscular irritability and nerve excitability are two types of distinct phenomena, and if they can be, at least theoretically, separated from each other and considered in isolation."[130] In his experiments on curarisation Claude Bernard compared the effects of galvanisation on a curarised frog and on a decapitated frog: in the latter, galvanisation of the lumbar nerves produced contractions in the limbs to which these nerves led, whereas not a single contraction occurred in the frog which was killed by curare; on the other hand, galvanisation applied directly to the muscles of both frogs produced contractions in both cases. The first experiment showed that "nerve excitability has [thus] been destroyed in the curarised frog," and the second one showed that, even in the same frog, "muscular contractility exists when nerve irritability has disappeared completely. These two phenomena are thus quite distinct, since they can exist independently of one another."[131] In his experiments carried out during 1854 and during the first lessons of 1856, Claude Bernard was preoccupied above all with showing that muscular contractility was increased in the curarised muscle. But in the 22nd and subsequent lessons, he was also interested in the mode of action and dispersal of curare, proving in particular that "under the influence of curare, motor nerves lose their properties from the periphery to the centre,"[132] which is the opposite of what occurs in deaths by asphyxiation or decapitation. The selective character of the action of curare and the direction in which it acts from the periphery to the centre, excluded its use as an anaesthetic: it only gave the appearance of producing anaesthesia because the animal under the influence of this poison could not express its feelings through movement. "Who knows," wondered Claude Bernard, "whether in certain cases of chloroform anaesthesia, one does not in fact feel, though one loses the memory of pain when one comes round?"[133]

The following stage, in his *Leçons* of 1856, consisted in undertaking a new dissociation in an attempt to establish that curare did not act on the whole nervous system, but only on the motor nerves while preserving sensory nerves. For that purpose, it was necessary to find a means of maintaining the centre from which reflex movements were initiated when the sensory nerves were irritated, *i.e.* the spinal cord, and at the same time protecting certain motor nerves from intoxication by curare. The very concept of the experiment was possible only because the work of Marshall Hall and Johannes Müller[134] had led to a new concept of the reflex arc, which implied that three distinct parts were active—sensory afferent pathways, intermediary locations for the conversion between sensations and movements, and efferent motor pathways. Through an ingenious experimental process, Claude Bernard ligated abdominal aorta of a frog to the lower body so as to intercept the circulation below it: curare introduced into the skin penetrated into all the areas above the ligature, but did not reach the posterior limbs; the motor nerves which controlled their movements were thus capable of producing signs of sensibility and of responding to a stimulation wherever it took place. The conclusion of the experiment was that curare was "a poison which not only produces the physiological isolation of nerves and muscles, but also the nervous system's two types of manifestations. It destroys movement but remains without effect on sensation; in this way, it seems to dissect the motor nervous system and at the same time to separate it from the blood stream, the muscular system and the sensory nervous system as well as other tissues."[135]

Thus the specific action of curare allowed physiological evidence to be obtained on the specialisation of the nerves' sensory and motor functions. Claude Bernard's experiments constituted a new step over and above Bell and Magendie's discovery by generalising the dual existence of the sensory and the motor activities, and his results were all the more remarkable as curare's effects left no anatomo-pathological lesions. In line with the standard Bernardian experimental process, there was only a functional modification, and this enabled an "uncoupling" between function and anatomical structure. However, Claude Bernard interpreted his experiments not simply as evidence of different functions but also as clear proof for two classes of distinct nerves which had different properties and functions. The immediate consequence was that the "chemical dissection" of the functions of nerves became a powerful means for diagnosing nervous-system disorders, and understanding the dissociation between motor paralysis and the maintenance of sensibility. The pathology of the nervous system was able to benefit from the teachings of experimental pathology which succeeded in artificially reproducing illnesses.

However, this concept of two classes of nerves being distinguished by their properties was soon the subject of controversy; veiled and indirect attacks came from one of Flourens' pupils, who stood in for him in his lessons at the Museum and who was to play a major role in the physiology of the nervous system—Alfred Edme Vulpian. In his *Leçons sur la physiologie générale et comparée du système nerveux,* published in 1866, without ever implicating Claude Bernard by name, he showed that there had been a confusion between properties and functions: "Contrary to the conclusion that was arrived at from the experiments carried out using curare, I fully trust that I will be able to demonstrate that the motor, sensory and sympathetic nerve fibres, indeed that all nerve fibres, may have the same physiological property, that they differ from one another only in their functions, and that these functions depend solely on the central and peripheral connections of these various fibres."[136] This served to clearly distinguish the question of functional specialisation from that of specificity, and to refocus research onto the field of nerve topography and of localising nerve fibres' points of origin and terminations using the same terms employed by Müller. Vulpian's subsequent lessons were dedicated to showing that curare acted at the level of the muscle's motor end-plate which had recently been described by Charles Rouget; this lay at the junction between the nerve and the muscle and curare paralysed its movements though it left the physiological properties of the nerves—particularly their electrophysiological characteristics—intact.[137] Only the physiological communication between nerve and muscle fibre was interrupted, although Vulpian was unable to explain the chemical mechanism for this. He suggested calling this intrinsic property of the nerve fibre neurility, in order to distinguish it both from the contractility of the muscle fibre and the excitability with which he credited all living tissue.

### Brown-Séquard, the Crossed Transmission of Sensations in the Cord and the Société de Biologie

The change in status of the spinal cord, from being a simple assembly point of the nerves into being a complete organ which was distinct from the encephalon and capable of playing the role of nerve centre itself, had prompted more exhaustive research into its structure and function. For a long time it had been noticed that the loss of sensibility in one half of the body was caused by a lesion in the opposite half, but how and when this crossover occurred remained a mystery. The physiologist, Longet, who had studied this question, thought that this crossed transmission took place not in the cord but in the brain at pons level, after the passage of sensory fibres through the restiform body and cerebellum. However, different pathological cases which were often of traumatic origin, collected by Serres and by Andral in his *Clinique médicale*,[138] showed that when there was a lesion of the pons or cerebellum not only was sensibility not suppressed but it was sometimes even increased and that, in any case, sensory transmission remained crossed, thus leading to the belief that this chiasma happened somewhere else. It was the physiologist Brown-Séquard who demonstrated experimentally that this crossed transmission took place in the cord; this assertion was met with a great deal of disbelief. Through his origins, as well as his endless comings and goings between France and the United States or Great Britain until he succeeded Claude Bernard at the Collège de France in 1878, Brown-Séquard epitomised that international community of experimental physiologists who were in the process of establishing modern medicine. However, originating with his thesis which was accepted in 1846, his research on the spinal cord found a particularly stimulating environment with the creation, in 1849, of the Société de biologie.

The Société was founded by the positivist Charles Robin[139] and by P. Rayer, with top-level foreign members (such as Du Bois-Reymond and Kolliker from Germany for example), and was attended by the best physiologists of the time, including Claude Bernard and Brown-Séquard. On the fringes of the Académie des sciences, it provided an intellectual environment where it was the rule that one should bring along new experimental findings, demonstrate them in front of colleagues who were at the top of their disciplines and submit a sober and concise interpretation of such findings which might later be validated through further experimentation. The incisive discussions were to the point and precise, and they were governed by a common quest for truth in which academic precedence and authority gained through previous work did not count. The founding members of the Société were guided by the principle of unity in the living world (vegetable, animal, human) within its different levels of organisation, and by the necessity of relying on established scientific disciplines which had developed earlier, such as physics and chemistry. They had greater reservations when it came to using statistics and numerical observations, denouncing "the vice of a method which tries to introduce a mathematical precision that the highly complex nature of their phenomena is quite beyond."[140] This was an attitude Claude Bernard expressed at a later date as he could see nothing more to statistics than "a generalised empiricism." It was a matter of establishing the need to undertake the physiological study of phenomena and of arriving at a precise comprehension of the functional mechanisms of living entities in order to build a sound basis from which to come to generalisations about observations. The pre-eminence of this criterion also had as a condition, and perhaps as a counterpart, the setting aside of medical practice. A long-term perverse consequence was the superiority gained by the theoretical and abstract scientific point of view over its applications as Auguste Comte had observed. This separation, which had been originally conceived as a temporary measure, was to influence the organisation of biological research in France in a lasting fashion. In his article "Sur la direction que se sont proposée en se réunissant les membres fondateurs de la Société de biologie" ("On the direction which the founding members of the Société de biologie, in joining together, proposed to take"), Charles Robin stated: "Anatomy, physiology and even pathology, considered as sciences, are today so advanced that henceforth we no longer have any need to be inspired by the study of pathological changes and the demands of the practice of medical art in order to study the normal state. Besides, too many examples have already shown us that by behaving in this way we have often described as pathological normal dispositions which have not yet been studied; we must therefore not hesitate to adopt as a goal the direct study of the normal state and only then descend to the observation of morbid conditions, leaving any medical applications to the intelligence of each individual."[141] Such was the general order of the relationship between the normal and the pathological which was structured according to a particular hierarchy. This conceptual tearing away from clinical medicine, which the founders had deliberately sought, would lead to research practices which were not only parallel, but often disjointed.

Thus in 1849, Brown-Séquard presented some experimental results to the Société de biologie; he described how, in a guinea pig, he had sectioned the right half of the spinal cord at the level of the tenth dorsal vertebra.[142] After a few minutes, it was noticed that voluntary movement had returned to the left hindlimb and that sensibility in this limb was nil, whereas in the right hindlimb, *i.e.* on the same side as had been sectioned, far from being abolished, sensibility had increased. This hyperaesthesia was, above all, met with astonishment and prompted new research. The series of experiments presented from 1849 onwards challenged Longet's ideas, which Brown-Séquard criticised as much from an experimental point of view as from an anatomical and clinical one.[143] He also disputed the criteria adopted for pain expressed by experimental animals, believing that the restlessness and cries which they made could have been interpreted not as manifestations of a preserved sensibility, as Longet thought, but as reflex actions.[144] However, he also had to refute another theory which had been developed in Germany, notably by Stilling, following the work of the Dutchman, Van Deen. Indeed Brown-Séquard himself had adopted this theory in his thesis in 1846. For these physiologists who had clearly observed "that sensibility is not lost posteriorly and on the sectioned side of a lateral half of the spinal cord, but [who] had not noticed the crucial fact that it diminished on the opposite side and was exaggerated on the side of the section,"[145] it was the posterior half of the cord's grey matter which was the exclusive seat for the transmission of sensations towards the brain. They also believed that the route of transmission could be achieved in any direction—in one lateral half or its opposite counterpart.

In his *Recherches* of 1855, Brown-Séquard performed an increasing number of hemisections at different levels (dorsal, lumbar, etc.); he applied the principle of providing a counter-proof in the Bernardian sense, by showing that, if an initial hemisection provoked a medullary hyperaesthesia, another hemisection carried out at a certain distance from the first, on the opposite side, very considerably reduced the sensibility of the parts which showed hyperaesthesia. For him, it was a matter of demonstrating that the observed phenomena could not be explained by a specific property of the grey substance, but indeed by the anatomical disposition of a crossing of the routes of conduction, and that this crossing was completely within the spinal cord. He even described its particular pathway, including the initial moment of conduction away from the brain which preceded decussation, *i.e.* the crossing of the median axis. To sum up these findings, one can refer back to the customary clarity of Claude Bernard and, in this case, to his caution: "The effects of the section are direct in the lower parts of the spinal cord, that is to say that they modify the sensibility of the side corresponding to the lesion; however, in the higher part of the cervical region they are crossed, *i.e.* they act on the opposite side."[146] A commission composed of Claude Bernard, Broca and Vulpian, amongst others, was formed under the auspices of the Société de biologie. The commission deemed in its 1855 report that Brown-Séquard's experiments were revolutionary and put an end to the era begun by Charles Bell, Marshall Hall and Longet; the experiment was qualified as "subversive" by Broca.[147] After verifying the experiments for more than a month, as much at the Veterinary School at Maisons-Alfort as on the premises of the Société de biologie itself, the commission concluded that the crossed transmission of sensations did indeed occur in the cord and that the phenomena could be observed no matter where the lateral hemisection had been performed. However, despite this success, many reservations remained. The attacks first came from Vulpian in France and if, during the 1880s, the "Brown-Séquard's syndrome" was clearly enough established within the anatomo-clinical framework, the significance of his hemisection experiments, as well as his hypothesis on the site of crossed transmission, were called into question not only in Great Britain, by Gowers, but also in Switzerland by Moritz Schiff.

In order to clarify this debate, it should be explained that there were a number of different issues involved: (*i*) Was crossed transmission in the spinal cord complete or incomplete, *i.e.* did it involve all the sensory fibres or was it necessary to acknowledge that some of them took a direct route? (*ii*) At which point in the cord did this crossing take place, which led back to the question of which tracts in the cord were the conduction pathways for sensibility; this apparently simple question concerning localisation in fact referred back to a series of hypotheses each with its own supporters: were the routes for sensibility in the grey or white substance? If they were in the white substance, were they in the posterior funiculi, the anterior or the anterolateral bundles? (*iii*) Was it possible to refer to sensibility in general, or should one distinguish between the sensations of touch, pressure, pain and temperature? The question as to whether there were specialised territories in the cord, each responsible for conducting a particular type of sensation, was certainly straightforward enough. However, because of the complexity of the problems, the question could not be resolved simply by means of experimental physiology. Clinical and anatomo-pathological observations were absolutely necessary in order to be able to provide an answer. Brown-Séquard was particularly fortunate in this respect in that he had practised medicine far and wide, especially in the United States and in Great Britain, which allowed him to understand the subtle shadings of certain positions.

The problem of delimiting sensory territories in the cord had already been brought up for a number of years. The German anatomist and physiologist Moritz Schiff, using the same type of cross-sections in the spinal cord as those made by Brown-Séquard, thought he had proved that the grey substance of the cord, just as the posterior funiculi of the white substance, had the power to transmit sensations. But he made a distinction between this ability and sensibility: "Thus the grey substance contains fibres which have the particular property of being able to transmit impressions to the brain which are communicated to them through the white substance; however, these impressions cannot originate there directly; thus these nerve fibres are not sensible, nor do they produce perceptions; their role is essentially only that of a conductor. As such nerve fibres were not yet known in physiology and it appears to me that it is necessary to distinguish them from the sensory fibres under a new name, I propose to call them aesthesiodic fibres (*odos*, path) and later I will prove that in the same substance there are analogous fibres for movements which I shall call kinesodical."[148] Such a concept, which was to be closely examined, quite clearly demonstrated the difficulties in interpreting experimental findings.

In his *Course of Lectures on the physiology and pathology of the central nervous system*, which was published in 1860, Brown-Séquard introduced the "solution, through pathological cases," of the problems of how sensations were transmitted in the cord. In 1863, he returned to some of these cases and developed them further in research in which he distinguished the sensations of touch, tickling, pain, temperature and contraction (in the muscular sense) in the spinal cord: this last sensation seemed to him to come about in direct fashion up to the level of the lower brainstem and he cited cases of haemorrhages at cord level in which sensibility was abolished on the opposite side of the lesion which showed up at autopsy, and in which the hyperaesthesia which he had described was confirmed; in other cases, however, especially stabbing wounds, the situation was more complex, as testified by one patient who became so sensitive on the left side that he was unable to put up with even the slightest contact or bedclothes, but who had conserved tactile sensibility on the right side though he had lost the sense of pain.[149] In more than twenty observations, it was noted that the sensibility was not totally abolished on the side opposite the supposed lesion, and that sometimes it was the senses of pain and temperature which had been lost while that of touch was retained, and sometimes other dissociations came to light. Within the category of touch, the sensation of contact sometimes remained though the patient was unable to recognise the shape of the object or the place which had been affected. Brown-Séquard restricted himself to asserting "that the conductors of certain sensory impressions cross in the spinal cord"[150] and he also acknowledged the participation in the transmission of sensations of certain bundles in the white substance. For the dorsal and the cervico-brachial zone, which he had studied in particular, he showed "that the conductors of different types of sensory impressions are distinct one from another and that these conductors are grouped in the cord, those of one type in one part, those of another in so many distinct parts."[151] If the conductors of hot and cold impressions seemed to be in the grey substance to him, those for pain seemed more dispersed than those of temperature and were to be found in the posterior and lateral parts of the grey substance; those for touch and tickling were in both the white and grey matter of the anterior parts. Thus, one could map out the different territories of the cord though these turned out to be considerably more intricate than would have been expected from the initial observations concerning localisation, not to mention the problem of crossed transmission. In the main, the broad outlines of sensory pathways in the cord were established at least as far as the question of differentiation of territories was concerned. However, clinical findings were insufficient without the additional input from pathological anatomy and microscopic examination; above all, because of the lack of adequate staining techniques for showing up the paths of different bundles, the discussions were unable to progress beyond coarse differentiations.

The attacks against Brown-Séquard's hypotheses came first of all from one of his closest colleagues, Vulpian, with whom he edited the *Journal de physiologie*. In the article on the spinal cord which Vulpian submitted to the *Dechambre Dictionnaire* in 1877, he proceeded with an orderly demolition, arguing first of all about the variability of the experimental results, and particularly about the persistence of a certain sensibility in the two halves of the body; according to him, this proved "that there is no complete crossing between the fibres of the nerve roots in the spinal cord" and "that the transmission of impressions does not occur in the cord through necessary, invariable routes."[152] The limited and variable persistence of sensibility in the side opposite the hemisection was, according to Brown-Séquard, an "apparent sign of survival of sensibility," which could be explained by a sort of reflex reaction. Vulpian challenged him with an experiment which he judged to be very decisive: "After having cut one half of the spinal cord transversely in the dorsal region of a guinea-pig, and after having established the extent of the sensibility persisting in the posterior limb on the opposite side, the sciatic nerve is sectioned on the side of the operation in the part which is entirely above the thigh. Thus the majority of reflex movements which can be evoked in the corresponding limb are suppressed. But in these conditions, the sensibility of the other limb suffers no new decrease, as should occur if there were grounds for M. Brown-Séquard's hypothesis."[153]

He also challenged Brown-Séquard's findings on the basis that there was a difference between the routes taken by fibres under normal conditions and those taken in pathological situations where a particular perturbation and circulatory disorders could force sensations to follow pathways other than their habitual route. However, these objections could not be dispelled by experimental methods. If none of Brown-Séquard's experiments appeared truly decisive to him, the analysis of the clinical findings seemed just as open to question: he expressed "doubts concerning the reality of the absolute abolition of all modes of tactile sensibility on the side where anaesthesia has been confirmed,"[154] perhaps for want of an adequate discrimination of the different sensations at the time of the clinical examination. In effect, the disagreement was as much about the problem of crossed transmission as about the existence of specific nerve fibres and medullary zones for each of the sensations of touch, temperature, pain, etc. For Vulpian, who also challenged the concept of "muscular sense" being a special sense, there was something of a paradox in the wish to systematise this specificity at an anatomical level since deeply situated organs innervated by the sympathetic system such as the intestine, liver and kidneys—which as a general rule did not give rise to pain—could, in certain circumstances, become the site of very acute pains: could one say that there exists therein painful "algaesthesic" fibres which accompany sensory fibres everywhere but are unemployed during most of our lives or, on the other hand, could one say that ordinary sensory fibres, under the influence of an increase in stimulation, are capable of provoking pain? Vulpian favoured the second hypothesis because he found the "anatomical cost" of the first unacceptable: "It would have to be accepted that special fibres, dedicated exclusively to transmitting painful impressions, were associated with fibres from all the sensory nerves of the body and, consequently, it would also undoubtedly have to be allowed that there are special elements, or particular cells found throughout the grey substance of the spinal cord responsible for gathering and translating these impressions. . . . A physiological theory which involves such an hypothesis is not admissible. Pain can only be conceived as a simple exaltation of the common, general sensibility . . . and the nerve fibres through which painful stimuli are conducted are the same as those which serve to conduct all perceptible and imperceptible impressions. And, as far as the skin is concerned, there can be no doubt that it is the same fibres which serve to transmit tactile and painful impressions."[155] The difference between the sensations and the finding that they were not all lost simultaneously could be explained by their having different modes of action on the nerves.

It is difficult to conceive of how better to illustrate the opposition to the theory of specificity and to shed light on the continuity between the criticisms which Vulpian directed implicitly at Claude Bernard and those which were aimed against the "territorialisation" of the different conduction pathways for sensations in the spinal cord. We might nonetheless mention a certain confusion—perhaps deliberate—on Vulpian's part concerning both the problem of the specificity of fibres and that of the specificity of their points of origin and destination as well as their pathways, which he rejected here in their entirety. In view of the clarity with which Vulpian himself had made this distinction with regard to Claude Bernard's work, it is hard to believe that he may have forgotten all about it some ten years later. It would in fact seem that beyond the criticism of the experimental and clinical results, we are really dealing with beliefs and depictions which do not entirely rest on experimental rigour and scientific wisdom, but on a type of stunned amazement and reticence at the implications of this hypothesis from the standpoint of the nervous system's complexity and of the data it must integrate. Need it be added that at that time, quite unlike Brown-Séquard, his experience as a hospital clinician left a great deal to be desired?

Thus, William Richard Gowers' criticism of the positions taken up by Brown-Séquard did not fall on virgin soil. He had at his disposal a vast experience of nervous disorders as a result of his work as a clinical professor at London University College, as a physician at the University Hospital and, above all, as a physician at the National Hospital for the paralysed and epileptic: his *Manual of Diseases of the Nervous System* is a classic which was reprinted a number of times. His opposition was not with regard to the general problem of crossed transmission, but the essential role which Brown-Séquard had assigned to the grey substance: "The earlier experiments of Brown-Séquard established one important fact, which has been fully confirmed by pathology, that the chief part of the sensory path crosses the median line shortly after its entrance into the spinal cord and continues towards the brain from the other side. These experiments also suggested that an ascending path is found in the intermediate grey substance. This possibility cannot be said to be entirely disproved, but later investigations have given it no confirmation and it is on the whole improbable."[156] According to Moritz Schiff and Woroschiloff's experiments,[157] the antero-lateral and posterior bundles in the white substance could be the sensory pathways for pain and tactile sensibility respectively but, in the latter case, without decussation. Besides, from experiments involving hemisections of the spinal cord, Gowers came to the conclusion that the routes for pain and touch were not close to one another. In 1878, he had carefully analyzed a case[158] in which a bullet had badly damaged a patient's spinal cord at the cervical level, with lesions of the lateral bundle and of the grey substance, whereas the posterior funiculi had only been caused to swell: the loss of sensibility to pain had been total on the side opposite the lesion, whereas tactile sensation had suffered no deterioration at all. It seemed certain to him that the antero-lateral bundle in the white substance was the sensory pathway for pain. On the other hand, the posterior funiculi seemed to play a role in the transmission of muscular sense (now known as proprioceptive or deep sensibility, coming from the muscles, tendons, bones, etc.) without decussating, and perhaps also in tactile sensibility: he felt that the characteristics of locomotor ataxia, still called tabes, which are the result of the nervous degenerative changes due to tertiary syphilis, seemed to confirm this hypothesis. In this illness which, among other things, manifests itself by a lack of co-ordination of movements because of a loss of proprioceptive sense and by the lack of tactile sensibility, the posterior funiculi of the spinal cord exhibit large lesions.[159] On the basis of other observations, Gowers thought that the pathway for thermal sensibility must be found close to that of pain sensibility, but he could not go any further. Gowers' contribution to the understanding of the different routes for sensory transmission in the spinal cord is thus important: if the experimental demonstration of crossed transmission belongs to Brown-Séquard as does the research into differentiating between the routes for different sensations in the spinal cord—a parallel line of research which was also carried out by Moritz Schiff—it is Gowers who deserves praise for a more precise localisation of these routes than that proposed by Brown-Séquard, and closer to our present-day knowledge. His critical reflection as a clinician, which combined experimental physiology, clinical medicine and microscopic anatomy, in all likelihood was one of the major reasons for his insight.

### *Von Frey and the Theory of Specificity*

At the very end of the century, the attention of anatomists and physiologists was drawn to undertaking research on receptors which were specific for pain and for each of the "primary sensations" at cutaneous level. Microscopic anatomy had allowed the identification of neural structures which were highly different from one another morphologically speaking. It was therefore logical to think that these structures subserved very different functions: the corpuscles of Pacini and Meissner-Wagner were already known of, as was the existence of free nerve endings; to these were added the bulbous terminal corpuscles of Krause which were identified in 1860 and those of Golgi-Mazzoni which were found a little later. At the same time, the question of an instrument for measuring sensibility—an aesthesiometer—was the order of the day. The reliability of such an instrument was a condition for improving neuro-pathological diagnoses and for arriving at less equivocal interpretations of the experimental results and clinical findings about disorders of the spinal cord which had been the subject of debate. Consequently, it seemed that the original studies carried out by Magnus Blix and by Goldscheider in 1881 on the specific energies of sensory organs in the skin,[160] and particularly those of Max von Frey, were the acme of research into the specificity of pain which had been initiated mainly by Johannes Müller. In truth, it seems that behind an apparent continuity the theory elaborated by von Frey between 1894 and 1896 had very different implications and led to a more restricted concept from which to understand pain. The underlying idea consisted in trying to find particular points in the skin which responded more specifically to one or another of the four fundamental cutaneous sensations—touch, heat, cold, pain—while all other sensations, especially tickling, stinging, etc., were considered to be derived or secondary sensations. However, in order to be in a position to determine the exact function of this mosaic of sensitive points which were found on the skin, an instrument of great precision was needed which would eliminate variations from one experiment to another, not only from the standpoint of individual reactions of the subjects being tested, but also from the standpoint of the actual conditions of stimulation. Von Frey invented an aesthesiometer in which the stimulus usually consisted of a hair, either from a man or a woman, or he sometimes used horsehair or a hog bristle. By determining with precision the diameter of the hair which was attached to a wooden stick, as well as its length and the weight which it was capable of supporting without breaking off, he was in possession of an instrument which made it possible to measure the force of the stimulus and to apply it in a very accurate fashion to a restricted area of skin. Even if sensations of touch, tickling and pressure did not seem really different, on the other hand, it was possible to clearly distinguish those points where a feeling of pressure was perceived from those where pain was felt. "This result did not depend solely on the force of the stimuli, but also on the location where they were applied."[161] The explanation which was generally given was that pain was perceived when the stimulation went beyond a certain threshold intensity or reached deeply located neural structures. However, von Frey observed that there was "no coincidence between the points where the sensibility to pressure was strongest and the points where pain dominated."[162] The fact that a threshold was high for a given stimulus did not prevent there being lower thresholds for other stimuli or for the same one applied in another way. The fact that the reaction time for pain was generally longer than that for pressure, and that it could last for some time after the stimulus had disappeared, constituted another discriminating factor:

pressure, for its part, provoked an instant reaction and disappeared when the stimulus was removed. Finally, when an electrical current produced by induction was used, its oscillations were not perceived on the "pain points" (a succinct expression used to describe a point which is sensitive to pain), thus giving an impression of continuous stimulation, while these oscillations were clearly distinguishable in the case of "pressure points." The consequence of these different observations was that "this dismisses the interpretation of pain as being an altered sense of pressure and leads to the unavoidable conclusion that pain results from the stimulation of special organs."[163] It became possible to establish a topography of "pain points" which revealed their extreme density as there were more than 100 points per square centimetre which were generally situated more superficially than pressure points and distributed fairly unevenly. Von Frey put forward a chemical explanation for pain, thinking that the latter did not appear directly, but as a result of the stimulus displacing the intercellular fluid and modifying it: the lag in reaction time could readily be explained by this mediation. He also sought to measure the stimulus threshold to determine the minimal point at which a stimulus applied to a pain point became perceptible, and to compare it to the stimulus threshold for pressure. With the help of an ingenious device a strand of hair attached to a metallic thread encased in a sliding sheath which provided light contact with a copper tube—he was able to push or pull the tube in such a way that the hair was lengthened or shortened: thus the maximum length corresponded to the weakest stimulus, while the shortest corresponded to the maximum force the hair thus being stretched and compressed as it were. The ability to feel pressure on a given point was, according to von Frey's calculations, about a thousand times greater than of the "pain points," but the ratio could change as a function of the area stimulated. Within this framework, pressure points corresponded to Meissner's corpuscles and, in zones where there were hairs, to their follicles, while the "pain points" corresponded to free nerve endings.[164]

This theory which based the mechanisms of pain on a specific neural apparatus—different from that used for pressure, heat or cold—was passionately debated and criticised. Even if the strict and rigid form in which von Frey presented the theory has been generally abandoned today, its theoretical contribution has been considerable. It opened the way for the notion of sensory receptors, which was later redefined by Sherrington in particular, as well as for research into measuring the intensities of stimuli and sensory thresholds. In addition, and it is a paradox which the history of pain theories fully brings to light, each tentative attempt to establish a notion of specificity whether at the level of the receptors, at the level of the fibres and transmission pathways, or at the level of the centres, turned out to be insufficient and surpassable; however, each of these attempts helped advance the understanding of the phenomenon. They did this by indicating particular conditions and modalities necessary for pain production which undoubtedly did not take the overall phenomenon entirely into account though a knowledge of these was nonetheless indispensable. This "methodological reductionism" proved to be extremely productive even if the results which it contributed have been not so much debated as permanently reinterpreted in relation to a collection of more complex data. What undoubtedly creates a problem is not due to having looked for a strict anatomical correlation between nerve terminals and sensations, but to having used this correlation as the basis of an explanatory principle and to having thought of pain being a simple response to a stimulus. It was also therein that von Frey's line of research distinguished itself radically from research into pain transmission pathways in the spinal cord: for the studies of Brown-Séquard, Moritz Schiff and Gowers did not claim to explain pain by the anatomical localisation of the transmission pathways, but only to say under which conditions in the spinal cord's anatomo-physiological integrity pain was felt and, on the other hand, under what conditions of disorganisation it was abolished. The same remark could be applied to the problem of cerebral localisations. Finally, von Frey's work seems to have neglected the difference between experimental pain and the pain caused by a wound or illness.

## *The Summation Theory According to Goldscheider*

Efforts undertaken in the first half of the 20th century to confirm this mosaic of specific points and find strict correspondences between histological structures and sensations did not generate positive results. However, as early as the end of the 19th century, Goldscheider—who himself had tried to localise specific points—showed the weakness of this explanatory model. He readily acknowledged the existence of a mosaical distribution of points corresponding to particular sensations—hot, cold, pressure and pain—which constituted a sort of network with high density zones where the points formed both aggregates and near-empty zones;[165] however, he proposed a completely different interpretation altogether. His attention had been drawn to three types of findings: firstly, a pathological aspect which we have already described in relation to tabes, namely the exacerbation of pain upon the repeated application of a stimulus, a thermal stimulus for example, producing a sensation which was always out of proportion with the intensity of the stimulus; it did not seem possible to explain this hyperalgesia—an increased sensibility to pain—at the level of the receptor alone; rather, one was led to consider a cumulative process which acted at the level of the central structures, the spinal cord and brain. Secondly, he had noticed that pressure applied to the skin with the head of a pin in certain places initially brought about the sensation of pressure then, after a short period, a sensation of pain; in his eyes, this delay in the appearance of pain, which was also reported by von Frey, could not be explained by the coincidence of two different types of receptors in the same place, but indicated a secondary effect which, once again, involved transmission through the central nervous system. Finally, in the small areas formed by a series of "pressure points" which inevitably included intermediary areas which were devoid of points, he had noticed strange phenomena involving the conversion of the sensation of pressure into that of pain, even though the increase in the degree of pressure was not distinguished, as well as certain transitional situations: aside from the so-called "pressure points," "there are other points where a granular sensation [linked to contact] is converted directly into pain, and others where the sensation of pressure persists up to a certain point and is then rapidly converted into pain. In these cases, a transition which goes from pressure to pain is observed."[166] What explanation can be given for these two different situations, on the one hand an impression of points specifically linked to a type of sensation, and on the other, clinical and experimental observations which made an interpretation in terms of specific receptors impossible? Goldscheider tried to find an explanation in terms of the intensity of the stimulus, all the more so after he had noticed some "eccentric anaesthesia" phenomena: when a very weak stimulus was applied to a point which is normally sensitive to pressure, the surrounding region became incapable of feeling a tickling sensation which had been felt there previously, and the same incapacity existed in the case of touch or pain, albeit to a lesser extent. The duration of such insensibility also depended on the duration of the stimulation. However, when the pinpricks on a given surface, for example a square, were markedly increased in number without modifying their intensity, the zone of insensibility stretched well beyond the nerve terminals or fibres which were affected by the stimulus. In order to explain these phenomena, whether of sensations being abolished, perhaps through "adaptation," or exacerbated, it was necessary to return to a theory of central summation.

In a small work published in 1920 entitled *Das Schmerzproblem,* Goldscheider put forward a synthesis of the ideas which he had already developed in 1898. In his opinion, "tactile" nerves could assume three different modalities—tickling, touch and pain; the sensation of tickling would be the normal mechanical response to the weakest stimulus, and the specific sensation of tactile nerve fibres could, with an increase in intensity, give rise to an impression of being pricked, and then of pain. The difference perceived among the sensations arose not from there being different receptors, but due to a bifurcation of the nerve fibres in their trajectories from the periphery to the centre—those conducting tickling or pricking sensations heading towards the posterior bundle in the white substance, those for pain being transmitted through the grey substance or the spinal cord, and more precisely through the posterior horns. The fact that the sensation of pressure was of a different order, and was not necessarily converted into a sensation of pain, even by increasing the intensity of the stimulus, had still to be explained. For this purpose, Goldscheider invoked the existence of special tactile fibres which, when faintly excited, would also give rise to a tickling sensation and, depending on the increased degree of intensity, to sensations of pressure and pain, following the same specialised trajectories in the spinal cord. Thus, there would be no specific fibres for pain, but special trajectories borrowed in a privileged fashion depending upon the stimulus intensity, the localisation of receptors in more or less exposed points, and the habitual manner in which these receptors reacted—it being understood that the more frequently a receptor or fibre reacted in a given direction, the more the effects of stimuli would tend to take this path.[167]

It is quite obvious that this theory of summation is more complex than von Frey's and it makes it possible to account for a set of facts which are difficult to explain through a simple specificity theory: for example, the exacerbations of pain due to the cumulative effect, the persistence of pain after the disappearance of its initial cause or the beneficial effects of therapeutic techniques involving electrical stimulation. In fact, Goldscheider's theory, which inspired a considerable number of theories on pain in the 20th century, stems from a compromise which includes the remnants of specificity on one hand, and a concept based simultaneously on the receptor-fibre couple and transmission pathways in the spinal cord towards the cortex which depended on a system selection and specialisation, on the other. The vitality of Müller's laws cannot on their own explain this attachment to specificity which Goldscheider stated he did not wish to violate. All the neurological work in the 19th century seems to have favoured this direction, as much for the spinal cord with its clearly defined territories responsible for fulfilling specialised transmission functions, as for cerebral localisations. The spectacular advances accomplished by Broca, Ferrier and Charcot on the delimitation of cerebral areas and the roles they played in certain specific disorders, such as those affecting language or tactile discrimination, could only reinforce the adherence to such a line of research. However, very refined electrical-stimulation techniques, as well as the contribution provided by embryological studies were needed in order to obtain more precise answers on the cortex's role in the production of pain. On an entirely different level, in contrast to the localisation trend, the whole question of whether pain was a sensation or an emotion remained to be explored.

## Pain and Illnesses in the 19th Century

A major leap forward occurred as a result of a better understanding of physical signs which allowed a more accurate differentiation of illnesses, their diagnosis, aetiology and their prognosis. This was thanks to the dual contribution from clinical work and pathological anatomy, which was increasingly being followed up by histological investigation using a microscope. The École de la Salpêtrière of which Charcot was the illustrious head in the 1880s, perfectly illustrated the success produced by the combination of these disciplines. As virtually all illnesses are accompanied by some form of pain, it would be impossible to review the entire field of pathology here. But in analysing different types of pains, it seems that two categories must initially be distinguished: on the one hand, the well-established pathologies which include the pains of gout or nephritic colic—but these are symptoms rather than illnesses proper—and of causalgia or the "burning pain" linked to war wounds; these can all be traced from Ancient times onwards, and it is necessary to unearth the evolution of their proposed aetiologies and the modifications in their treatments; on the other hand, a whole series of disorders which began to be classified individually during the 19th century thanks to the new methods previously mentioned and also to the new investigation trends concerned with exploring the central and peripheral nervous system: they were not new illnesses in the true sense but were distinguished as a result of newly-established nosological categories, based on new ways of grouping symptoms and of determining, within the previously accepted pathological realities, new categories which now made sense because they could be associated with precise aetiologies or specific localisations. Such was the case, for example, of neuralgia, of the Brown-Séquard syndrome mentioned earlier, of multiple sclerosis, of syringomyelia or progressive locomotor ataxia *(tabes dorsalis)*, whose syphilitic origin was only recognised quite late thanks largely to Fourier who gave a description of it in 1876.

### Neuralgia and the Temptation of Suicide

The appearance of the term "neuralgia" in classifications of diseases is relatively recent, since it was only after the publication of the Table synoptique de la névralgie ("Summary Table of Neuralgia") by François Chaussier in the year XI (1802), that this entity was recognised by the uniformity of its symptoms and the tissues which it affected. It is true that a number of previous works had described symptoms which were analogous but these were usually lost either in unrelated collections, such as in André's treatise of 1756 on illnesses of the urethra, or listed within the framework of humoural explanations by the corruption and acrimony of the humours, as was the case in the work of the English physician, Fothergill. Nonetheless, in a work called *De ischiade nervosa,* the Italian Cotugno had, as early as 1767, described a neuralgia of the femoro-popliteal (sciatic) nerve, whose sole characteristic was the nature of the sharp, excruciating and irradiating pain within the projection of the nerve.[168] However, it was Chaussier—professor of anatomy and physiology in Dijon and then later at the École de Santé in Paris from the time of its establishment in 1795 until 1832—who was the first to define it with perfect clarity, no matter which nerve was affected. His definition was completely in line with the idea that the type of tissue involved provided the basis for identifying an illness rather than the localisation of the organ. He concentrated particularly on trifacial neuralgia which affected the face, following the distribution of the three sensory branches of the trigeminal nerve, namely the ophthalmic nerve of Willis, the superior maxillary and the inferior maxillary (mandibular) nerves. Chaussier characterised neuralgia: "1. by the nature of the pain which is simultaneously sharp, excruciating, and sometimes, especially at the onset, is accompanied by torpor or tingling, more often with successive throbbing, shooting pains and twinges, without flushing, without burning, without tension or apparent swelling in the part, which returns by fits and starts for shorter or longer attacks, closely spaced, often irregular and sometimes periodic; . . . 2. by the site of the pain, which is always associated with a nerve trunk, or on a branch of the nerve and which, during paroxysms, propagates itself and surges up from the originally affected point to all its ramifications; it rapidly travels along them like a streak of lightning as far as their terminal extremities, following them in their diverse connections and affecting them successively one after another at times and sometimes all together or, at other times, restraining itself more particularly to one or two of its filaments."[169] With this authoritative description, Chaussier effectively gave birth to the nosological category of neuralgia, setting it apart from both the vague, undifferentiated class of neuroses or nervous disorders, and from that constituted by inflammation of the nerves, or neuritis. Under the heading "anomalous neuralgias," Chaussier grouped together neuralgias caused by tumours which were compressing nerves in their trajectories, and those which were the consequences of wounds inflicted on the nerve—as a result of blood-letting, for instance—even a long time after the accident occurred and healing was complete. He noted, almost as a distinctive sign of neuralgia, that sectioning the affected nerve stopped the pain for a certain length of time, though it inevitably returned, sometimes in a more severe form.

However, the idea of a radical separation between neuralgia and neuritis was not readily accepted as, in 1819, Montfalcon—the author of the article "Névralgie" in Adelon's *Dictionnaire des sciences médicales*—wrote: "The natural place for neuralgias within a nosological framework is among the phlegmasiae (*i.e.* inflammations); one day they will be allotted that place and will retain it."[170] At the same time, the neuroses class, which was particularly abundantly filled in Pinel's work, included all nervous illnesses for which it was not possible to reveal any organic lesion, for example. Thus neuralgia was caught between two interpretations, one which wanted to identify it as being due to a transient inflammation of unknown causes even though this did not agree with most observations, and the other which, in view of the failure of pathological anatomy to find any lesions, was likely to identify it as a mental illness, that occasionally was of hysterical origin. This debate was pursued throughout the 19th century since, in 1840[171], physicians were still asking themselves whether neuralgia was a simple "functional disorder" or whether it was caused by some irritation of either the nerve sheath, the neurilemma, or of the nerve core: during that period, the histology of the nervous system and the use of the microscope did not seem to have penetrated into the world of hospital clinicians and physicians. At the beginning of the 19th century, neuralgia was an enigma to medicine—a collection of symptoms with a very diverse range of proposed aetiologies (syphilis, wounds, gout and rheumatism, toothache, a predisposition likely to be set off by the slightest breeze). At times, neuralgia was treated by anti-inflammatories (local blood-letting, diet) or irritants (vesicatories, rubefacients), at others, by surgical operations which were generally not very successful and were virtually abandoned by the middle of the century, or by cautery (a red-hot iron was applied to the inferior maxillary nerve in order to destroy it), or by moxibustion. All these therapeutic means were aimed at stimulating the sick nerve. In this therapeutic arsenal, which was heterogenous enough, the particular place of the antispasmodics should be noted (diethyl ether, opium alone or associated with camphor, orange flower) and, among these, Méglin's famous pills composed of zinc oxide, black henbane extracts and wild valerian root—these plants having sedating and analgesic properties. Méglin's pills, which were the object of controversy in the medical press, constitute a good example of pharmaceutical advertising at the beginning of the century. But the results were not always satisfactory and the clinical description of the pains of trigeminal neuralgia, with its intermittences and paroxysms which could be triggered simply by eating or talking, concluded by stating that medicine was powerless: "They are so sharp that patients willingly submit to the most painful operations without flinching and they themselves ask for vesicatories, the burning moxas, and the cruellest cauterisations; in the hope of putting an end to their ills, they desperately wish for the succour of the scalpel which is generally so dreaded. We have even seen several sufferers who, driven to despair due to maxillary neuralgia, did not hesitate to have all their teeth pulled out one after another . . .; others, having unsuccessfully tried a multitude of different treatments, and no longer believing in either the power of medical craft or of nature, have finally put an end to a miserable existence which had become the most awful of tortures for them."[172] The doctor who wrote these words had nothing condemnable to say against suicide in this case. Despair when confronted with intractable pain inevitably led to such a solution and the physician's only comforting words were to emphasise that not all neuralgias were so merciless.

Despite the clarity with which the pains were described, the prevalent concept of neuralgia amongst clinicians around the middle of the century attributed trigeminal neuralgia, or essential neuralgia, to the same category as some visceralgias which were generally more extensive, "saturnine" neuralgias caused by lead poisoning, and even the "marsh fever" neuralgias which were linked to malaria: rather than being an illness, neuralgia was seen as a symptom. Among other physicians of the period, Valleix tried, in his *Traité des névralgies,* published in 1841, to update the classification of essential neuralgia by proposing new criteria: the superficial location of the nerves attacked by neuralgia and the sensitivity to pressure which could unleash attacks, already mentioned by Chaussier, took on considerable importance. At the same time, he was carrying out research to determine the particularly painful points along the trajectory of the nerve, which were the true foci of sensibility from which pain irradiated in an ascending or descending direction. Valleix, who was a physician at the Central Office of Civilian Hospitals and Hospices and a member of the Société médicale d'observation and of the Société anatomique, had accumulated a considerable amount of experience on all sorts of neuralgias, based on hundreds of cases of each type. These "Valleix's points," which remain in medical terminology, corresponded particularly to points of emergence of nerve trunks where they were sometimes compressed, and also occurred at the level of nerve terminals as well as along the nerve fibres at those points where they tended to come close to the skin surface.[173] Thus Valleix made the distribution of his painful points one component to be considered in a differential diagnosis and equated them to tactile-sensibility disorders. Trousseau, in his medical clinic at the Hôtel-Dieu, accepted the localisation of painful points at spinal-process level where the nerve emerges from the intervertebral foramen, the sensitivity to pressure, the hyperaesthesia as well as the nature of the pain; he used these signs to characterise "*tic douloureux*," "trigeminal neuralgia," which is also named after him, and which simultaneously involves motor (spasms, convulsions) and sensory disorders. However, he raised a major question which was going to fuel the debate about whether the origin of neuralgia was central (in the cord) or peripheral.[174]

The reappearance of pain after the section of the affected nerve did not prove to be as decisive an argument in favour of a central origin for neuralgia as was at first thought. In fact, as there was an initial relief, the supporters in favour of the peripheral origin of neuralgia were in an advantageous position in maintaining that the neural lesion extended above the point of the section. Two other explanations for this return of pain after the interruption of the communication pathways were put forward: the presence of recurrent fibres, which were conceived along the lines of the model of recurrent sensibility, which had just been put forward by Magendie, and the questioning of the functional independence of each nerve—a concept which had been dogma since Müller's laws and which had led some physicians to refer to sensibility "supplemented" by anastomosis. The observations which were made notably by Axenfeld[175] seemed to indicate that pain did not restrict itself to the trajectory of the nerve, but extended to the sensory fibres and to the multiple connections which they maintained. Experiments carried out by Arloing and Tripier[176] showed that the sensibility of a finger, for example, did not depend on a single branch of a nerve, but that at least four such branches had some involvement in the sensory innervation of any one given territory. As a result, one was never sure that all the branches apt to produce pain had been sectioned. Moreover, this type of analysis was only possible thanks to the use of electricity as a means for diagnosis rather than just for treatment. Finally, recurrent pain brought another distinction into play which was particularly important: it was a matter of determining whether neuralgia was "real," that is whether it was truly located in the spot where the patient thought he felt it, or whether it was only "virtual," in other words referred by a reflex action.

Few painful symptoms have led to so many contradictory interpretations as did neuralgia in the 19th century—it was one of those questions for which clinical medicine could provide no irrefutable answers one way or another. The supporters of the "central theory of neuralgia," in addition to drawing analogies with the sharp, "lightning" pains of tabes which, in this case, originated in the cord, based their convictions on the absence of any lesions under microscopic examination: Anstie in Great Britain, Romberg in Berlin, and Vulpian in France proposed that there was "a modification in the intrinsic nutrition of the neural tissue" and thought "that this dynamic modification, imperceptible to our available means of analysis, is the cause of neuralgia."[177] However, when it came to nutritional (or trophic) disorders, it was clear that it would be necessary to look for their cause at the level of the nerves' points of origin—whether these arose directly from the cord or from the posterior root ganglion—which was their "trophic centre"; according to Vulpian, in particular, such a central origin allowed for a better understanding of a number of complaints associated with neuralgia: muscle spasms or paralysis, zones of anaesthesia, vasomotor problems, etc. Thus, even when the origin of the neuralgia seemed to be linked to a lesion of a peripheral nerve (neuralgia following a blow or a wound for example), Vulpian thought that the persistence of pain after its original cause was eliminated could be explained only by the fact that the disturbance had altered the nerve centres. Some, such as Ouspensky, basing his reasoning on the hypothesis of nutritional problems, foresaw chemical metamorphoses—"modifications in the molecular exchanges"—but the available means of analysis in no way allowed a decision to be made between these two theories. Neither clinical observations, nor any microscopic evidence from histological studies, nor the reasoned analysis of the results provided by different therapeutic methods led to any satisfactory explanation for such an unyielding, cruel pain with which a variety of factors seemed to be associated; moreover, it was highly difficult to untangle the web of interdependent causes and effects: for instance, the flushing and burning which might occur in certain neurological attacks were indeed due to the paralysis of the vaso-motor nerves, but this was not the cause of pain but rather a consequence of reflex action.[178] From the therapeutic standpoint, in view of the failure of neurotomies whose use was advised only as a last resort, only two methods were considered to be relatively effective: on the one hand, hypodermic injections of morphine at the painful points and, on the other, the use of electrisation, a method which was used for almost all pains and to which we will return in a more composite fashion. The very results of these therapeutic techniques provided a mass of contradictory clues as to the causes of neuralgia and the debate remained open.

### *Weir Mitchell: Pains of Traumatic Origin and Causalgia*

The American Civil War was responsible for a considerable number of wounds of traumatic origin, and these ultimately resulted in either amputation or, when it had been possible to save the affected limb, in neuritis due to the presence of a bullet or the partial destruction of the nerves. In both cases, as is true with fractures, dislocations and different forms of nerve compressions, the soldiers continued to feel awful pains for months and sometimes years. The most common of these pains were neuralgia, phantom-limb pain, and causalgia, or the impression of intolerable, intense, burning pain. All surgeons who have had to practice in times of war have become acquainted with these chronic, frequently inexplicable pains, for which medicine was then singularly ineffective. Leriche, during the First World War, was confronted with the same type of cases as Weir Mitchell, and military surgery up until the present day has had to resolve the terrible after-effects of nerve wounds. At the time when, during 1863, Weir Mitchell began to put together his observations acquired at the hospital in Philadelphia to which—at the initiative of W. A. Hammond, the Surgeon General of the American army—all nervous illnesses were referred, there was still no substantial body of observations available for reference. Furthermore, it was still not always clear how to distinguish the psychological after-effects of the traumas of war (anxiety, hallucinations about the circumstances under which a wound was inflicted, changes in character) from the existence of a very real persistent pain, which lasted for a long time after "recovery," and which tended to be interpreted as a psychological problem, or an imaginary pain, rather than as a real pain.

The work of Weir Mitchell, by the number and the quality of his observations, and by his permanent concern for linking them to anatomo-physiological knowledge and for advancing the aetiology and semiology of nervous-system disorders, constitutes a classic reference which is still in use today. He brought all the available methods of investigation together and sometimes tried to test hypotheses derived from his clinical observations by experimenting on animals. The way in which he set out the differential description for neuritis and neuralgia deserves to be recalled: "Subacute neuritis is difficult to diagnose when it is of average intensity and does not result from a traumatic accident . . . The affected nerve is sensitive for a considerable part of its trajectory, the points of emergence from the bones or aponeuroses are the most painful. By contrast, in genuine neuralgia, the Valleix's points are the only ones which produce acute pain in response to pressure. Contrary to neuralgic hyperaesthesia, that of neuritis is uniform and constant at all times."[179] If, in both illnesses, exacerbation of the paroxysms might occur at night, on the other hand, the impression of sclerosis of the nerve trunks, which became hard and swollen, and the disorder's regular ascending propagation allowed them to be distinguished easily: "Everything which excites the circulation heightens the pain and exacerbates the suffering. This last fact is so constant that absolute rest is a vital part of the treatment."[180] In addition to this, Weir Mitchell prescribed enveloping the painful limb in an ice pack all day long—this was at the time when rubber bags appeared in hospitals—and since the pains occasioned by the ice were initially difficult to endure and supplemented those of the neuritis, he also administered hypodermic injections of quite high doses of atropine sulphate and morphine every four hours. It was on the basis of observations made following treatment with morphine for chronic pains due to war wounds that the first questions concerning morphine dependency were posed: Weir Mitchell indicated the routine usefulness of the remedy "without forgetting that by using it unnecessarily, one engenders very unfavourable physical and moral conditions."[181] One can gain quite a precise idea of the importance of its use in the United States in the hospital run by Weir Mitchell by the fact that he notes that 40,000 doses of narcotics were distributed in one year, and mentions that one patient received up to 500 such doses. "This mode of treatment with narcotics has come into current use today and one cannot have too great a confidence in it. In our work at the military hospital for nervous diseases, resident surgeons went into the wards with the injection apparatus two or three times per day, finding themselves faced with . . . anguish and pain, and afterwards they left behind them well-being and sometimes a smile. This picture is not exaggerated, since few hospitals have seen as much suffering and torture as ours. There were times when each assistant gave between 60 and 80 hypodermic injections per day and per night."[182] There was consequently no reticence at all in using morphine, which was administered directly by a surgeon, and not by a junior intermediary; Mitchell stated that it was superior in its effectiveness to anything else which had been proposed—aconite, atropine, veratrine, etc. Thus it was after the 1870s that the limitations of opiate remedies began to be questioned by the medical world which, up until then, was not aware of the problem.

Among the painful ailments which were best analysed by Weir Mitchell was causalgia, whose etymology itself evokes a sensation of burning, a very common symptom of this strange illness, shared by the *causos* mentioned in the *Hippocratic Collection*. Appearing much later than the wound itself, causalgia tends to be localised in the hand or the foot.

Those afflicted by this condition are sensitive to the smallest external stimulus—a breath of air, the lightest caress, vibrations caused by walking—and, in order to avoid these, they must take all sorts of precautions such as enveloping the affected limb in bandages, or keeping it in water all the time. This particular pain is a veritable torture, which compromises the health and psychic equilibrium of the patient. Mitchell left a classical description of it: "In our medical practice, we have frequently encountered patients who complained of very acute pains, which they themselves compared to a burn, or to the action of a very hot mustard plaster, or to the effect of a red-hot file abrading their skin,"[183] which often takes on a glossy appearance. The treatments used to relieve causalgia were the same as those used for neuritis, but this was the only condition for which Weir Mitchell extolled the value of an injection of morphine directly into the painful part—while for all other ailments the site of injection was unimportant, and the result identical. To this were added wet compresses which were renewed as often as possible, and vesicatories. The problem of causalgia rekindled the debate on the cause of pain and resulted in different observations which suggested that the causes were multifactorial: circulatory disorders due to the action of sympathetic vasomotor nerves, nutritional problems, reflexes mediated through the cord which brought about "subjective," but real, pains remote from the original lesion, hyperaesthesia of the skin, exacerbation of pain due to the anxiety of the slightest contact and the continual attention paid to the pain or to the means of preventing it. Thus one was dealing with a vicious circle in which different factors accumulated and mutually reinforced one another. In addition, Weir Mitchell, being very critical of the notion that there was a specific sense of pain, was fully aware of the complexity of the pain phenomena which he considered in its entirety, and he remained cautious of unequivocal explanations, such as those put forward by Vulpian. Very curiously, the preface which the latter wrote for the French translation of Mitchell's work was ambiguous—it paid homage to the work accomplished but was fundamentally critical, particularly of its explanations for pain.

*Charcot, the Clinique de la Salpêtrière, and Nervous Illnesses:*
*The Example of Progressive Locomotor Ataxia (Tabes Dorsalis)*

During the last third of the 19th century, Charcot's work dominated the pathological study of nervous and mental diseases to such an extent that we are only able to provide a brief overview of his research here. Even if one limited oneself strictly to his contribution to the study of cerebral localisations or the different illnesses named after him, or his contribution to the definition of hysteria, in each case there would be enough material for substantial discussion.[184] We have therefore chosen to describe an illness—*tabes dorsalis* or progressive locomotor ataxia— which is slipping from our memories but which was very prevalent in the 19th century. The clinical signs of this illness had been described by Duchenne de Boulogne in conjunction with the electrisation treatment, while its histological signs had been studied by Charcot. The tertiary stage of syphilis, ataxia is also remembered through some remarkable works of literature, the best-known being *Le Docteur Pascal* by Émile Zola[185] and the intimate notebooks of Alphonse Daudet, who suffered from the illness, and left us a testimony of rare intensity under the title of *La Doulou*. To describe the spirit of the Salpêtrière briefly, one can do no better than let Charcot himself speak: during Vulpian's funeral, he referred to this critical moment in the history of medicine when morbid microscopical anatomy had had its day, when Cruveilhier had "extracted" everything one could expect of it and when medicine would henceforth walk on two feet: "Without the backing of experimentation, pure observation is often useless, while, on the other hand, experimental findings, at least when they are concerned with human pathology, have virtually no legitimate application unless they are constantly subjected to the supreme test of clinical medicine."[186] Before the description of the problems of ataxia by Duchenne de Boulogne in 1858, what was in fact a lack of motor coordination without any decrease in muscle strength was diagnosed as being paralysis or a loss of muscle strength.[187] A few sketchy descriptions had been given by Benjamin Olivier Todd in Great Britain and by Romberg who failed to make a distinction between ataxia and paraplegia, both of which he included under the title of tabes. The classical description of this "primitive chronic illness which simultaneously attacks different parts of the nervous system and which, as a rule, progresses degeneratively and often has a fatal issue" included three phases. The first involved fulgurant or shooting pain, sometimes accompanied by partial loss of sight. In this phase, the attacks often occurred at night and lasted for several days; the patients then described a sensation of the body being stabbed by a dagger which was then driven in and twisted—or what was called a boring pain—which was then followed by an increased sensibility; this phase also included pains which gave the patient an impression of being compressed in an armour or a vice,[188] and this feeling lasted throughout the intervals between the shooting pains. It was in the second phase that the lack of motor coordination appeared which led to a hesitant, hopping type of gait like that of what Daudet described as a "hobbling cripple"; thus going down stairs or crossing a road became quite a challenge. Finally, the last phase was one of paralysis. The patients remained completely lucid throughout the illness which was also characterised by a progressive loss of tactile sensibility. Histological analysis showed a deterioration of the posterior funiculi in the cord. Daudet described what his life was like during his illness, his growing isolation from the rest of the world, the joy obtained from injections of increasing doses of morphine in order to calm the suffering, the bromides, the time spent, uselessly, at Lamalou-les-Bains or Neyris observing the impending demise of patients at a more advanced stage of the illness than himself, the sessions of elongation of the cord when he was suspended for hours by a hook with Seyre's apparatus and from which he came away feeling shattered. Above all, there was the certainty of the relentless progress of the illness which he had learned as a result of a meeting with Charcot, a keen "observer" whom he likened to himself by the succinct and sharp nature of his remarks.

No doubt it is significant enough that the medicine of nervous diseases at the end of the 19th century achieved so much in localising the lesions which brought about the pains of tabes, the attacks of multiple sclerosis or of infantile paralysis (poliomyelitis), although it did not progress to the same extent where aetiologies and treatments were concerned. In the case of tabes, the time-lag is striking: Charcot described in minute detail "the sclerosis of the fine lateral fibres of the posterior funiculi" of the cord which were "the only constant lesion of locomotor ataxia"[189] and spoke of radicular pains. However, no treatment against syphilis was yet envisaged, although several authors had described the link between these nervous symptoms and syphilis. This situation points up the problem concerning the limits of whatever model is dominating medicine at a given moment in time, as well as that of the rift between the disciplines at the very heart of "scientific medicine": in this instance, for example, the absence of links between microbiology on the one hand and neuro-histopathology on the other was shocking. In terms of the politics of scientific research, such a situation in which each discipline works to their own accepted set of norms, but without any direct communication with other branches, may provide food for thought.

### *Pains for the Rich, Pains for the Poor: A Myth to be Demolished?*

Among the oldest pains described in medical literature are those of podagra or gout, which was often presented as being an illness of the wealthy who were healthy and too well fed. This criterion was even used to distinguish it from rheumatism: gout "respects hot countries virtually entirely, does not affect poor populations nor nomadic tribes: it is an illness of temperate climates, civilised countries and large cities. Rheumatism in contrast is rife in every zone and every latitude, and at the same time befalls the peasant, the city dweller, isolated individuals and those in crowded urban areas. One is the aristocratic ailment which generally prefers to afflict the rich classes: a generous table, intellectual activities and inactivity predispose one to it. Rheumatism attacks everyone and if it has a preference, it is for the sick who lead a poor, tiring life of toil and who are exposed to inclement weather without having sufficient resources to afford the reparative nourishment needed to withstand it."[190]

Gout, which was an hereditary ailment according to the majority of 19th-century doctors, thus necessitated a particularly sober lifestyle if one were to have any hope of escaping the transition to a constitutional diathesis brought about by this "morbid constitution." An attack of gout was always characterised by violent pain in the big toe, which might occur in the middle of the night, although the sufferer believed him- or herself to be in perfect health. Trousseau reiterated the main elements of Sydenham's description, noting the nocturnal exacerbation, the impossibility of putting up with blankets and the awful suffering caused by the vibrations of the floor when a vehicle went past in the street: "Those who have endured it compare it to the sensation of a nail being driven into their joints, to the tearing of the flesh with the jaws of powerful pliers, to the bite of a dog whose teeth gnaw their bones, to an intense pressure from a vice, to the agony of the boot when the torturer squeezed the legs of the unfortunate victim between oak planks and hammered wedges into the space separating them. In a word, the person with gout used the most excruciating images to express the infernal suffering which he endured."[191] If the clinical description was well-established, it was virtually only at the very end of the 18th century, however, that a link was established between gout and uric acid: "The day when, in 1797, Wollaston and Tennant discovered that gouty concretions were made up of 'urate of soda' [monosodium urate], considerable progress was made in the history of the disease."[192] In the middle of the 19th century, discussions continued as to whether gout was a localised disorder or a more generalised predisposition, not only producing articulation pains, but affecting the whole organism; this was not without relevance to the ultimate treatment. It was the work of the British doctor, Garrod,[193] which was authoritative, however, because he took into account all the symptoms and explained gout both in terms of the quantity of uric acid in the blood and by a malfunctioning of the kidney. He also established criteria other than sociological ones for clearly distinguishing chronic gout from rheumatism: the mode of invasion of the joints was less selective in the case of rheumatism, the pain was also less intense, there was a lesser degree of inflammation, and heart complications were more frequent.

The evolution of treatments and the way pain was dealt with were progressing: to Cullen's old precept, "Patience and flannel" were substituted remedies which brought greater relief, nonetheless always accompanied by the persistent attitude that the painful attack was a salutary crisis which had to be allowed to run its course: "Without recommending an abortive treatment of the attack, it does not seem to us indispensable that patients should be allowed to suffer, without trying to alleviate their distress; we believe, in fact, that the length of their attack can be reduced without entailing future troublesome consequences. . . . Managed with prudence, the medicines can greatly relieve the sufferers' pains and even induce the crisis to disappear progressively without provoking the slightest visceral repercussion."[194] The caution of this remark, though it had Charcot's wholehearted support, reveals to what an extent the vestiges of humoural pathology were hanging on, and the apprehension that gout could "return," "revive" or "retrocede" and therefore become all the more difficult to cure, even though its chemical aetiology seemed well established.

This opinion still prevailed at the beginning of the 20th century in the work of Georges Dieulafoy, for example, who in his *Clinique de l'Hôtel-Dieu,* reprinted numerous times, still asked whether one should "let nature act" or whether intervening exposed the patient "to the terrible hazards of revived gout." And this was despite the fact that efficacious remedies were known of from the end of the 1870s: quinine sulphate to relieve febrile and inflammatory symptoms of the illness, and above all sodium salicylate, aspirin's predecessor, whose efficacy against acute articulatory rheumatism came to be appreciated. Sée, in his two communications to the Académie de médecine,[195] extolled it unreservedly in spite of the hesitations of his colleagues. Considered to be a first-class analgesic, sodium salicylate was administered in large doses—6 grams daily, then, in the case of Dieulafoy, 8 to 10 g. However, others limited themselves to 2 to 3 grams daily, fearing that this drug was not properly eliminated. These posological variations in the use of a new remedy had numerous reasons: fear of stopping the inflammation which was considered to be beneficial, fear of provoking a toxic accumulation of the medication, the lack of hindsight in an area which smacked of therapeutic experimentation, as well as plurisecular beliefs in a particular essential sequence in the course of the illness. Thus, the other great remedy, whose action was evident in "the very rapid disappearance of articular swelling, by the easing of the pain and the sedation of the pulse,"[196] that is meadow-saffron *(colchicum autumnale),* was not given at the onset of the attack, but only after about a week when the patient started to weaken. This cautious attitude, due to various factors, was, to some extent, also in contrast to the audacity of the therapeutic trials in the first half of the century and up until the 1860s. It was as if physicians confronted by "the chemical explosion" which characterised therapeutics at the end of the century—and in view of the established toxicity of certain products (as is eloquently obvious in the case of veratrine)—became more acutely aware than they had been in the past of the necessity of testing certain products thoroughly before being able to use them properly; the problem of dosage, for instance, was not a simple matter.

At the opposite end of the sociological spectrum—though gout was only referred to as the "aristocratic disease" discreetly by the beginning of the 20th century—was a particular class of especially painful neurites. It might well be asked to what an extent these did not constitute "a nosological artefact." What we are referring to here is multiple neuritis (polyneuritis) of toxic origin, often caused by alcoholism among other factors, which was particularly reputed to affect the working classes.[197] Many of the symptoms of polyneurites, which had previously been considered to be disorders arising in the spinal cord, were, at the beginning of the 1880s, interpreted as peripheral afflictions of infectious or toxi-infectious origin.

Recent histories of alcoholism have shown to what an extent the definition of a type of pathology may be linked, at a given moment, to a set of political and cultural circumstances. It is therefore essential, beyond the widespread discussions on the subject, to arrive at some critical evaluation as to their actual degree of reality. The medical historian knows that at certain points in time, outside of the well-defined framework of epidemics, a particular pathology can come to the forefront, not necessarily because it was particularly ravaging but because the attention of the medical world was focused on it—be it from an aetiological point of view which allows a new coherence to be given to a group of symptoms and enables them to be organised differently, or from the standpoint of possible treatments. For example, this was emphasised by Gowers when he presented the three possible forms of impairment to the nerves in neuritis, *i.e.* impairment to the exterior sheath or perineuritis, impairment of the interstitial tissue of the fibres, or a degenerative deterioration of the fibres themselves: "Our knowledge about this latter type is recent, but whether it is acute or chronic, this process has proved to be much more frequent than was thought some years ago."[198] Polyneuritis of alcoholic origin affected young adults between the ages of 30 and 50, and particularly women, although men were reputed to be more addicted to drink. This disorder took on the magnitude of a full-blown cultural and social myth, partly as a result of literature which adopted the topic as did the work of Zola, for example, and thus fuelled its prominence. In constituting this "end-of-the-century" myth which crystallised many social fears were a collection of pseudo-scientific "reasons," including Morel's theme of inherited degeneracy, political fears such as the growing organisation of the working classes in urban areas, and economic and social ones like the development of work for women in workshops and factories, and their effect on challenging the traditional role model of women as keeper of the house and family. But even if this myth did not actually reflect reality, it nonetheless did, in its own way, express a truth which casts light on fundamental issues in times of crises.

## Between Diagnosis and Treatment: Promising Techniques in the Fight Against Pain

By considering a certain number of particularly painful illnesses, we have shown the diverse methods of treatment in which a central role was played by the isolation of new substances under the impulse of pharmaceutical chemistry which at the time was being industrialised.

However, particular attention must be paid to two techniques, which were not unknown in the 18th century, but which took on greater importance in the following century: on the one hand, the use of electricity in medicine and on the other, the techniques of suggestion and hypnosis. It might appear surprising to consider these together in this study since, at first glance, everything concerning electricity, a material agent, seems quite the opposite from anything concerning hypnosis. At the beginning of the 19th century, it had become perfectly clear that the latter did not act as the result of the flow of whatever imponderable fluid the supporters of animal magnetism in the previous century had thought. However, three reasons may be put forward for the reconciliation of these two techniques: both bring up the issue of actual innovations in medicine; both were conceived initially for therapeutic ends, then became instruments of diagnosis; and finally each had uses which, particularly in the case of hypnosis, preceded the elucidation of their mechanisms of action, and these two techniques may ultimately hold promises which have not yet been totally explored.

### Electricity and Medicine

The use of medical electricity dates back a very long time as, even in Ancient times, the properties of the torpedo fish were known, and it was used to ease rheumatic pains. However, in the middle of the 18th century, with the invention of the Leyden jar, the use of static electricity in medicine took a great leap forward, above all in the treatment of paralysis and nervous illnesses, particularly hysteria. The physicians who took an interest in it, Jallabert in Geneva, Sauvages in Montpellier, the Abbot Bertholon, and most particularly Mauduyt, or Marat, among many others,[199] defined the different modes of application of this electricity by means of sparks, or in an "electric bath," etc. They also defined the duration, intensity and principal illnesses in which it could be of use. A second stage was reached with the discovery of "animal electricity" by Galvani:[200] this electricity was produced by a "metallic arc" that placed a muscle and a nerve in contact via the intermediary of a conducting metallic body. In these decisive experiments which were performed in 1786, then published in 1791, although the frog was dead, its muscles contracted energetically though all other stimulants tested turned out to be powerless. These experiments were to form a basis for all of future electrophysiology and were soon famous through-

## ⟜ PHARMACEUTICALS AND CHEMICALS AT THE END OF THE 19TH CENTURY: FROM SIMPLE LABORATORY TO MAJOR ENTERPRISE ⟜

Throughout the period when anaesthetics were first developed, and morphine was originally isolated and prepared, questions arose about ensuring the necessary supply of raw materials, the required purification processes, and the subsequent conservation of the end products on a much larger commercial scale than simply that of the pharmacy dispensary. But a new era opened up completely with the advent of synthetic medications which began to compete with the natural substances obtained from the animal and plant kingdoms.

The first international pharmaceutical congresses, particularly the 1865 Congress in Brunswick and the one in Paris in 1867, reveal that initial attempts were being made to conciliate pharmacists' commercial and economic interests along with the official status they sought to acquire for their "sanitarian" mission. A great deal of discussion focused on the definition of the pharmaceutical "speciality," because this field conferred the right to exploit therapeutic remedies; however, this guarantee often came up against opposition when secret formulas were involved. The Union of French Pharmacists was established in 1877 with the aim of "assembling, under the Pharmaceutical Society's patronage and with their help, all isolated French pharmacists who have maintained the scientific aptitudes they developed in school and wish to cultivate them further and preserve them for the benefit of French society as a whole." At the same time, a number of publications were also consecrated to the link between chemistry and pharmaceuticals, and particularly the *Journal de pharmacie et de chimie,* founded in 1809. For its part, the *Bulletin de thérapeutique,* providing information on the latest methods and products as well as devoting considerable space to the most recent discoveries from abroad, was particularly interested in publishing clinical results. Keen interest in therapeutic research also accounted for the establishment of the Société thérapeutique and for the numerous discussions on all these questions at the Académie de médecine.

It is quite common to contrast the powerful German chemical industry (as well as the Swiss) in the 1880s to the more artisanal nature of the small family firms producing pharmaceutical products in France at that time. It is true that chemical dyes had earned considerable industrial importance in Germany which influenced research substantially. However, as P. Boussel, H. Bonnemaison and F. Bové emphasise in their *Histoire de la pharmacie et de l'industrie* published in 1982, there were also small family concerns in Germany just as there were in France, who were producing pharmaceutical products and expanding dynamically. The first such pharmaceutical factory was established by Genevoix, under the "July Monarchy"; the Étienne Poulenc factory was founded in 1859 in Ivry, and specialised in mineral compounds for photographic and therapeutic uses; the Usines du Rhône corporation was later founded in 1885; and this was followed by the Roussel firm at the turn of the century. Aside from the different economic potential in each country, it is worth examining their different approaches to pharmaceutical research both in their universities, on the one hand, and through the industrial world on the other.

Thanks to *L'Officine ou Répertoire général de pharmacie pratique*, published by Dorvault throughout the 19th century and continually updated, it is possible to keep track of the progressive appearance of multiple pain remedies with suggestively feminine-sounding names (in French at least!) such as: Migrainine, Anesthesine, or Duboisine. They can be generally classified as "encephalic agents (cerebrospinal)" which included tetanic convulsants (nux vomica, strychnine, upas, curare, etc.), stupefactive convulsants (bitter almond, cherry laurel), stupefactive narcotics, both deliriant and nauseant, and "inebriants" (hashish, mandrake, chloroform, nitrous oxide), as well as "metallic encephalics" (lead, zinc, silver, gold, bismuth). In 1867, for instance, the *Officine* listed aconitine, the active principle in aconite, discovered by Brandes in 1833 and used to relieve rheumatismal pains, gout, cancers, and neuralgias; veratrine, extracted from white hellebore roots or from autumn crocus, was used to relieve the same ailments.

Willow bark, whose properties had been known since Antiquity, was studied by the French pharmacist H. Leroux, who was able to extract its active principle in 1829; this paved the way for research on salicylic acid, soon followed by work on the salicylates, substances whose antiseptic properties were recognised as much as were their analgesic properties; Sée recommended sodium salicylate for acute rheumatoid arthritis. The 1880s witnessed an explosion of research aimed at discovering new therapeutic chemical agents, and there was evidence then of the problems to come due to the many various names given to products which contained substantially the same active ingredients, or some which were composed of dubious combinations with more toxic than beneficial effects. The *Journal de pharmacie et de chimie* published its "Information Sheets" every month, providing details on newly produced remedies, their secondary effects, and their withdrawals from the market. In 1883, Ludwig Knorr prepared a synthetic product, anti-pyrine, introduced on the market the following year to reduce fever and relieve pain. The salicylates, which had already been known for a while, were supplemented at that point by a product which was to revolutionise the relationship human beings had to pain, a product called aspirin and marketed by the German firm Bayer. In actual fact, aspirin had already been synthesised by Charles Frederic Gerhard in 1852 (acetylsalicylic acid), but this Strasbourg inhabitant, who also synthesised acetanilid, had come to work in Paris but had been unable to gain proper recognition from the all-powerful *protégé* of Napoleon III, the chemist Dumas. In 1897, however, the pharmacist and chemist Felix Hoffmann, who had begun working for Bayer in 1894, and had an ambition to discover a substance that could relieve pain without the unpleasant side effects of sodium salicylate (nausea, stomach cramps, edema), rediscovered aspirin's composition and the product was finally commercialised in 1899. This medication's success and the industrial and financial stakes involved gave rise to what were known as "the Aspirin Wars" (*The Aspirin Wars. Money, medicine and 100 years of rampant competition*—a study by C. C. Mann and M. L. Plummer).

Where acetanilid is concerned, this precursor of paracetamol also enjoyed a rebirth in 1888 thanks to the Frankfurt pharmacist E. Ritsert, who managed to remove all its impurities; it was listed in the *Officine* in 1898 for its analgesic and antipyretic properties. In the *Journal de pharmacie et de chimie,* the first adverts were beginning to appear and the distinction between information and publicity was rather nebulous. After all, wasn't Adrian, who was in charge of the regular column, "Note sur les nouveaux remèdes" (Note on the latest remedies), also a corporate head promoting the benefits of a composite medication made up of aspirin, caffeine, and citric acid, marketed under the quaint name of Migrainine? Under the "Anaesthetics, hypnotics and narcotics" category (the term analgesic was not then currently in use), nearly thirty new products were itemised each year, providing an ample supply of remedies for therapeutic experimentation. Even if morphine remained the fundamental medication at the turn of the century, the aspirin and acetanilid groups were beginning to modify life substantially.

out Europe. However, they came up against a different, purely physical interpretation such as was proposed by Volta, which explained the contractions as being due to two heterogenous bodies being put into contact, rather than being due to electricity which was inherent to the human or animal body. The controversy which arose as a result has already been analysed many times[201] and if one wanted to provide a quick assessment of it, one could say that, on the one hand, it opened up the way for the development of electrophysiological studies which were very actively undertaken in Italy by Nobili, Matteucci and Marianini, and then in Germany,[202] and this permitted the concept of electrotonus to be described in detail; on the other hand, it led to the use of galvanic current, which was obtained with the help of Volta's battery.

However, at the therapeutic level, galvanisation did not produce the expected results and, in the years 1820–1840, it had ceased to be used on any large scale; the disenchantment was proportionate to the confidence which had been placed in it. Its use proved to be painful for the patient, sometimes with tragic consequences, such as that reported by Duchenne de Boulogne in the treatment of a case of amaurosis in which his patient lost his sight.[203] The apparatus used was hardly manageable, the current was quickly used up and its intensity was not constant. In addition, as had been shown in the case of Golding Bird in England, the practice of using electricity was sometimes considered to be practiced by quacks and, in any case, required no professional qualification.[204] In the second age of electricity, which was characterised by galvanism, a few apprehensive trials were carried out by Jean-Baptiste Sarlandière, for example, who in 1825 had published his *Mémoires sur l'électropuncture*. He had tried to combine galvanism with acupuncture needles placed where they would normally be inserted; he tried this technique to cure neuralgia, but the process was too painful to be acceptable to patients and was not met with the approval of his colleagues. Moreover, electricity remained more of a stimulation therapy for treating motor disorders than as a means for relieving pain. Thus, when Magendie presented a paper at the Académie des sciences, on an unyielding case of trigeminal neuralgia which had been cured with electricity,[205] he attracted new attention to a practice which was running the risk of falling into disuse, despite the interest which Rayer and Andral showed in it.

A shift was taking place in France from the 1850s onwards thanks to the research of Duchenne de Boulogne and also on the other side of the Rhine, under the impetus of Remak, Du Bois-Reymond, then Helmholtz. However, while Remak vigorously defended galvanisation, Duchenne chose to administer electricity obtained through induction. In the medical hierarchy of that time, Duchenne was regarded as being somewhat different, a sort of outsider who had begun to be interested in the medical use of electricity and had trained himself in it. Following family problems, he had left Boulogne where he had first settled, then returned to Paris where he had studied. At a time when the title of Hospital Doctor or Professor was a prerequisite for professional and social recognition, Duchenne earned a living from his private practice;

however, thanks to links which he had kept with his old teacher Cruveilhier, who was a professor of pathological anatomy and a specialist
in the study of the nervous system, and with Briquet who was at la
Charité, he was gradually able to gain access to every hospital service,
to become the expert to whom difficult cases were referred and thus
to gather together an enormous series of clinical observations which
allowed him to establish a theory of "localised electrisation." Without
these links, the technical innovations which he had thought up would
have been of little value; in fact, he received a prize from the Académie
de médecine for his double-reeled magneto-faradic apparatus, and for
perfecting the current interruptor and the maintenance of a current of
constant intensity. The use of the intermittent pulses of induced current
was important because Duchenne had observed that the effects of the
current were apparent not while the current was being conducted, but
when the circuit opened and closed. In memoirs presented to the
Académie des sciences in 1849 and to the Académie de médecine in
1852, he undertook a critical examination of the different types of
electrisation apparatus and the different types of electricity.

However, his innovations relied less on the choice of a new type of
apparatus or a new form of electricity which, since the time of Oersted
and Faraday, was well-known and had been used by certain physicians,
notably by Marshall Hall in Great Britain. Rather, they were based on
the conviction that one could master this particularly powerful agent
completely and direct it exclusively to where one wanted it to act, thus
sparing healthy tissues: "To direct and to limit the electrical power in
each of the organs, without puncturing nor incising the skin, such is
the aim of this new method. . . . It seemed to us that it would be
possible to obtain results which would perhaps be more dramatic and
more regular if it was possible either to restrict the electricity to the
skin without stimulating the organs which it protects, or to cross this
tissue without affecting it in order to concentrate this power in a nerve
or a muscle or, finally, to direct the electric agent so it would penetrate
into organs which are deeply situated."[206] These variations could be
obtained by modifying the state of the metallic excitors (electrodes)
which were connected to the poles of the battery as well as the surface
to which they were applied: when both it and the electrodes were very
wet, a very sharp, deep sensation ensued which, in turn, allowed Duchenne
to discover the "muscular sense," or a contraction was obtained depending on where the electrode had been placed.

Duchenne conceived the idea of "faradisation"—it was he who suggested substituting this word for "electrisation" or "galvanism" which were too vague or inexact—to describe a precise, selective instrument which acted as much on the cutaneous surface, as it did directly on tissues or nerves situated deep within the organism. It was a kind of analysis of the living being, an *in vivo* dissection which was aimed at breaking down the specific action of each nerve and muscle involved in physiological movements, facial expressions and pathological processes.

The vast analytical semiology that Duchenne called on allowed the exact targeting of the action of a current to a strictly localised area; his work was also based on the conviction that all types of current were not equally apt to fulfil this objective, but that they had different physiological effects. Thus in the hands of Duchenne, faradisation became a powerful means of diagnosis which, on the basis of the continued electro-muscular contractility which he detected, allowed him to foresee the possibilities of recuperation in certain cases of paralyses and to distinguish serious lesions of the cord from curable cases: thus he distinguished progressive (spinal) muscular atrophy, infantile paralysis due to poliomyelitis and ataxia; he separated these, among others, from paralysis due to lead poisoning or hysteria.

Duchenne used faradisation for quite an extensive range of illnesses, including sciatica, rheumatic pains, trigeminal neuralgia, hysterical pains, and angina pectoris. He considered it as a stimulant in the same class as vesicatories or sinapisms, but infinitely more powerful, to be applied directly to the painful location and, if possible, during the attack. According to him, this "electrical fustigation" or "sound thrashing" acted like a perturbational method, and because it could be applied in measured quantities, faradisation did not produce any of the serious sores caused by cauterisation and could be reapplied several times until the cure was complete.

In one of his most important works, *De l'électrisation localisée et de son application* which was published in 1855, Duchenne gathered together hundreds of detailed observations on the results of electrical treatments which he had carried out, whether for pains of many different origins, or against anaesthesias.[207] Contrary to some of his predecessors in electrotherapy, Duchenne did not claim to be able to cure everything, and from the failures or very limited improvements which he obtained, he was able to add elements to the knowledge of pathology, while medical treatment and the study of symptoms through faradisation then represented two aspects of the same technique. It was because of this that Duchenne de Boulogne has been considered one of the founding fathers of neurology, and Charcot, Vulpian and many others worked alongside him and recognised the merits of someone who, with no university qualification other than a diploma in medicine, was in the process of creating a new medical specialty.

While Duchenne wanted to restrict the use of galvanism to surgical uses (cauterisation of wounds, treatment of aneurisms, tumours, blood coagulation), Robert Remak in Germany claimed that induced currents caused debilitant "hypostheniant" effects, and that only galvanisation could be efficacious. A long dispute ensued. Remak accused Duchenne of keeping his method secret, and on the latter's death, Lasègue, who wrote his obituary in the *Archives générales de médecine* where much of his work had been published, reproached him for his exclusive preference for induced current. Remak also produced some observations on patients with rheumatism, sciatica, and trigeminal neuralgia whom he had cured in some dozen sessions or so, but his work, from the clinical and semiological point of view, did not have the same rigour and precision as Duchenne's, while its physiological bases and his explanation for the mechanisms of currents were all different, and undoubtedly sounder. He relied on the ideas and work of Du Bois-Reymond (1816–1896) concerning the existence of a current special to the nerve, which the effects of the galvanic current stopped; this was what he called "the negative oscillation of a nerve" and he attributed it to the relationship between the direction of the electric current and that of the nervous current, the latter being increased when the excitatory current was flowing in the same direction, and lessened when it was flowing in the opposite direction.[208]

For Du Bois-Reymond, a professor in Berlin and pupil of Johannes Müller who had sent him to study under Matteucci in Florence, the bio-electrical phenomena were strikingly confirmed in a posthumous revenge of Galvani's ideas over those of Volta; from them, he deduced a certain number of laws concerning the direction of nerve currents, from the nerve's longitudinal cross section to its transversal cross section, on which he hoped to found electrotherapy.

In order to explain the disappearance of pain following galvano-therapy, Remak in particular introduced the concept of "a catalytic action," which produced two series of phenomena: " *1.* Dilation of the blood and lymphatic vessels and, consecutive to this dilation, disgorgement of swollen cells within the blood and the lymph, reabsorption of exudates by exciting a flux of liquids inside the tissues; 2. electro-chemical mutation in the tissues, accompanied by an electrodynamic transport of liquids."[209] These two processes resulted in modifying the osmotic conditions of the muscle and the medium in which it was to be found, and corresponded well enough to the first explanations which Dutrochet had given for the phenomena of osmosis and endosmosis; moreover, they were in line with the humoural aetiology which Remak adhered to for rheumatism: "If the stimulation of the sensory nerves can, by reflex action, produce a diminution and dilation in the calibre of the vessels, and if it has an influence on the skin's secretion, the neuropathological theory of rheumatism is not devoid of all foundation."[210] According to Remak, two other actions—an antiparalytic and an antispasmodic one—could be added to this catalytic effect: the treatment for neuralgias (in the widest sense of the word) depended on the antispasmodic one, as the lessening of the pain was obtained through a loss of excitability on the part of the nerve.

In his *Manuel d'électrothérapie*, Tripier noted the ambivalence of electrical phenomena as far as pain was concerned: the passage of an intermittent current in a nerve produced a sensation of pain whose intensity was directly related to the intensity and frequency of the current; however, the passage of a continuous or intermittent current of a given duration brought about a state of numbness and finally anaesthesia, as had been observed by the physician, Julius Althaus,[211] when he was investigating this decrease in sensibility by experimenting on the cubital nerve in his own arm. More than ever, the therapeutic rationale for using electricity against pain was thoroughly constrained by the problem of whether it was stimulating or enfeebling, although electrophysiological studies brought out the importance of the direction of the current: "Pain is thus produced by both the unexpected modifications brought about in the electrical state of the living tissue which is a part of the circuit, and by the amount of the current. These two conditions tend to cause the making and the breaking of the electric circuit to be painful; it is only with the latter, *i.e.* the strength of the current, that painful phenomena are observed during its passage. Current's influence is not confined to the previous effects; there is yet another, less obvious, but sometimes perceptible phenomenon which again demonstrates the complex influence of electricity on the sensitive element: we are referring to anaesthesia,"[212] which reduces sensibility to pain, without suppressing the other "sensibilities" (touch, muscle sense and temperature).

Electrotherapy from 1860 until 1880, be it galvanic or induced, effectively calmed a fair number of pains, above all those of a neuralgic type, even though the mechanisms which made them effective remained uncertain. A considerable number of books and manuals on electrotherapy were published in different countries;[213] portable apparatuses and special instrument cases such as those of Trouvé were produced although, despite all that, medical electricity did not become a distinct, institutionally recognised medical specialty. Due to its origins, it depended on a certain technical knowhow which, within French medical education, had less prestige than the fundamental, theoretical and abstract disciplines, and it could not in any case have provided a fitting career.

Electrophysiology, which might have represented the "noble route" in this research on the links between electricity and medicine, was less developed in France than in the Germanic countries and there was a tremendous lack of experimental physiological laboratories according to Claude Bernard, in his *Rapport sur les progrès de la physiologie en France* of 1872. One must link to these structural problems the question of the relationship between medical faculties and science faculties, in which the distribution of chairs and disciplines reflected the old division between science and art: the whole issue of the emergence of biophysics and biochemistry and of their institutional position was also at stake through this question concerning electrophysiology and electrotherapy, although, traditionally, the "medical physics" chair could have encompassed the teaching of these subjects.

A fourth stage in this history of medical electricity was reached with the discovery of high frequency currents and the work of Jacques Arsène d'Arsonval. He had a medical background and was one of Claude Bernard's last pupils; he first audited his lessons at the Collège de France and soon thereafter became his assistant and helped him in his research on animal warmth. From the beginning, he showed a very active interest in the industrial applications of electricity. He notably demonstrated this interest during the Universal Exhibition of 1881 and the First International Congress for Electricity which were held together. He immediately linked this interest in electricity to his preoccupations as a physiologist. Thanks to Paul Bert who became Minister for Public Education, he obtained a laboratory of biological physics at the Collège de France in 1882, and, in 1894, succeeded Brown-Séquard in the Claude Bernard Chair.

We cannot go into detail here on all the apparatus for producing, measuring, and recording the effects of electrical currents for whose invention science is indebted to d'Arsonval, but we must highlight the two preoccupations which guided his research into medical electricity. Firstly, in pursuing a project which had previously been initiated by Duchenne, one of his objectives was to find an electrical current which could stimulate only the nerve, or only the muscle, selectively; secondly, he sought to understand how it was that very high frequency currents could pass through a living organism, without any doubt producing a physiological effect but without causing it any damage, whereas currents of much lower frequency would have killed it. From 1876 onwards, he had taken an interest in the physiological effects of alternating currents. He showed that they activated the metabolism and that their action could be useful in nutritional problems. One of the problems which had to be resolved in order to achieve a better use of electrotherapy was that of finding a standard measure for whichever source of electricity was being used: d'Arsonval showed that "the effects of these different electrical sources could be foreseen, measured and reduced to a common unit through a knowledge of the electrical wave which arises from each of these sources. It is this electrical wave which I have called the characteristic of excitation. When this curve is the same, the physiological results are the same, no matter what the electrical source is or where it comes from."[214]

The most important distinction necessary for this research concerned whether the current was of a constant or a variable nature. A state of constant current produced effects which depended on its intensity and which were translated in physiological terms inside the tissues, by chemical decompositions produced by electrolysis; that is why it has been used in surgery for cauterisations and to destroy living tissues. The local action of the constant current, at the point where the electrodes were applied, depended on the density (intensity per unit of surface). By contrast, a variable current did not depend on the intensity, but on the difference in potential at the excited point, which d'Arsonval summed up by the following law: "The intensity of the motor or sensory reaction is proportionate to the variation of the potential at the excited point."[215] Once the measure of the "physiological power" of the variable currents had been calculated and its relationship with the electrical wave had been grasped, it became possible to dream of therapeutic applications for "sinusoidal currents" and to differentiate the desired effects on the muscle or on the nerve by varying the electric potential and the application time.

Three types of effects could be obtained: an action on the motor nerves, producing a muscular contraction whose energy was a function "of the speed of variation of the potential difference between the two excited points of the nerve, all else being equal";[216] an action on the muscle which necessitated a less abrupt variation and an activation of metabolic exchanges (an increase in the exchange of respiratory gases and, above all, the production of warmth). D'Arsonval wrote about the first two: "As a result of these findings, when using very brief, high potential discharges, one must excite above all the motor and sensory nerve terminals . . . as a consequence, by sufficiently lengthening this discharge, we must succeed in virtually no longer exciting the nerve tissue but, on the contrary, in strongly exciting the muscle tissue."[217] But with very high-frequency currents—and following Tesla's experiments in the United States, d'Arsonval used apparatus capable of generating up to a thousand million electrical vibrations per second—he observed that there was no muscle reaction and no sensation, which he explained by the fact that our nerve terminals are not in harmony with this vibratory frequency. However, he also noted that, quite on the contrary, one could observe a persistent cutaneous insensibility for up until half an hour later, phenomena of vasodilation, and sweating.

Electrotherapy using high frequency currents was given the name "d'Arsonvalisation" until the wider term of "diathermy" came into use around 1920. The experimental use of electrotherapy began in gynaecology at the Hôtel-Dieu in 1892 under Dr. Georges Apostoli, and was used on 34 patients, 12 of whom had fibromas. Even if the high-frequency currents did not seem to make the fibromas regress, he was nonetheless able to draw attention to the fact that "it exerts a very clear action on the symptom of pain; this action is confirmed as early as the first sessions and more often than not immediately after the end of the session." He concluded: "Up until the present, it is the medication *par excellence* for pain, and as such, if it in no way supplants faradic and galvanic applications which have proved themselves, it is nonetheless one more weapon, and conservative gynaecology can only accept everything which serves to enlarge and strengthen its domain."[218] The validity of this means of fighting pain soon went beyond the field of gynaecology in order to treat a majority of "painful illnesses."

The dream of mastering a particularly powerful and formidable material agent, which had been passionately pursued by Duchenne de Boulogne, became a brilliant, concrete accomplishment with the work of d'Arsonval; with it, the divorce between electrophysiology and electrotherapy became a thing of the past, as was the division between industrial technical applications and fundamental research; through his work, as well as that of his contemporary, Étienne Jules Marey—both successors of Claude Bernard—the physical phenomena of life came of age in France, but for how long?

### Hypnosis, a Poorly Mastered Weapon

Without doubt, an 18th century intellectual who was a disciple of Mesmer would have thought it was quite normal to go from electrical fluid to magnetic fluid, even if the latter were "animal magnetism." The whole subsequent history of magnetic or somnambulistic sleep, and hypnosis has been tarnished by that first, original slur, which led to the condemnation of Mesmer as a quack and an imposter, and the judgement delivered by the Commission de la Société royale de médecine in 1784. The judgement was motivated not only, as has sometimes been said, by the scandals, sexual indiscretions and secrecy which surrounded the sessions around the magnetic tub. Rather, the reasons were really because the sessions were not reproducible by doctors who wished to duplicate the experience, and also that the members of the commission refused to accept, without substantial proof, the existence of a fluid—a material agent—circulating from the "magnetiser" to his patients, who were joined to one another as well as in contact with the tub through iron strands. What the members of the commission, and particularly Bally, objected to, was the actual performance itself which tended to substantiate hypotheses which they were unable to ascertain themselves. Jussieu and other physicians interested in somnambulism, such as Ménuret de Chambaud from Montpellier,[219] were readily willing to admit that something did happen and that the sessions undoubtedly had a beneficial and often timely though limited effect in the treatment of nervous illnesses, especially hysteria. However, they could not agree that the resulting improvements, which were fewer and less spectacular than Mesmer claimed, were due to the effects of a mysterious fluid. Instead, they tended to favour an explanation where the psyche or "morale" (*"le moral"*)—which at that time had nothing to do with what might be considered "moral" (*"la morale"*)—played its part more effectively than did "the physical."

If, consequently, there was an original blemish, it was because all subsequent trials to induce magnetic sleep for therapeutic purposes, such as those carried out by the Marquis of Puysegur, Abbot Faria and others, were interpreted with reference to the continuity of Mesmerism whereas, from our standpoint, their parting of ways should have been clear. It was the confusion between magnetism and hypnotism, brought about by the permanence of the earlier term—and perhaps maintained deliberately—that weighed against this practice. This was soon relayed by the fear of seeing morality undermined by the singular relationship involving a suggestible young girl or young woman (there was always an element of the female sex) being subjected to her "magnetiser," and it was finally condemned by the Congrégation du Saint-Office in 1851.

A cursory study of sources from the first half of the 19th century shows that "magnetism" was never abandoned totally despite a checkered history, and that it was practised in order to relieve suffering and to cure certain pains. To show its popularity prior to 1845, one need only point out the importance which Julien-Joseph Virey conferred to it in the article "Magnétisme" in the *Dictionnaire des sciences médicales* which was published in 1819, and the numerous histories of magnetism which flourished, such as the one by Deleuze. But more simply, within an eleven-year period, the Académie de médecine was apprised of the performance of an operation without pain while under the effects of magnetic sleep: the first time was in 1826, when Hippolyte Cloquet proceeded with the excision of the breast of a young girl who, though still able to speak, felt tinglings but no pain at all, although the surgeon took his time with the operation and blood flowed. The finding appeared so extraordinary that, during the meeting of 28th February 1826, a commission was appointed, composed of Double, Itard, Guéneau de Mussy, Guersent, Magendie and Leroux among others. After an enquiry lasting five years, the commission came to the conclusion that the findings were authentic.

Despite the conclusions of the report which were recalled in 1837 by Husson, and despite the testimony of Cloquet who refused to give an interpretation but confirmed the facts, disbelief persisted and explanations such as stoical courage, will, etc. were suggested. However,

Cloquet affirmed that he had not noticed any of those disturbances—such as changes in pulse rate, palpitations, etc.—which are visible even among the bravest. As proof of the confusion which was brought to the debate, it was noted that the patient claimed to have seen a red plaque on her pleura which was not present at autopsy, and this was used to raise suspicion.

During the meeting of the Académie de médecine on 24th January 1837, the story was reported of an extraction of teeth carried out on a young, 25-year-old woman who was rendered insensible by magnetic sleep: it was Oudet, a member of the Académie who had performed the operation and he, too, wanted to stick resolutely to the facts, not being able to put forward any other explanation. Neither the event nor the epistemology of the bare facts were enough to convince the Academicians, some of whom at best even accused him of having let himself be duped, while others such as Roux were more vociferous in their disapproval: "Thus, gentlemen, let us get straight to the point: quackery will find enough voices outside these walls to defend it; it must find none amongst us."[220]

Another affair, which had exploded a little while previously, concerned a man who claimed to be able, while under magnetic influence, to read a book which he did not know, with his eyes shut. Georget, an alienist physician and nervous-system specialist, had declared himself in favour of magnetism as a result although those who had performed the demonstration in front of him had apparently stated after his death that they had tricked him. Thus it was in a passionate atmosphere that a session with the academician Roux was organised on 3rd March 1837, in order to confirm a young girl's insensibility to being pricked by a pin while "magnetised," *i.e.* hypnotised. The session took place before the members of the new commission appointed by the Académie (Bouillaud, Cloquet, Caventou, Emery, Pelletier, Oudet, Dubois d'Amiens and Roux), and after some time, the "magnetiser," Berna, was even placed behind a door in order to avoid all possibility of fraud.

In view of the stakes involved, the session's protocol went beyond the problem of insensibility to pain and described the phenomena of hypnosis studied by Charcot. The points which were considered as important by the commission were as follows: (*i*) somnambulisation; (*ii*) corroboration that there was insensibility to pricking and tickling; (*iii*) reinstatement of sensibility at will; (*iv*) obeying the mental instruction to lose the ability to move; (*v*) obeying the mental instruction to stop replying during a conversation; . . . (*viii*) after a mental instruction added by the "magnetiser," the persistence, on awaking, of insensibility and also persistence of the faculty to lose or recover sensibility at the will of the "magnetiser"; (*ix*) the same confirmation regarding the loss of movement, which could pass from one side of the body to the other at will.[221]

In spite of the detailed report of the findings, their authenticity was challenged because the basic issue of the insensibility to pain was eclipsed by the debate on the possibility "of seeing without the help of the eyes" which Berna claimed to be able to prove. This particular matter was, of course, a fiasco but had serious repercussions on the rest of the case. Some of the fears about the morality of ventures in hypnosis and their consequences on the indirect accomplishment of crimes and murders became firmly rooted in this very episode. Its effects were long-lasting, since in the *Dictionnaire Dechambre*, in 1873, the article which was attributed to Dechambre himself, on the subject of "Mesmérisme" (and the choice of this word to describe the subject was not insignificant), tended toward "nothing less than the pure and simple negation of the truth of these hypnotic phenomena."[222] It was this which obliged Paul Richer and Gilles de la Tourette to draft a new article, called "Hypnotisme," in 1889, in the fourth series of the Dechambre, in order to re-establish a certain number of facts, and retrace the evolution of knowledge on this subject.

It was in Great Britain that research into hypnotism first resumed, thanks to James Braid, a doctor in Manchester who, in 1843, published his *Neurypnology or the Rationale of Nervous Sleep in Relation with Animal Magnetism;* the work was translated into French only in 1883, with a preface written by Brown-Séquard; this had become virtually indispensable because of the experiments of Preyer, Charcot, Dumontpallier and a few others.

The previous year, in 1882, Brown-Séquard had published a work called *Recherches expérimentales et cliniques sur l'inhibition et la dynamogénie* ("Experimental and clinical research into inhibition and facilitation") in which he defined inhibition as "the momentary or permanent disappearance of a (normal or morbid) property or activity in a nerve centre or a muscle, occurring without visible organic alteration (at least in the state of the blood vessels) happening immediately or at least shortly after the production of an irritation at a point in the nervous system, more or less distant to the place where the effect is observed";[223] dynamogenesis (facilitation), on the contrary, was a sudden increase in activity. Was it possible to compare the phenomena of inhibition, which had no discernible organic basis, with insensibility to pain while, on the other hand, facilitation would be comparable to hyperaesthesia? Was it not surprising, and of great importance, to see one of the principal neurophysiologists of his day, who had advanced the understanding of sensory pathways in the cord, become interested in these processes in which, no doubt, psycho-physiological data would have to be included?

However, Braid's work had been recognised before that date in France: it was disclosed through Todd and Carpenter's *Encyclopaedia;* and revealed to the public at large in 1852 in *La Presse* by Victor Meunier, a man who championed all difficult causes and who defended Pouchet against Pasteur; it was described in the 10th edition of Nysten-Littré-Robin's *Dictionnaire de médecine;* it was used to advantage by Dr. Azam from Bordeaux, and endorsed by Velpeau who presented a paper by Azam to the Académie des sciences on "L'anesthésie chirurgicale hypnotique" ("Hypnotic surgical anaesthesia") during the meeting of 5th December 1859; it was soon thereafter relayed to the Société de chirurgie by Broca.

Was hypnotism to compete with chloroform? Nothing ever came of it for reasons which were quite different from those which had previously been put forward. Braid's work had, in fact, brought hypnotism into a more acceptable arena through science, insofar as he had eliminated the atmosphere of mystery and the ceremony surrounding hypnotism, by demonstrating a relatively simple technique to induce it, and by recalling that he had first attended demonstrations in Manchester by a French doctor, Dr. Lafontaine, as a sceptic and even an "unbeliever" himself. "Take any shiny object (I generally use my scalpel-holder) between the thumb, index and middle finger of the left hand; hold it at a distance of between 25 and 45 centimetres from the eyes, in such a position above the forehead as to induce the greatest necessary effort on the part of the eyes and eyelids of the subject to stare at the object. The patient must be made to understand that he must keep his eyes continually on the object and his mind must remain totally attached to the idea of this sole object. One will observe that because of the synergistic action of the eyes, the pupils will firstly contract; shortly afterwards they will begin to dilate, and after having become thus considerably dilated and after having made an oscillating movement, if the pointing and middle finger of the right hand, extended and slightly separated, are taken from the object towards the eyes, it is highly probable that the eyelids will close. . . . After an interval of about ten to fifteen seconds, by gently lifting the arms and legs, one will find that the patient, if he is strongly affected, is disposed to keep them in the position in which they have been placed."[224] He made a careful note of a rise in pulse rate during a first excitatory phase, then of a switch into a kind of torpor and a stiffness of the muscles which were, however, capable of going from being rigid and insensible to being supple and hypersensitive.

Braid concluded that these experiments clearly showed that there were psycho-physiological phenomena therein, which were explicable—although he could not provide a satisfactory rational explanation—and quite independent of the communication of any fluid whatsoever. Braid thought that there were modifications "of cerebro-spinal centres, and the circulatory, respiratory and muscular systems; a disturbance brought about, as I have explained, by the concentration of vision, the total relaxation of the body, the fixedness of attention and a suppression of respiration which accompanies this fixedness of attention"[225] and as a consequence completely excluded the hypnotist's personal volition. Moreover, he was convinced that this state could not be induced without the consent of the hypnotised person, which was of great importance from the medicolegal point of view; excluding any form of simulation, he also asserted that, when in this state, the subject was physically incapable of opening the eyes. What intrigued him particularly was the psychic dualism which was brought out by hypnotism: there was no memory on waking of what had happened during the hypnotic sleep but, on the other hand, it was possible to reproduce images, emotions or thoughts which had arisen during this state, if one only placed the patient into identical conditions again, as if there were natural connections between certain gestures, certain areas which had been touched and certain psychic manifestations. This splitting of the consciousness manifested itself during one of his very first observations of a young girl who admitted to theft in the hypnotic state, without having the least recollection of it on waking.

However, among the reasons which spurred him to persevere with his research were certain astonishing therapeutic results as in cases of acute tetanic (spasmophilic) crises and obstinate pains. In his *Neuryp-nology*, he reported that several cases of *tic douloureux*, some of which had lasted for months, had been cured and eliminated during one session of hypnosis lasting a few minutes; other successful cases dealt with dreadful headaches, epileptic attacks, rachialgias (spinal irritations), cardialgias, without forgetting the easing or suppression of pain which occurred in surgical operations. Braid clearly concluded that the method was efficient: "Hypnotism can be used, very successfully, in the treatment of numerous sprains arising from the weakness of some muscles or from an excess in the strength of contraction of their antagonists; I am certain that, by this method, we can correct a good number of cases which until now have been treated by sectioning the tendons or the muscles";[226] all the cases of lateral curvature of the spine, which brought about very painful contractures, could, according to Braid, be cured in those cases where the problem was recent and very well relieved in the longer-standing cases.

The number of observations gathered by Braid and his wish to exclude all supernatural or irrational factors made of his work a major document on the therapeutic possibilities of hypnosis, even if he was incapable of giving a precise explanation for the effects on the nerve centres of this psycho-physiological state of attention. In addition, it seems necessary to emphasise that Braid's research was not prompted by a vague concern with the effects of the mind over the body, or simply by the power of the imagination; rather, it really had a psycho-physiological perspective, albeit that such a term did not yet exist. Such an approach recognised in psychic and physiological phenomena the two sides of human behaviour which were both governed by a scientific rationale.

However, if this work did not directly produce the research which, by right, it should have inspired, and if Charcot's initiatives at the Salpêtrière from 1878 onwards gave rise to an excessive reaction, notably from the École de Nancy (principally on the part of Bernheim), it was due to a new problem which came to be posed: what was the real basis of hypnosis and, consequently, of its power to ease pains in patients, aside from those who were hysterical? Were not the spasms, contractures and pains cured by Braid also of a hysterical nature? In brief, were healthy subjects capable of being hypnotised, and was hypnotism itself not a pathological state or a phenomenon which always brought such a state to the surface in predisposed subjects?

Not only was Charcot's work at the Salpêtrière unable to answer these questions but to some extent actually prompted them. In the clinical study of nervous illnesses, hypnosis could be of use in bringing to light the hysterical origin of muscular or sensory disorders; this was true notably in the case of hysterical brachial monoplegia of traumatic origin, *i.e.* deprivation of movement in one arm after a shock or accident without any lesions of the nerve centres, and also in hemianaesthesia (insensibility of only one side of the body).[227] With the help of hypnosis, Charcot dedicated himself to showing that patients, above all during the third somnambulic phase,[228] were capable of exactly mimicking the clinical signs of paralysis or anaesthesia which Charcot had witnessed in a group of "control" patients whose illnesses he suspected of having an hysterical cause. Thus, by suggestion, he was able at will to induce patients into and release them from psychic paralyses or insensibilities, sometimes segment by segment—the wrist, forearm, shoulder, etc.

This type of analysis was corroborated by the results of therapeutic tests in which the use of a dynamometer provided a means for measuring the development of muscular energy. These showed that baths and electrisation acted above all psychically, not directly, but by suppressing the inhibition attached to the production of the necessary mental picture for accomplishing a gesture. "Thus, if I am not wrong, we behave above all psychically. It is known that the production of an image or, to put it another way, of a mental picture—however concise and rudimentary it may be—of the movement to be carried out, is a preliminary, indispensable, condition for the voluntary accomplishment of that movement. But it is very likely that in our two patients the organic conditions which are normally in charge of the representation of this mental picture have been deeply disturbed, to the point of making this image impossible or, at the very least, very difficult; this is a consequence of an inhibitory action exerted on the cortical nerve centres through the fixed idea of motor powerlessness, and it would be to this circumstance, above all, or at least to a large extent, that one would attribute the objective realisation of paralysis."[229]

The principal explanation for these states, in which the psychical and the physiological are connected, is easily transposable to painful sensations: these could arise from the phenomena opposite that of inhibition, through a sort of increase in dynamogenesis (facilitation) of the sensory centres. A demonstration of this extension of the principle was given in relation to a case of coxalgia, a pain in the hip, tending to reach other joints (the knee), sometimes with shooting pains and always a handicap to the patient when walking. It concerned a railway worker, the victim of an accident which had thrown him into the air to a height of about 3 metres; this brought on a sharp pain accompanied by a feeling of numbness and swelling of a limb, and thereafter, he was able to walk only with the help of a stick and could not work. By carefully following the criteria provided by the English doctor Brodie who, in 1837, had been the first to describe this illness under the name "hysterical pathology of the joints"[230] and particularly by distinguishing the pains felt by his patient from those of organic coxalgia which awaken the patient at night, Charcot was able to identify the illness as an hysterical coxalgia, which in no way lessened the reality of the pain but did allow the treatment to be planned accordingly. Had the patient agreed, Charcot would have liked to use hypnosis as a means of both diagnosing and treating the disorder.

The very act of explaining the theory for these pains or any other hysterical incidents in front of the patient might, in certain cases, hasten recovery. But it might also—and this was very much the criticism levelled by Bernheim and also by Babinski—artificially create pathological manifestations; amongst those were certain recurring conditions such as a narrowing of the visual field and loss of focus or anaesthesia of one part of the body which could have been directly induced by the knowledge which the patient could have obtained during previous medical consultations, or through reading, about the "stigmata" of hysteria: "I am convinced that these phenomena are the product of autosuggestion or rather of the doctor's unconscious suggestion as, moreover, has already been suggested by Dr. Bernheim from Nancy; the interrogation of the patient and the questions which one would usually ask him to determine his state of sensibility could be enough to arouse the idea in his mind of a hemi-anaesthesia or of a visual problem; it is therefore essential to take every precaution in order to avoid this danger," warned Babinski.[231]

Thus, the problem of pretence, which had been posed from the beginning of the century by physicians such as Double who were trying to define the valid conditions for medical interrogations, cropped up again with astonishing validity at the end of the 19th century. As the discrimination among the various different forms of sensibility became increasingly perfected, so, too, was the patient's participation increasingly required. It was therefore necessary to forestall the risks of mistakes by taking care not to introduce elements of bias into the wording of questions and by testing the sensibility of one part, not only because of what the patient said, but in order to observe reflex phenomena.

The identification of pricking sensations, or the sense of contact, or the ability to recognise objects by palpation (stereognosis) had to be introduced by questions of the type "What do you feel, what have I just done?" and never by "Can you feel what I am doing?" or even "Can you feel it as much on one side as on the other?"[232] in such a way as to avoid any risk of suggestion.

By taking such precautions, most of the so-called classical manifestations of hysteria, especially those of a speculative nature, did not occur. The possibility open to the physician of making them appear at will and even of making them disappear thus became a means of arriving at the differential diagnosis of hysteria, and resulted in discarding many cases considered to be cases of hysteria by Charcot into the category of nervous illnesses of organic origin. In fact, Babinski really proceeded to take traditional interpretations of hysteria apart. By placing the various manifestations of hysteria into a historical context, he also explained why these changed according to the times and cultural variations. In an attempt to define the attributes exclusive to hysteria, Babinski settled on the possibility of reproducing its manifestations through suggestion, and making them "disappear under the exclusive influence of persuasion."[233]

After a fashion, this conclusion certainly was an accurate reflection of the views of Charcot—whom Babinski identified as his master—who had shown the possibility of inducing a psychic paralysis in subjects and then releasing them from it. However, he disagreed radically where the interpretation was concerned and, rigorously distinguishing between persuasion and suggestion, finished by questioning the therapeutic usefulness of hypnotism. Indeed, if none of the habitual characteristics used to define hypnotism (unconsciousness of one's lethargy, amnesia on awakening, abandonment of free will to the somnambulic state) seemed to be fully validated nor unassailable, then there was no particular reason for using hypnotism rather than psychotherapy or anything else which, after gaining the patient's confidence, one could use on him.

Babinski, on the basis of his authoritative background, earned as a result of his close association with Charcot and his experience of hypnotism at the Salpêtrière, explained his discontinuation of the practice in 1910 in the following way: "It will be easy for me to explain the reasons why I hardly use hypnotism any longer for therapeutic purposes during this talk. If hypnotism neither creates nor increases suggestibility, it seems pointless. In the past, it is true, a large number of hysterical patients were cured after having been hypnotised; but the cure was not the result of hypnotism, it arose from the fact that patients were suggestible or capable of falling under the influence of persuasion. The practice of psychotherapy in a state of wakefulness would have released them from their pain just as well."[234]

Babinski's condemnation was full of consequences for the practice of hypnosis as a therapeutic instrument and all the more so since the neurophysiological basis of this particular state had not been elucidated: as a result, there was far less research to explore it in as much as Babinski had cast doubts on the specificity of its somatic manifestations (neuro-muscular hyperexcitability and plasticity of the muscles when in a state of catalepsy). By reducing the hypnotic state to one of semiawareness, it was the whole problem of the dissociation of the psyche which Babinski had dismissed, but which Braid had noted in his time. Though the criticism levelled at Charcot on the role of medical suggestion in the production of the manifestations of hysteria were fully substantiated by a large number of observations and arguments, the opposite was true when it came to the unverified assertions that the hypnotised somnambulic person did not completely lose voluntary control over his actions or his critical faculties and that he was not unconscious.

As often happens in scientific debates, particularly when, as was the case here, they are not fully backed up by experimental evidence, the strength of certain arguments or of certain objections wins total adherence in favour of either accepting or rejecting a hypothesis, often on the basis of those aspects which appear to be least convincing. Babinski's position was reassuring from the medico-legal position since the free-will of the hypnotised person was no longer in question. However, this undermined the very essence of the therapeutic value of hypnosis by questioning its usefulness. In one way or another, the implications of this position have continued to make themselves felt, despite the efforts undertaken since the end of the 1950s to re-evaluate the problem of hypnosis.

In spite of these criticisms and the eclipse of some of the achievements of the École de la Salpêtrière which was perceptible from the time of Charcot's death in 1893, the influence of his work had been immense, as much for psychoanalysis as for psychophysiology. It is not for us to analyse it, nor to recall all the debates between the École de la Salpêtrière and the École de Nancy which were prompted by the practice of hypnosis. However, the importance of this practice must be emphasised, not only as a means of diagnosing or treating disorders but also as an instrument for understanding pain and the complexity of the processes which produce it.

If there is a medical experience which can shatter the dogma of specificity, it certainly may be considered to be pain of hysterical origin, within an infinite circle of physical and psychical causes, each echoing the other. Even if one believes that hypnosis affects only actual or potential hysterical individuals—"neuropaths who have yet to know it" according to the vocabulary of the end of the 19th century—and even if one asserts that hypnosis constitutes a process which exaggerates hysterical phenomena, this very practice could have and could still provide information concerning the "normal" functioning of the nervous system, the role of the unconscious and involuntary functions, the emotional effects which cause the appearance of pain and sometimes make it persist for many years. "The experiments into hypnosis . . . thus represent the best demonstration of the automatic functioning of one part of the brain, a function which has already been studied by physiologists and which has been called cerebral automatism or unconscious cerebration."[235] Though unconscious, automatic, and emotional as much as sensory, the pain phenomenon was nonetheless the province of the physiological branch of experimental study; if, moreover, at the closing of this century—which had seen considerable advances in neurology—pain clearly encouraged adopting a monist approach to mankind, it was up to neurophysiology, endocrinology and biochemistry to shed light on its processes and the conditions of its integration.

# 7

## Communication Strategies
### *The Approach to Pain during the First Half of the 20th Century*

> "What is a man capable of? I will fight anything beyond a certain magnitude, except bodily suffering. That's where I should undoubtedly begin, however."
>
> Paul Valéry, *The Evening with Mr. Teste*

When an historian analyses our present century, so near and yet already remote, he faces two major dilemmas: one concerns the legitimacy of the time divisions used to delineate a past which has yet to be fully unravelled or elucidated, while the other involves the type of criteria to adopt in order to uncover rational patterns and rhythms governing such a broad spectrum of experiences. These questions are evidently pertinent to the work of historians everywhere and for every period of time, but they seem all the more pertinent here, as the trauma of two world wars has brought society and science more closely together than ever before and, as a result, has profoundly affected our understanding of pain, the systems for evaluating it and, to a lesser extent, the available remedies.

The passage from one century to the next does not necessarily imply there has been any shift in general knowledge nor in practical methods; any break actually coinciding with the turn of the century would, more likely than not, be quite artificial. Indeed, if among the possible sources of change we examine investigative methods, philosophical concepts, or the kinds of problems being examined by physiologists in relation to pain, we realize that new developments had already come about as of the 1880–1890s. The methods for investigating the nervous system (histological staining techniques, for instance, and the electrical stimulation of the brain had taken on as much importance as analysing the effects of surgical ablation) were developing considerably. A number of major advances took place before 1900: Golgi's discovery of the silver-stain method, later adapted by Ramon y Cajal to prove the discontinuity between neurons,[1] occurred about twenty years before the start of the century, while Fritsch and Hitzig explored the cortex using electrical stimulation as early as 1870.[2] The study of the nervous system's embryonic development to chart cortical areas and pinpoint their functions goes back to Flechsig,[3] and efforts to recapitulate early man's evolution through the brain's architecture were directly linked to the widespread dissemination of Darwin's theory in the second half of the 19th century. The pharmaceutical field contributed a great number of synthetic substances (antipyrine, aspirin, and paracetamol's precursors) which were produced and quickly sold on the open market, while local anaesthesia and electro-therapeutics were also progressing at a rapid pace. Does this mean, then, that continuity prevailed, and that the only meaningful changes were brought about by the political and social upheavals? This had an indirect effect on the scientific field, either because of power redistributions occurring in certain countries, for instance, or by forcing the development of the neurosurgical branch—still in its infancy in the previous century—because war had made it indispensable.

Another problem, linked to the complexity of the subject under study and uncovered gradually in the course of experimentation, calls into question the simpler original concept of a peripheral receptor connected to a central terminal. If we carefully analyse the various earlier hypotheses and the groundwork carried out, it seems clear that within any field of discipline such as the physiology of pain, related knowledge does not all develop concomitantly. The idea that peripheral receptors were structurally diversified and that relative specialisation existed in the production of sensations had been accepted, though the existence of specific pain receptors had not yet been definitively admitted. The conviction was more certain, however, that the concept of sensibility needed to be broken down into its various heterogeneous components, as several types of sensibility actually existed; it was also uncontested that their conduction in the spinal cord corresponded to different body regions, and that the anterolateral spinothalamic tract was the main pathway followed by impulses produced by painful stimuli. Finally, it was accepted that certain functions could be localised in the brain despite the fact that functional diagrams drawn up on the basis of experimental studies of pathological lesions stirred considerable debate. Could the common denominator behind all these accepted views have been the actual somatic establishment of these functions, and their territorial limits and localisation, even if the precise level on which each function occurred became increasingly specific and less clearly visible, from organ to tissue, tissue to cell, and from each overall function down to the mosaic of cellular activities?

If we accept this assessment, the broad outlines as well as the gaps in the general knowledge acquired at that time about pain are more obvious. The primary preoccupation concerned the pathway connecting one point to another, the time it took to travel along it, and the various stages and transmission relays along the way. Research on pain was dominated by problems of communication, speed, and efficiency. There were two dominant traits which characterised pain research during the first half of the 20th century (and later), even though more traditional approaches persisted and continued to provide answers. The first was the introduction of a temporal notion in studying the mechanisms of sensation; this shifted the focus from the localisation of receptors, transmission pathways, and central organs, and concentrated attention on closely related time and space factors: the conduction rate of nervous impulses, the study of temporal "summation" and also, on an entirely different level, the possible connection between the nervous system's embryonic development, the origins of the human species, and the individual's mental processes (both intellectual and emotional) as illustrated by a comparison of earlier brain charts and later stratigraphies. Though much of this research was undertaken during the last thirty years of the 19th century, it only provided answers much later when the problem of synaptic transmission began to be elucidated. The second major characteristic came about as a result of introducing Darwin's theory of evolution into the general research on the mechanics of pain, not from the standpoint of individual behaviour, as this had already been done, but in terms of the neurons. The neuronal theory appeared in both cases as the pivotal concept and necessary condition for any further research. Such a view reinterpreted the question of pain not from the standpoint of an organism's struggle for survival within its environment, but on the level of actual cell activity within an internal environment; it introduced the notion of competition and adaptation in the transmission of sensations, and viewed the nervous system as a whole integrated system endeavouring for the best possible defence against every type of aggression. This perspective maintained a clear relationship to the temporal question, as it situated the response to a noxious stimulus, at whatever level it occurred, within a framework defined by spatial and temporal factors.

### Classical and More Recent Methods of Investigating the Nervous System

*Déjerine, or the Golden Age and Limitations of Clinical Examination*

A close relationship exists between the investigative methods used and the results obtained, as well as the types of questions under scrutiny and the notions scientists had already formulated about the way nervous impulses were transmitted. As long as the problems being investigated remained limited to topographical regions and to the detailed histology of gray and white matter, three types of methods were used, either separately or in combination. The oldest, which served as the foundation of experimental physiology throughout the 19th century, consisted in the surgical ablation of carefully selected tissue in the central nervous system, or in severing certain nerves to examine the resultant effects. The problem with this kind of exploration remained that the extent of the damaged areas could never be fully guaranteed with any degree of certainty. Pathological conditions also provided similar cases to the ones artificially produced by the mutilations carried out by researchers on test animals; these provided the opportunity to verify their conclusions on human beings. The semiology of nervous-system disorders and animal experimentation were normally used in combination to achieve results.

Déjerine's work at the Salpêtrière Hospital in Paris, which ultimately pinpointed the role played by the thalamus in sensory functions, illustrates this approach. As early as 1903, in a paper written for the Neurological Society, Jules-Joseph Déjerine presented two "thalamic syndrome" cases—which had yet to acquire that name—in which the location of the lesions was confirmed by autopsy. The clinical symptoms, as described in the 1906 medical report, were as follows: (*i*) a mild hemiplegia, with no contracture, which tended to disappear rapidly; (*ii*) superficial hemianesthesia which, for its part, continued and was accompanied by deep-sensibility disturbances; (*iii*) coordination disorders in one-half the body and inability to distinguish the volume and shape of objects when handling them (astereognosis); (*iv*) sharp persistent pains with paroxysms on the hemiplegic side, unrelieved by analgesics; (*v*) again on the affected side, abrupt and irregular, involuntary movements, particularly of limb extremities.[4] These symptoms as a whole, and particularly the intensity of the pains which were of central origin, in contrast to the moderate degree of paralysis and the absence of any other specific disorders affecting areas adjacent to the thalamus (ocular paralysis for example) permitted a differential diagnosis.

A micropathological examination undertaken by making a series of transverse cross sections made it possible to determine the exact location and extent of the lesion. The dissociation of sensory and motor disorders prompted a search for an area where the sensory and motor fibres were quite distinct from one another, rather than intermingled. Microscopic examination of the lesion showed clearly that the thalamus had been damaged where the ascending sensory fibres transmit, exactly level with the lateral nucleus. Such rigorous clinical observation and close analysis of the lesions allowed a much broader conclusion on the exact role played by the thalamus in general sensibility. "We say that such a lesion severs the ascending centripetal neurons of the general sensory tracts which all converge in the thalamus, but we cannot go so far as to describe these various tracts as we now accept . . . that the median fillet represents only one important stage in relaying sensory impulses. This sensory conduction includes relaying all the superficial peripheral sensations (pressure, pain, and temperature) which, along with deep-sensibility messages, are also implicated in these cases though to varying degrees."[5]

The certainty that the thalamus represented a major relay station for sensory tracts, as opposed to Charcot's conviction that it was the internal capsule (level with the cerebral peduncle) which served as the junction for the sensory pathways,[6] became accepted as a direct result of Charcot's method. Just as he had stressed in his *Inaugural Lecture* in 1911, Charcot's main contributions were his use of Laennec's method of clinical examination and the concept of cerebral localisations. "It is thanks to him, in fact, that we gained the key concept that, where nervous disorders are concerned, resultant symptoms are entirely dependent on the exact location, rather than the nature, of the lesions."[7] In evaluating the progress made in neurology since Charcot's death,

Déjerine firmly upheld the validity of Laennec's clinical-examination method which was benefitting from important new investigative aids: the new technique of examining series of micrographic cross sections of tissue—a method we mentioned above and which is sometimes called Stilling's method—Golgi and Ramon y Cajal's silver-stain method which made it possible to observe the absence of continuity from one nerve cell to another with unequalled precision, and Marchi's method which allowed the examination of nerve-fibre degeneration by staining the myelin nodules caused by recent degeneration with osmium. His *Semiology of Nervous Disorders,* published in 1914, provided a remarkable synthesis of all the method's achievements and suggested further discoveries it could provide. The work's entire section devoted to sensory disorders, the study of sensibility syndromes, and the analysis of "radicular" pain, *i.e.* stemming from the posterior spinal nerve roots[8] demonstrated the amazing advantages of this method. For instance, in the case of syringomyelia,[9] a disorder in which a cavity forms in the cervical region of the spinal cord, interrupting the fibres in the spinothalamic bundle and causing a dissociation of sensory impressions (loss of temperature or pain sensations, but conservation of the tactile sense and muscular awareness), the value of analysing such pathological conditions in this way was obvious, as the pathways conducting sensory impulses could be clearly understood; at the same time, the precise demarcation of the anatomical areas suffering a loss of sensitivity due to these lesions made it possible to determine the exact extent of their radicular topography.

However, Déjerine was just as lucid in 1911 when he indicated the limits of this approach and the possibilities other modes of investigation could offer. After devoting time to deductive "physiological" synthesis, a mode of exploration he felt was far from having been completed in every field, he suggested efforts should be consecrated to undertaking "a pathogenic synthesis" or, in other words, providing an overall explanation about the genesis of health disorders, without which no therapeutic approach could be successful. He also stressed the dangers of a neurological branch turned in upon itself, specialised to such an extent that it ignored findings related to general pathological disorders. He terminated his lecture by mentioning the role played by the emotions in nervous disorders. He admonished with foresight, "the clinical-examination method has permitted remarkable developments in the field of neurology. But limiting the scope of future research to strictly those areas which it can elucidate would be, to my way of thinking, to reduce neurology to a narrow, confined science."[10] These comments may not have pleased every listener at the Salpêtrière.

Nonetheless, such an exact description of the thalamic syndrome cleared the way for new research on the functions of the thalamus and its relationship to the cerebral cortex. The basic research on the functions of what was formerly known as the optic thalamus was carried out by Gustave Roussy and aroused considerable interest when it was published in 1907. After giving a brief history of what was known about the thalamus until his own research, and after recalling Nothnagel's opinion that it was the relay centre for motor impulses, Roussy definitively excluded the possibility that the thalamus played a role in motor functions on the basis of his own laboratory work with monkeys, dogs, and cats. He discarded the hypothesis that the thalamus served as a coordinating centre for reflexes or involuntary expressions, or as the centre which integrated the sympathetic nervous system; he stressed rather this organ's role in processing sensory impressions. "At the core of the cerebral hemispheres, the optic thalamus represents the relay station for the tracts transmitting peripheral sensory impressions to the higher cortical centres. Both the anatomy and physiology of this organ demonstrate the intimate link it has with sensory functions."[11] He meticulously described the fibres connecting the thalamus to the cortex and the cortex to the thalamus, but submitted no hypotheses concerning the functional connection between the two. It was Head who suggested a possible explanation for this relationship.

The hypersensitivity of patients suffering from this syndrome gave rise to two hypotheses. Either the thalamic lesion produced a permanent source of irritation and transformed even the mildest, otherwise quite bearable, sensation into a painful stimulus or, as Head conjectured, the lateral nucleus of the injured thalamus was in fact a center which was controlled and regulated by the cerebral cortex. This control mechanism would therefore seem to have become inhibited as a result of the lesion. A closer look at these questions necessitates a prior description of the brain regions which had been localised at that time.

*Cerebral Localisation and the Localisation of Pain Centres:*
*Trial and Error with Numerous Dead Ends*

The theory that pain had a distinct relationship to consciousness and that it supposed a transition from a sensory state to a perceptual state, as well as calling on complex individual emotional reactions associated with prior experiences, invariably came up against the debate centered on the actual localisation of sensory, emotional, and intellectual functions. The search for an ultimate pain centre was part of the more complex and long standing issue about the very feasibility of locating specific functional areas in the brain. The scientists in favour of localisation came up against a "unitary" concept applied to the brain, jealously defended in France by Flourens and by an important spiritualist movement that foresaw the downfall of religion and the relegation of man to animal status if the attempt to assign a specific location to the soul's functions succeeded.

However, partisans for the existence of functional correlation in the brain such as Brown-Séquard continued to put forward rigorous scientific arguments to back their views. These included accounts of compensation and recuperation processes witnessed after injuries to carefully identified zones, or evidence of overlapping sensory-motor phenomena suggesting the intermingling or interlacing of zones rather than strict demarcations. Though a number of different "localisationist" schools existed, any discussions attempting to establish the correlation between brain structure and function in relation to the intellect immediately precipitated sharp clashes between materialists and spiritualists, particularly in France; this conflict was further exacerbated by the antagonism which existed between science and religion at the time. Despite the efforts of Bouillaud, and then Broca, the battle was fiercer in France than in Great Britain due to a fundamentally different philosophical stance; in Great Britain, the sensualist tradition could be said to have laid the philosophical foundations required to establish the connection between the sensory centres and intellectual functions. The political implications of these issues were enormous insofar as the discovery of specific regions, particularly the role of the left hemisphere in what are called the higher functions, tended to establish a correlation between such characteristics as cranial volume and mental development to the degree of evolution witnessed in specific individuals or races.

Before any attempt was made to distinguish an hypothetical pain centre in the cortex, neurologists had tried to define sensory and motor areas in the brain. This had necessitated prior rejection of the notion that the brain was an homogeneous whole. The concept of cerebral localisation, developed by Charcot and Broca in the final third of the 19th century in France, and by David Ferrier in Great Britain, was soon criticised as being a basely material approach. It was all the more vigorously condemned as medical research, aside from straightforward clinical observation. They were now carrying out experiments using a faradic current to stimulate the brain of monkeys, a procedure which had led to the inference that these animals were being equated to man. It was only in 1909 that the American surgeon, Cushing, was able to apply this investigative technique directly to human beings by operating on fully conscious epileptics.[12]

Gall's phrenology doctrines were denounced all the more severely due to the analogy being made between human beings and anthropoids, particularly as a result of violent reactions against the implications of Darwin's theory. These were all the more vehement as there had been a number of initial successes, including Broca's work[13] which, in 1861, localised the seat of aphasia in the region of the Sylvian fissure at the base of the 3rd left frontal convolution, a discovery which aroused considerable hostility due to the close link which exists between language and thought. A further polemic opposed Déjerine and Pierre Marie, in 1907, as the latter contested the role played by the 3rd frontal gyrus in the concept of language imagery, and insisted in addition that all aphasic patients were intellectually deficient.[14]

However, if we examine the general knowledge acquired at the very start of the 20th century about the brain's sensory zones, and particularly about the cerebral cortex, we must admit that facts were scant. David Ferrier, to whom we undoubtedly owe the generalisation of the terms afferent and efferent rather than sensory and motor,[15] drew up a clear diagram of the cerebrospinal nerve centres providing a systematic view of the sensory tracts graduating up from the spinal cord towards the thalamus and onto the cerebral hemispheres. Ferrier depicted sensory centres, themselves dependent upon sites located in the frontal region, responsible for moderating and coordinating complex actions between the sensory and motor centres of the lower levels, which he identified as the seat of attention and consciousness and termed, for these reasons, the "katanoetico-kinetic" centre.[16] Would it be possible, however, to distinguish the same type of sensibility discrimination in the brain's hemispheres as was being discovered in the spinal cord? By destroying the hippocampal region in a monkey, Ferrier found that its sense of touch had been suppressed on the side opposite the lesion and that the animal could no longer even use its limbs on the insensible side, though a reaction could be elicited by a very intense stimulus.[17] In piecemeal fashion, thanks to happenstance pathological conditions presenting extremely localised brain lesions due to tumours or hemorrhages, and through the selective-destruction method, a body of knowledge was growing, though more readily in the case of motor functions than sensory functions. Based on the study of an epilepsy case which included the sensation of an "aura" around the thumb, John Hughlings Jackson[18] localised the sensory zones in the posterior portion of the brain. Others such as Schiff or Munk envisioned an extension of the sensory zones on either side of Rolando's fissure and considered that the two types of functions actually overlapped, with the existence of a sensimotor or sensorimotor area. Nothnagel's work (1873) concluding that sensibility was localised in the parietal lobes seemed contradicted by the excision of these lobes, however, as no sensory disorders were provoked in the dog, while the removal of the so-called motor zone resulted in a diffuse reduction in sensibility to pain and a clear alteration in the tactile sense.

To summarise the uncertain information available around 1900 on the cortical localisation of sensory disorders, Bechterew's testimonial is worth analysing as he was quite in favour of non-exclusive specialisation and tended to explain the different results obtained by searching for the centres at deeper levels of superimposed strata rather than those spread over the surface. "For the vast majority of physiologists, including Schiff, Munk, and Bastian, the Rolando's area in monkeys and human beings, and the sigmoid gyrus in dogs, are exclusively sensory and represent the zone which perceives and registers muscular, tactile, painful and temperature sensations, an awareness of which precipitates the appropriate response. The destruction of this zone causes only sensory disorders and, in consequence, motor disorders. For others such as Ferrier, Horsley, and Schäffer, the destruction of the motor zone under no circumstance ever causes sensory disorders, not even muscle-sensibility disorders. The localisation of the sense of touch varies according to the author: some settle on the hippocampal convolution or the callosal gyrus, while others designate the parietal lobe. Between these two apparently opposed concepts, a mixed theory going back to Hitzig should be mentioned as it holds that the cortical centres are actually sensorimotor and represent either a junction of the motor centres and sensory centres, which overlap but remain independent to some degree (Luciani, Seppilli, Tripier, Tonnini, Mott and Goltz), or are composed of similar elements, translating sensations into movements."[19] Whatever the hypothesis, they all stressed the importance of awareness factors in the development of sensory disorders. The generally accepted conclusion was that more extensive, deeper cerebral lesions were necessary to produce sensory disorders contrary to those which produced motor disorders; in addition, the inhibition of pain and temperature sensibility was rarer than the loss of "extensity" or spatial awareness of objects, and compensation occurred more rapidly in the case of sensory than motor disorders. These various theories readily agreed with the notion that sensibility was an essential condition for survival and adaptation to the environment, and that the cerebral zones involved must therefore be highly protected and were perhaps less specialised.

All sorts of problems arose in attempting to interpret these initial findings. For instance, muscular sensibility as well as deep sensibility in general seemed essential for movement and its coordination to occur. As a result, it became excessively difficult to differentiate the parts of the cortex truly specialised in the projection of sensations by using the selective-destruction method. Even between the two wars, both cerebral localisation and the inheritance-of-acquired-characteristics theories smacked equally of brimstone as Alajouanine testified: ". . . alarmed as if by hellfire, many fought the cerebral localisation doctrine fiercely; we saw the same assaults which had been launched against Lamarckism a century earlier, and also witnessed redoubled onslaughts against the theory of transformation in general. Adversaries and critical arguments were much alike as the cerebral-localisation concept is ultimately just an aspect of 'transformism' and the victory of the first would definitively ensure the triumph of the latter."[20]

At the onset of the 20th century, cerebral localisation research was oriented in a different direction linked to the influence of the cellular theory. Indeed, as a result of the studies carried out by the Viennese "alienist" Meynert[21] on the structural composition of the cortex made up of five layers, each represented by a specific type and distribution of cells, the question arose as to whether this disposition might not have a functional significance. It was only a short time later that an attempt was made to map out the different cerebral areas based on the detailed studies of the Spanish histologist, Ramon y Cajal. A kind of architectonic diagram of the cortex was drawn up, both in England by Campbell in 1905, and also by Vogt and Brodmann in Germany in 1909,[22] based on the types of cells present in each area and the major anatomical divisions.

The difficulties involved can easily be measured if we remember that, in his *Physiology of the Nervous System* published in 1938 and translated into French in 1947, Fulton states: "A complete map of the human or primate brain has yet to be drawn up."[23] At the start of the century, knowledge remained limited to the major regional divisions, though a more precise histological understanding had been acquired along with a gradual confirmation that the parietal lobe was involved in sensory functions, particularly in the somatic sensory functions, *i.e.* sensations from the body, while other zones were held responsible for distinct senses (the occipital lobe for vision for instance). However, the precise confines of the somesthesial zones in the cortex (the zones responsible for bodily sensations) left the question of localisation of the psyche or consciousness quite unanswered. For one, there were a considerable number of cases recorded in which frontal lesions were responsible for mental disorders that, according to Alajouanine's classification system, affected recent-event retention, emotional feelings, activity and initiative, as well as spatial orientation.[24] In addition, it seemed impossible to locate the psyche in any one specific delimited zone, and a number of researchers stressed the risk of making false assumptions or, more specifically, of holding the frontal lobe responsible for disorders originating in the corpus callosum. In short, the more the cortex's extreme differentiation became clear, the harder it became to associate any one area with a specific function. Alajouanine, who suggested that a fairly narrow localisation probably applied to space-related functions, pointed out that "localisations are above all functional localisations, and the more a function calls on the higher association mechanism, the less precise is its localisation."[25]

The quest for a hypothetical, strictly delimited pain centre in the cortex seemed therefore to be an increasingly illusory goal, not because investigative methods were inadequate but because the question was not properly posed. The experiments in which electrical stimulation was applied directly to the cortex of human patients—during operations on epileptics or to remove brain tumours—provoked no pain though attempts to surgically remove the somatic sensory regions to alleviate the pain of phantom limbs brought no relief. It would have been possible to search for another pain centre, such as the thalamus for instance, whose importance we have seen; however, it was becoming increasingly clear that there could be no "cartographic" answer to this problem and that a more dynamic investigation of the multiple connections linking the various ascending tracts should be carried out. The conscious perception of pain was dependent upon the associations between the system's different stations or relays, and particularly between the thalamus and sensitive cortex and the links between the different areas of the cortex. Among other things, pain awareness pointed up the need to distinguish the projection areas linking cortical centres to the periphery as well as the association zones between neurons inside the cortex.

In 1911, Henry Head and Gordon Holmes published a joint study entitled "Sensory Disturbances from Cerebral Lesions," in which they produced a model showing the relationships between the cortex and the thalamus based on their many observations of the thalamic syndrome. In contrast to Déjerine and Roussy's work, they underlined the emotional dimension involved in hyperesthesia and hyperalgesia to a greater degree: "The characteristic thalamic response does not consist in an excessive reaction to painful stimuli only. In suitable cases, we have shown that the response to pleasurable stimuli, such as warmth, is also greater on the affected side. Moreover, the manifestations of general mental states of pleasure or discomfort may be more pronounced on the abnormal half of the body."[26] While lesions affecting the sensory cortex provoked changes in the aptitude to distinguish the weight, size, shape, position of objects in space, or everything which required a fine-tuned discrimination between sensations, thalamic lesions seemed to modify the emotional or affective tone with which any sensation was perceived, transforming even the lightest touch into a source of extreme pleasure or pain. The thalamus therefore seemed to behave as an amplifier of all the sensations transmitted by the afferent tracts, and had the capacity to produce unbearably loud and painful sounds if the thalamus no longer fulfilled its role as a conductor, controlling and regulating the lower level.

Head and Holmes concluded, on the basis of precise anatomical studies carried out by Roussy on the functions relayed from the thalamus to the cortex and vice versa, that the thalamus had three major characteristics: it was the terminus for all the afferent sensory pathways which were then redistributed in two directions: to the cortex on one hand, and to the body of gray matter in the thalamus itself on the other; this gray matter represented the essential centre for certain sensations and complemented the sensory cortex; and finally, that the lateral portion of the thalamus served as a receiving station for fibres from the cortex and was the region through which this organ exercised its

regulatory functions because damage to this area was always evident in
the thalamic syndrome. Head and Holmes continued, "Thus we believe
that the essential organ of the optic thalamus is the center of consciousness for certain elements of sensation. It responds to all stimuli capable
of evoking either pleasure or discomfort, or consciousness of a change
of state. The feeling tone of a somatic or visceral sensation is the product of thalamic activity, and the fact that a sensation is devoid of feeling
tone shows that the impulses which underlie its production make no
thalamic appeal."[27]

In this hierarchical model, in which the cortex would control the
thalamus under normal circumstances, we are once again confronted
by the disparity between the two forms of sensitivity studied by Head,
one diffuse and the other discriminatory and, in addition, by the
contrasting emotional functions and intellectual functions assigned to
the cortex. This organ of conscious attention to sensations would have
served as a type of general storehouse of past sensations from which
bodily impressions were elaborated, as well as the representation individuals formed of their own body and its activities; it also made it
possible to interpret the cortex from an evolutionary point of view. This
division of roles accounted for the different reactions: "All stimuli which
appeal to the thalamic center have a high threshold. They must reach
a high intensity before they can enter consciousness, but once they have
risen above the threshold, they tend to produce a change of excessive
amount and duration, and this it is the business of the cortical mechanisms to control. The low intensity of the stimuli that can arouse the
sensory cortex and its quick reaction-period, enables it to control the
activity of the cumbersome mechanisms of the thalamic center."[28]

Two major concepts emerged from Head and Holmes' research: the
sensation of pain could not be explained by a strictly linear theory of
transmission, with only intermediary relay stations between departure
and arrival. These relays in actual fact reorganised and modified the
information being transmitted, while various interactions took place
between the different stages. We can therefore understand in what way,
though not strictly from a localisational standpoint, the thalamus could
be considered if not "the pain centre," then certainly a decisive stage
in its production. Such a model has never really been abandoned,
though its functional principles have been considerably expanded to
reflect the dynamic connections between the various levels. The other
aspect of Head's work that should be remembered is the marked

significance of its evolutionary basis, which also covers its comportmental and ethical implications as he considered that man's evolutionary goal was ultimately to control his emotions and instincts through such mental processes as judgement and discrimination.

### Head: Epicritic Sensibility and Protopathic Sensibility

The utility of pain as an alarm signal takes on a special significance in relation to Darwin's theory because individuals lacking this capacity would receive no warning of imminent threats from the outside world and would therefore be an easy prey for their aggressors. Indispensable, therefore, according to this scheme of things for the protection of the organism and the survival of the species, sensibility to pain must be a universally shared attribute among living beings (even the simplest forms). It must be particularly well protected, and should at once be the trait spared the most despite wounds inflicted on the nerves, as well as the sense to recover fastest when these have been badly damaged.

These ideas were undoubtedly current when Henry Head began his research on various pain sensations related to visceral disorders.[29] His research was closely linked to work being carried out by Gaskell and Langley in Cambridge during that period and, in addition to this scientific circle, Victor Horsley, the leading practitioner in the field of neurosurgery, was providing useful insights; Head was also gaining invaluable experience from his work at the London Hospital and his observation of a substantial number of interesting cases. As he wrote in 1905 in his introduction to an article in which he distinguishes three kinds of sensibility: deep or visceral, protopathic, and epicritic (the one which allows fine distinctions); it was the absence of correlation between the dominant ideas about the physiology of sensations and clinical observations which prompted him to study the problem, focusing on the nerves in the hand and forearm for his research.

Surgeons and neurologists knew that lesions to the median nerve caused a loss of sensibility in the index and middle fingers, and that sensibility in the area of the palm which it normally innervated was reduced though not entirely suppressed. The same observations applied to the ulnar nerve when it was injured and to its resultant effects on the sensibility of a portion of the palm and the annular finger. In order to explain the absence of correlation between the severed nerve and the loss of sensibility in the zone dependent upon it, the overlapping quality of nerves in a given territory was ordinarily brought up as justification. Using such logic, however, it should have been possible to demonstrate variations in sensibility which had depended upon the severed nerve, a fact unsubstantiated by clinical observation. "If each nerve occupies another's territory over a sufficient range to prevent the total loss of sensation in such a large area of the palm, it's obvious that the destruction of the ulnar nerve must cause a reduction in sensibility over half the median area. This loss of sensibility should vary in proportion to the amount of sensation retained after the destruction of the median nerve. However, even the closest examination of the hand does not make it possible to establish the least reduction of sensation in the median half of the hand as a result of the ulnar nerve's destruction. What has always been termed a reduction in sensibility caused by severing the nerve amounts, in fact, to a loss of certain forms of sensibility while others are preserved."[30]

Could this difference be explained by the existence of specific receptors for each kind of sensibility such as they had been described by von Frey? This proved untrue, however, when the process of recuperating sensibility in the areas in question was studied and attempts were made to distinguish what had been lost and what remained. A close examination showed that the tactile sense had disappeared, as well as the ability to determine points of contact or appreciate average temperatures. During the recuperation phase, sensibility to pain and sensibility to heat and cold returned rapidly, while the sense of touch remained absent in exactly the same area as at the start. Moreover, sensitivity to a pinprick (the test for pain sensibility) produced a sensation which, instead of being precisely localised, radiated widely and produced an unpleasant and abnormal sensation, "an unusual discomfort" which Head set apart from pain. Finally, the ability to discriminate between temperatures was altered because warmth could be detected beyond a certain threshold, but temperature differences within an average range could not be distinguished.

In order to shed light on these aspects, Head decided to undergo an auto-mutilation procedure on April 25th, 1903, by severing, then suturing back together, the cutaneous branch of his radial nerve near the elbow. He subsequently observed the rate at which healing occurred. He began by noting the absence of any sensibility in the area subserved by this branch of the radial nerve, though he could clearly perceive a pinprick, which he could only explain by the presence of afferent fibres alongside the motor nerves of the same type as those found in the muscles or tendons; this deep sensibility could detect pressure and pain when firm or intense. However, the experiment's real value lay elsewhere, in the two types of sensibility which were recuperated successively. After seven weeks, a pinprick was perceived on the forearm at first, then on the hand but with the same diffuse quality of pain which made strict localisation impossible. Fine tactile discrimination returned much later, but had not been fully recovered nearly two years after the mutilation.

In the initial phases of recuperation, temperature and pain perception required much higher threshold stimuli than prior to the lesion, while sensation was generally felt to be more painful and often seemed referred to a distant point from where the stimulus had been applied. In addition, simultaneous stimulation of multiple hot and cold points, as well as stimulation of pain-sensibility points to a lesser degree, considerably lowered the threshold of perception, while temperature gradations did not correspond to the sensation gradations. Head was thus confronted with a new form of sensibility, a previously unknown, diffuse, radiating form of sensibility whose exact localisation could not be determined; it was a form of sensibility to pain and temperature extremes which triggered a reflex system that reacted rapidly and indiscriminately to painful stimuli. "To this form of sensibility we propose to give the name 'protopathic.' . . . With the gradual return of sensation, it again becomes possible to discriminate two points touching the skin at distances more nearly normal, and the widespread radiation, so characteristic of the first stage of recovery, ceases and is replaced by an increasing accuracy of localisation. To this form of sensibility we propose to give the name 'epicritic' since it is peculiarly associated with the localisation and discrimination of cutaneous stimuli."[31] This latter form of sensibility seemed more highly developed and limited to the skin, while protopathic sensibility was to be found everywhere within the body, including the skin, and could be equated with the confused sensibility of the viscera. These two systems naturally led to postulating that there existed two types of fibres, with the protopathic ones regenerating more rapidly.

A considerable number of similar experiments have been carried out to verify Head's results. Trotter and Davies[32] also underwent auto-mu-

tilation procedures on various types of nerves and observed a number
of differences, particularly with regard to hyperalgia. If, in fact, the
recuperation process was not entirely uniform, it was nonetheless impossible to infer that two types of fibres existed. Trotter and Davies did
not exclude the possibility of chemical reactions occurring at the lesion
site or of the central nervous system intervening. Weddell,[33] for his part,
and as a result of the incomplete sensory perception in the protopathic
stage, suggested that the number of nervous fibres which had regener-
ated themselves were initially less numerous than before severance, and
only in unique locations, thereby preventing any summation process
and explaining what appeared to be an "all-or-nothing" phenomenon,
a subject we will examine later. Head's research clearly showed that the
skin's specific receptors could not account for his observations, as these
related to a time frame rather than a specific localisation. In France,
Déjerine immediately adopted Head's distinction between the two
forms of sensibility and his experiments were mentioned in most physiological treatises devoted to the nervous system. Nevertheless, Roger
and Binet's *Manual,* the classic physiology reference between the two
wars, preferred to limit itself to the differences between deep sensibility
and superficial sensibility. Could this have been due to a reticence
concerning the link Head established between protopathic or basic
sensibility and phylogenesis, or due to the fact that his approach encompassed the nervous system's integrated functions rather than staying confined to traditional clinical observations?

In the rather biased commentary Jean Lhermitte makes about Head's
ideas, there are a number of reservations: "In his highly valuable work,
Henry Head once again stipulates the characteristics of so-called thalamic pains and underlines their emotional or affective tone by establishing their link to functional states and their dependence on certain
specific stimulations, originating not in the cerebrospinal system but
the autonomic nervous system. The thalamus is once again described
among the centres where stimuli from the viscera terminate and are
mirrored. As far as superficial sensibility is concerned, H. Head demonstrates that thalamic destruction does not inhibit sensation so much
as the discrimination between stimuli and, above all, replaces their
specific qualities by a painful or disquieting, highly uncomfortable and
uniform sensation. H. Head has described the patient as 'integrating
all his sensations in the pain,' but with regards to the cortex's inhibitory
role on the thalamus, he felt the idea was tempting, though not founded

on any precise anatomical data."[34] Head's work prompted the need for a new interpretation of pain which was provided by Sherrington.

## Pain's Space-Time Continuum

### *Sherrington and the Integrative Action of the Nervous System*

The work of Sherrington (1857–1952), a neurophysiologist who studied at Cambridge in Michael Foster's laboratory under Langley and Gaskell's supervision,[35] really belongs to contemporary science due to the nature of his theories. Sherrington, who received the Nobel Prize for Medicine in 1932, rounded off his education by doing a number of internships abroad, especially in Germany where he worked with Virchow and Koch in particular. His experience was highly diversified by the time he obtained a physiology professorship, first in Liverpool, then in Oxford, as he had worked in an animal hospital in London among other activities before writing his best known work, *The Integrative Action of the Nervous System*, published in 1908.

He regarded this opus as having been written under the double influence of both Virchow's cellular theory and Darwin's theory of evolution; it could almost be said that he transposed the theory of evolution to the level of the neuron and the synapse, a term he coined for the junction separating two neurons.[36] The originality of his methodological approach lay in his combination of analysis and synthesis. In order to study the integration achieved by the nervous system in transforming separate organs into a single individual whose parts all work together, he focused first on the simple reflex arc which called on a receptor, a conductor, and an effector, and then the composite actions resulting from several coordinated reflexes. He called the simple reflex an "artificial abstraction," but he needed to examine this simple unit's construction, the predecessor of the motor-unit concept which earned him the Nobel Prize, in order to demonstrate that the nervous system did indeed operate as a single integrated whole. To do so, the simple reflex had to be treated as the basic first step in the coordination process. This was a legitimate approach if one accepted the premise that the activity produced by the effector was the appropriate response to the stimulus transmitted by the receptor, and that this process called several neurons into play.

If an animal was able to present itself as a single entity, it was because at each moment of its existence reciprocal simple-reflex adjustment processes were occurring simultaneously, ultimately forming a pattern through the summation of its reflexes, on both a spatial and a temporal level. In this system, a receptor's adaptation to specific stimuli and the appropriateness of responses were conceived along Darwinian lines since selection made it possible to adjust the response to the stimulus. This concept agreed with a non-transcendent view of the final teleological purpose of nature, and was reflected by the most efficient conduction possible. It explained the selective nature of a stimulus capable of exciting a nervous receptor, not on the basis of structure but in terms of adaptation and specialisation which could, even if only in part, be the result of evolution. "The main function of the receptor is . . . to lower the excitability threshold of the arc for one kind of stimulus, and to heighten it for all others."[37]

This theoretical framework made it possible to account for the results obtained from the different types of receptors, and to resolve the problem posed by the existence of specific pain points, as well as by the pain produced by excessive heat, cold, pressure, etc. The concept of nociception put forward by Sherrington as early as 1903 to explain reactions to a set of painful stimuli was obviously broader than the idea of specific pain receptors, and it provided the answer to a major dilemma: how was it possible to reconcile the idea of a minimum threshold, specifically adapted to a certain type of stimulus on one hand, and the fact that cutaneous nerve endings were capable of responding not only to painful stimuli, but also to any other type of stimulus, whether mechanical, thermal, etc., which seemed opposed to the selection principle, on the other? The solution to this problem was actually supplied by the very definition of nociception: all the stimuli mentioned had the common characteristic of being potentially damaging to the skin. Where nerve endings were concerned, "Instead of but one kind of stimulus being their adequate excitant, they may be regarded as adapted to a whole group of excitants, a group of excitants which has in relation to the organism one feature common to all its components, namely, a *nocuous* character. . . . It does not seem improbable, therefore, that there should under selective adaptation attach to the skin a so-to-say-specific sense of its own injuries."[38] This adaptation would not have depended upon any minimum threshold because, by definition, all the stimuli which caused pain were intense; it would, however, have needed to respond to a broad range of stimuli, *i.e.* all those which presented the individual with any risk of aggression in his environment. It was therefore more sensible, instead of referring to "pain nerves," particularly from the standpoint of the reflexes in question, to speak of "nociceptor nerves."

Research on combined reflexes, in situations where two stimuli compete to win "the final common path" through which an effector is reached, for example, has shown that it is always the reflex in response to nociception which wins. The predominance of the nociceptor reflex over all others does not depend on the intensity of the stimulus, but on the presence of an emotional quality which strongly affects the subject's experience. The association of sensation with an emotional reaction may have originally been introduced in order to deeply ingrain past events which gravely imperilled the individual or the species in the memory. If, in the struggle to use the common path—which all stimuli must take though they cannot do so simultaneously—it is the nociceptive defense mechanism which gains exclusivity, this is due to its importance for the survival of the species and the frequency with which animals are thus exposed. The definition which describes this link between basic reflexes and emotional tone is as follows: "Physical pain is thus the psychological adjunct of an imperative protective reflex."[39] We have therefore described a special class of stimuli which affect the skin's nerve endings and whose common characteristic is their tendency to endanger the skin. Four traits describe the reflex they excite which: "*(i)* is *prepotent; (ii)* tends to *protect* the threatened part by escape or defence; *(iii)* is *imperative; (iv)* and, if we include psychical evidence and judge by analogy from introspection, is accompanied by *pain*."[40]

However simple it may seem, this definition still requires a detailed study of simple reflexes and their coordination. "The common path principle," for which competition between the responses from different receptors arises, is based on close observation of what occurs when several receptors are called on simultaneously. "When two receptors are stimulated simultaneously, each of the receptors tending to evoke reflex action that for its end-effect employs the same final common path but employs it in a different way from the other, one reflex appears without the other. The result is this reflex or that reflex, but not the two together."[41] A selective principle intervenes which supervises the bottleneck that might otherwise occur, so to speak. In actual fact, the organism is constantly subject to multiple stimuli and the simple reflex-arc in question is more the exception than the rule.

Sherrington classified stimuli on the basis of their origins into three categories:[42] "proprioceptive," or arising from organs deeply situated within the body, particularly the muscles, tendons, joints, and blood vessels; in contrast to this category in which stimuli originate in the body itself, there are the "exteroceptive" receptors found over the body's entire surface which are in direct contact with the environment; and finally, there is another type of sensibility which is involved in such functions as digestion, absorption, etc., called "interoception" and known also as the visceral sense, that we are not generally conscious of. The ordinary functioning of a living entity implies coordination among all the reflexes arising from these different stimuli. Certain reflexes are "allied" and reinforce one another in consequence, with the first stimulus "easing the way" for the second in a sense; others, on the contrary, are antagonistic and the outcome of any conflict results in the inhibition of one reflex.

If we observe the dog's scratching reflex, which Sherrington studied closely, two stimuli applied to the skin a certain distance apart but in the same exteroceptive field—in this case the area which provokes the scratching reflex—mutually reinforce the scratching reaction. This situation is identical to the one we examined in relation to the simple reflex action describing the summation process. One strong stimulus is in fact not enough to set off the scratching reflex, while two weak stimuli, each by itself below the minimum threshold, reinforce one another by having a cumulative effect if administered at sufficiently close intervals. In the case of antagonistic reflexes, which sometimes elicit incompatible sensations (the sensation of pain repressing the tactile sense, for instance), one group suppresses the other, producing an inhibitory phenomenon. Here again, the simple-reflex situation serves as a model on which to base our interpretation since we may consider the refractory stage to new stimulation, *i.e.* the stage immediately following an initial stimulus, as a special case of inhibition.

In this conception of the organism's way of functioning, we are not simply dealing with the trajectory from a point of departure to a terminus, but with a set of dynamic exchanges and interrelationships in which nervous impulses from a variety of origins are competing, at every moment, to take over the path which will allow them to have an effect. The hiatus introduced by this concept can perhaps be more readily understood if we examine Head's metaphor to describe the central nervous system in 1905: he compared it to a post office receiving missives from all sides and, after centralising them, redistributing them in every direction to the right addresses. This analogy does not represent Head's ultimate vision of the scheme of things but it did provide useful insights for analysing the brain's way of functioning, while Sherrington's view, for its part, emphasised the integration concept as well as functional complexity.

Two questions can be raised in relation to this model: the first concerns the relationship between reflexes and sensations, particularly where pain is concerned, and the second, the general utility of the different reflexes, particularly in response to nociception. Without wishing to delve into their ontological nature, but only to examine the parallels between mental and physiological phenomena, Sherrington underlined the close resemblance between nervous reactions originating in the muscles and those of a clearly sensory nature, with the first serving to study the second. Next, he pointed up the clear utility of reflex actions and sensations which introspection makes patently obvious in man and whose self evidence can be inferred down through the animal kingdom. The usefulness of reflexes is certainly manifest and represents an evolutionary progress which has ensured the survival of the species. "Pure reflexes are admirably adapted to certain ends. They are reactions which have long proved advantageous in the phylum, of which the existent individual is a representative embodiment. Perfected during the course of ages, they have attained a stability, a certainty, and an ease of performance during that course beside which the stability and facility of the most ingrained habit acquired during an individual life is presumably small."[43]

In the most highly evolved animals, though, reflexes are partly dependent on the brain and within certain limits can be suppressed, modified, or generated. This can be seen in the writing skill, for instance, since once the apprenticeship is over, the ability becomes automatic and we no longer need to pay attention to the way we form the letters. There is therefore no abrupt break between reflexes and consciousness, whose primary function is precisely one of control. The development of organs which ensure this control even constitutes "the best criterion to the success of an animal form in the competition which lies at the root of animal evolution."[44] The emotional or affective tone (pleasure or pain) which accompanies a certain number of reflexes and calls on mental associations and memories reinforces the adaptative behaviour of individuals to the situations which confront them.

In Sherrington's view, the principle underlying the three main aspects of comportmental behaviour, including the motor reflex, the emotions, and consciousness, appeared to be the search for utility destined to benefit the individual. "These nervous organs of control therefore form a special instrument of adaptation and of readjustment of reaction to better suit requirements which may be new. New adaptations whence the individual may reap benefit are thus attained."[45] From the synapses between neurons to the brain's most elaborate functions, the natural selection principle which accounts for the reaction mechanisms best suited to the survival of the individual is obviously at work and explains behaviour. Should we therefore reproach Sherrington for conferring a "psychological dimension" to neuronal activities and their synaptical relationships by introducing an unjustified finality principle? That would amount to making the same unwarranted accusations against him as those made against Darwin's theory itself. Sherrington did not transfer the utility principle to the neurons themselves, but saw in such efficiency a clearly intelligible underlying principle directing the reaction mechanisms of the integrated whole making up each living individual, and he always interpreted such efficiency or advantages from an all-embracing viewpoint.

The inroads of this principle—based not on psychological analysis but on a careful study of the nervous system, focused on the mechanisms of nociception in particular—gave the "utility-of-pain" concept a theoretical foundation it had previously lacked and which some may have deemed forgotten. This concept goes beyond the issue of pain's value as an alarm signal, as it also comprises the utility of the self-protection mechanisms and physical strengthening provoked in organisms as a result of pain, as Cannon's research clearly brought to light. But the heuristic and explanatory value of Sherrington's work and the research which followed dominated Western culture's view of pain to such an extent that its fundamental significance was gradually overshadowed; its findings were used in other fields of research and for treatises or discussions which had neither the same objectives nor the same methodological rigour.

### Herbert Spencer's Legacy: The Theory of Evolution and the Language of Pain

If research on pain has been so closely linked with the theory of evolution, in Anglo-Saxon countries in particular, this may be attributed at least as much to the widespread dissemination of Darwinism as to the importance Herbert Spencer in his *Principles of Psychology* and John Hughlings Jackson placed in the psycho-physiology and brain pathology field. Here, two approaches converged and identified pain as a mode of emotional reaction shared by both man and animals, stemming from the most basic cerebral structures, anterior even to the development of the earliest hominids, and retained due to their adaptability value. One approach, which based its findings on embryological studies, recognised in man's present brain an encapsulation of more primitive structures, while the highly developed telencephalon appeared later in the course of evolution to ensure more highly elaborate functions. The other approach studied psychological phenomena from the standpoint of evolutionary influences. This perspective was adopted by Spencer in particular, who maintained: "If the evolutionary theory is true, then the mind can only be interpreted in relation to its evolution. If the highest animals only acquired their well-integrated, clearly defined, and highly-diverse organisation after untold eons of accumulated modifications, and if the nervous system of these animals only gradually attained its complex structure and functions, then it clearly follows that the complicated forms or correlative consciousness made possible by these structures and complex functions must have emerged by degrees. And since it is truly impossible to understand the body's general organisation and particularly that of the nervous system without following its successive stages of elaboration, it is also impossible to comprehend mental organisation without studying those periods. We will therefore begin our study of the mind by examining its objective manifestations in the ascending stages of various types of sentient beings."[46]

If we consider that every feeling, which includes both sensation and emotion, corresponds to a certain type of excitement of the nervous fibres from a material standpoint, then the most automatic reactions would be ones which have occurred most frequently in the course of evolution and which have encountered the least resistance to stimulation of the nerves and muscles. The most frequent situation arising in the animal kingdom has in fact been one of antagonism towards other members of the species, with its resulting highly unpleasant feelings. From an evolutionary point of view, pain, anger, and combat have always been closely associated, which accounts for descriptions of emotional states in terms of the shouts, trembling, and fits common to those states. "Though man experiences many situations other than

antagonistic ones which arouse unpleasant feelings, and though antagonism only leads to actual combat in extreme conditions, nonetheless as
antagonism was usually the most common, dominant adjunct to unpleasant feelings in the lower animals from which we descend, and as it
has generally remained such an adjunct in mankind, a correlation established itself at the outset between unpleasant feelings and the muscular reactions usually elicited by antagonism."[47] According to Spencer,
an expression such as a frown, which has the dual significance of either
annoyance or pain, must have played a role in combat situations; an
ingenious hypothesis (reminiscent, though in an entirely different context, of some of Bernardin de Saint-Pierre's "providentialist" explanations) suggests that vigorously contracting, or "knitting," the muscles
above the brow may have been an advantage for an animal when the
sun was shining in its eyes. Don't certain monkeys have a bony crest
covered with fur in that spot, for instance, which serves to shade their eyes?

Underlying principles of association are therefore at work among the
various behavioural patterns inherited in the course of evolution. "Since,
in the lower animals as well as in man, both physical and psychological
pain are inseparably linked to the various facets of combat, their physiological effects are intimately linked to the physiological manifestations
of combat posture. This is evident to the extent that pain, no less than
anger, sets off a considerable number of muscular reactions that originally occurred spontaneously to ensure the successful issue of combat."[48] This psycho-physiological evolutionary view prompted certain
scientists such as Cannon to wonder what useful purpose the physiological, outward display of pain might have today.

### Defence Mechanism to Avoid Pain

Aside from the most obvious motor reflexes designed to avoid or reduce
pain (flight reflex, or scratch reflex in response to local irritation), doctors
had long observed that violent emotions had considerable physical
impact: modifications in heart rhythm and blood circulation, pallor,
feelings of intense cold, sweat, or even fainting at times. These symptoms, initiated automatically and difficult to control through will power,
and usually linked to fright or passionate love, had become as much
literary or pictorial subjects as medical ones; before the end of the 19th
century, however, there were no satisfactory explanations for these
phenomena which largely involved vascular changes in blood circula-

tion and the flow of fluids. The concept of internal secretions, closely linked to the concept of the organism's internal environment, both subjects developed by Claude Bernard and further elaborated by Brown-Séquard,[49] provided the necessary theoretical basis for experimental research on internal bodily modifications triggered by emotional states and, chief among them, the sensation of pain.

Thus Walter B. Cannon's *Bodily Changes in Pain, Hunger, Fear and Rage,* published in 1915, can be situated at the crossroads between experimental physiology and the newly emerging endocrinology. This study of somatic reactions was prefaced by observations about digestive disorders provoked by such emotions. In 1911, in his *Mechanical Factors of Digestion,* Cannon demonstrated that anger or pain interrupted gastric secretions and hindered digestion, while happiness and satisfaction improved it. His work had benefitted from Pavlov's findings on conditioned reflexes, and it substantiated the similarities between human and animal behaviour. A cat subjected to a dog's menacing, repeated barking adopts a certain behaviour (raised hackles, claws extended, strength increased tenfold, sound effects) that clearly manifest anger. Cannon's research focused on the internal changes occurring simultaneously and he then analysed their utility in terms, above all, of Darwin's theory and of evolutionary psychology, both of interest to physiologists and psychologists. "During the past four years, there has been conducted in the Harvard Physiology Laboratory, a series of investigations concerned with the bodily changes which occur in conjunction with pain, hunger and the major emotions. A group of remarkable alterations in the bodily economy have been discovered, all of which can reasonably be regarded as responses that are nicely adapted to the individual's welfare and preservation."[50]

Pain's newly acquired status, from a simple sensation to a sensation accompanied by emotion, underpins Cannon's entire approach. In his view, however, pain cannot be reduced to a set of behavioural reactions which produce secretions and other reflexes, that neither lessen its distressing reality nor its significance. These reactions, linked to the sympathetic nervous system's activities, undoubtedly always accompany pain, but pain is not only this set of emotional reactions. Cannon disagreed with another tradition on this subject, upheld in the United States by W. James who, in his 1884 article, equated emotion with its physical expression.[51] Despite Cannon's critical comments refuting James' reductionism and describing a number of distinct emotions occurring alongside various physical changes, research on the importance of emotions in relation to pain, in conjunction with research on the cerebral localisation of emotion—a subject which roused little interest until the end of the 19th century—led to a new line of investigation which emphasised the complexity of the many factors involved (sensory, emotional, and psychological) in the sensation of pain.

Cannon first demonstrated that pain and other strong emotions were accompanied by increased adrenal secretions, which depended on the sympathetic nervous system, as did the other alterations observed under similar conditions (inhibition of digestion, hair on end, dilation of pupils, etc.). Using a moderately strong intermittent current, the researcher applied what would have been a painful stimulus if the animal (a cat) had not been anaesthetised. Dilation of the pupils occurred, the breathing rate augmented and deepened, and increased adrenalin was present in the blood stream due to the increased secretions from the adrenal glands dependent on a splanchnic nerve. The physiological effects caused by the increased secretions produced by the adrenals were threefold: blood-sugar levels increased; the smooth muscles in the bronchioles relaxed; blood circulation was altered, with increased flow to the heart, lungs, central nervous system, and extremities at the expense of blood to the viscera. The adrenalin helped stave off muscle fatigue and coagulation time decreased.[52] All these effects were confirmed by experiments in which a certain amount of adrenalin was injected into the blood stream.

Cannon stressed first that these were reflexes and, as such, fulfilled a final adaptive purpose which had proved successful in the fight for survival throughout evolution. A "phylogenetic association" had come about in which, to each type of situation, there was an appropriate corresponding physical reaction to ensure the survival of the individual. These reflexes mirrored the primitive relations between prey and predator in which only two attitudes are possible—flight or fight—summarised by Mac Dougall as the "flight or fight principle"[53] corresponding to fear or to anger. The utility of the effects produced by increased adrenal secretion in emergency situations was clearcut: either it served as a form of anticipation with regard to the outcome of a conflict, or it served to multiply available energy in order to withstand external aggression. This is exactly the case, as the activated secretion increases muscle strength and contraction duration, providing an antidote to fatigue and reducing the risks of hemorrhage during combat. These are all highly useful means of ensuring the preservation of the individual.

The immense value of Cannon's work and of those who followed was in taking a fresh look at the mind-versus-body quandary, or "body-mind-problem," from a physiological and experimental point of view as much as a philosophical one. It provided a reasonable argument for a monistic view of man which, while not doing away with the distinction between man and animal, did not exaggerate it nor claim it was due to an intrinsic basic difference but only to a higher degree of evolution.

A less convincing aspect of this research may be the result of likening pain to anger or to fright, as just one emotion among others. The experiments carried out by Cannon on the physical alterations occasioned by pain are indisputable, but nothing proves that pain which is not actually felt by a subject has the same repercussions, including physically, as pain which reaches consciousness. Finally, the utility of physiological reactions brought about by fright or anger seems more obvious than the usefulness of reactions to pain. Cannon's explanations certainly follow the lines of Spencerian psychology and the behaviourist approach to pain, where the analogy to other emotions is inferred rather than actually demonstrated. "These bodily changes are so much like those which occur in pain and fierce struggle that, as early writers suggested, the emotions may be considered as foreshadowing the suffering and intensity of actual strife. On this general basis, therefore, the bodily alterations attending violent emotional states would, as organic preparations for fighting and possible injury, naturally involve the effects which pain itself would produce. And increased blood sugar, increased adrenalin, an adapted circulation and rapid clotting would all be favourable to the preservation of the organism that could best produce them."[54]

The problem posed by pain and the reflexes it elicits can be approached from differing angles which are not without ambivalence. Are the physiological reactions provoked by pain useful, or is it the pain itself which has the utility? A shift from one view to the next within the theoretical framework of "survival of the fittest" as it applies to the psycho-physiological field undoubtedly oversimplifies the issue and has elicited reservations. Two main schools of research evolved during the first half of the 20th century, in fact, which opposed Anglo-Saxon versus French scientific views. Once again, the acceptance of Darwin's theory was placed in question. In France, evolutionary psychology and evolutionism in general were under the crossfires of both spiritualists and rationalists who felt Darwin left too much to chance and not enough to determinism. They also contested the non-transcendent view of the ultimate teleological purpose of the theory of adaptation, particularly as exemplified by the "purposive reflex" concept. These ideas therefore had little success in edifying French research in the field encompassing the nervous system's physiology and the psycho-physiology of pain or, for that matter, where fundamental social behaviour was concerned.

### *Experimental Embryology and its Consequences on Pain Analysis in Children*

The development of experimental embryology had major consequences on the understanding of the nervous system and the conditions necessary for its functioning. In some respects, experimental embryological research could be considered to be the opposite of the Wallerian degeneration approach. Paul Emil Flechsig, its main practitioner, had noticed while working in Wagner's laboratory in Leipzig, in 1872, that the myelination of nerve fibres occurred at different rates during development and that even in the newborn baby, mature myelinated fibres co-existed with nonmyelinated fibres. Only the first were deemed fully functional. Each nerve bundle and tract in the nervous system becomes myelinated at a different stage of development: first the peripheral nervous system, then the reflex tracts of the spinal cord and medulla oblongata, then the white fibres of the cerebellum, followed by those joining the cerebral cortex to the gray matter in the spinal cord, the pons Varolli, and the cerebellum, and finally the fibres ensuring intra-cerebral connections, particularly to the telencephalon, which are the last to acquire a myelin sheath.

This research method[55] made it possible to distinguish the different cortical areas according to the period at which the axons become myelinated; Flechsig made it clear that physiologically similar areas become myelinated at the same time, and that the first to do so are the most important. The sequence of events corresponds to a functional hierarchy which is also apparent in terms of individual development (ontogenesis) as well as from a phylogenetic standpoint. Weren't the first areas to become myelinated the same as those to become myelinated in the less developed species and groups, as well as those areas which ensured the essential functions of animal life? Haeckel's understanding of Darwinism and his formulation of the "basic biogenetic law," in which ontogenesis recapitulates phylogenesis, further validated comparative physiological research and the conclusions inferred from anthropoids. This view was also virtually contemporary with Herbert Spencer's *Principles of Psychology,* an evolutionary psychological approach translated into French by T. Ribot in 1875. For many years to come, the psycho physiological analysis of sensibility and pain was based on this dual influence.

Analytical embryology proved to be a powerful aid for investigating the cortical regions, but it had certain indirect consequences in the 1930s on the medical appreciation of the reality of pain in the newborn and young infant. If the perception of pain was closely linked to awareness, an absence of maturation in the intra-cortical association areas was thought to make the perception of pain impossible in infants. As the field of nervous symptomatology of the newborn child gradually developed through the study of reflexes, of head motion, etc., it became increasingly clear that the nervous system of the newborn child was rudimentary and that, at birth, "the functional centres are not all fully connected to one another yet."[56] Binet, in 1933, stressed the persistent disagreements as to the degree of maturity of the central gray nuclei (corpus striatum and thalamus) at birth, whose development would only be completed, according to certain authors, by the 6th or 7th month. In addition, mental development could only get underway once sensory stimuli reached the brain, including sight, hearing, and touch which developed progressively. A number of years later, André-Thomas in his *Neurological Research on the Newborn Child and Infant* summarised the requisite conditions for consciousness: "At birth, the various parts of the cerebrospinal nervous system have not all reached the same degree of maturation. The cortex lags far behind in comparison to the developmental stages undergone by the subcortical centres. It is even further behind physiologically and biologically if we consider that to function fully the cortex must first register a considerable number of afferent stimuli from the outside world (exteroception), while at the same time responding to afferent stimuli originating from within the organism (interoception and proprioception) in order to react in its best interest, and to gradually learn where its own domain and boundaries begin and end, to distinguish its individual being and property from that of others and, in other words, to at last acquire a general sense of awareness and of self."[57]

Finally, just as the cerebral fibres only acquire their myelin sheath progressively (not before the 4th month for intra-cortical fibres), the same is true of the peripheral nerve fibres which are only slightly myelinated, are of small size, and undergo a particularly active development phase around the 3rd to the 6th week after birth which is only completed by the age of two. Experimentation showed that their response to electrical stimulation also seemed less pronounced than in the adult. A link was often made between the absence of a myelin sheath and inactivity at that time, though André-Thomas advocated prudence with regard to the exact function of the myelin sheath, particularly since Remak's nonmyelinated nerve fibres were functional nonetheless; the fibres could also be excited before myelination was complete as was shown in studies of young animals, contrary to the reactions observed in adults. The question seemed, in fact, more complex than was first thought because the sufficient degree of myelination necessary for satisfactory functioning was unknown; whether the physiological activity of neurons actually depended upon myelination was also in question.

Despite these unanswered questions, doctors and surgeons generally considered that the subjective state we term painful could not be compared in the infant and the adult during the first year of life, although a certain number of emotional reactions seemed similar. Other lines of reasoning were also used to back this theory, including the general absence of childhood memories, or the difference between an emotion and the conscious perception of a sensation, or between behaviour interpreted as expressing pain and the actual awareness the subject has of it. The newborn infant's cries of pain, his facial expressions, and the various gestures which could possibly express suffering were thus taken little into account for two reasons which mutually reinforced one another: first, the general conviction that the tracts linking the thalamus to the cortex were not all in place, nor were the association areas, and therefore that neither consciousness nor pain nor memory were possible yet; and secondly, because of the results of experiments carried out to reproduce expressions of anger, joy, pain, and emotional reactions of every kind in "thalamic" animals, *i.e.* deprived of their cortex, or with anencephalus subjects, and whose reactions were all interpreted as simple reflexes which bore no relation to consciousness.

We should refer to Darwin on this point as, in his work on *The Expression of Emotions in Man and Animals,* he refused the idea of a teleological purpose in the anatomical disposition of the facial muscles and stressed that "when the sensorium is highly excited, an excess nervous energy is generated which is transmitted in certain predetermined directions dependent on nerve cell connections and, in part, on habit."[58] It was thus possible to find many animals with facial muscles which could be prompted to express anger or pain. In his chapter devoted to special expressions in human beings, Darwin analysed children's facial expressions, their cries and tears, convulsive movements, vascular and breathing changes in terms of just so many reflex actions, reinforced by habit. As a result, expressions of pain or other emotions commonly observable in that frequently grouped-together tribe comprised of "animals, children, savages, and the insane" could under no circumstances imply the awareness of pain. Any spontaneous, uncontrolled show of emotion in certain types of patients was explained away as a form of developmental arrest or as a kind of retrogression due to pathological factors, undoubtedly in an analogy to developmental arrests witnessed in embryology.

It is thus at the crossroads between experimental embryology and Darwin's interpretation of physiognomy and the emotions that we may probably situate the origins of the practice of performing surgical operations on newborn children and infants without anaesthesia. In the previous century, a time when infant mortality was high in any case, the question hardly arose since few operations were performed on young children; when surgery was performed, however, the dose of anaesthetics was simply reduced according to weight and surveillance increased, though this did little to put an end to the numerous accidents. The main preoccupation for a long time had, in fact, been concerned with finding ways of restraining children in order to operate quite peacefully. It was first and foremost the wish to reduce post-operative infant mortality and the reassurances provided by studies of the nervous system's embryogenesis which prompted such practices. It was quite common in the 1950s to practice paracenteses, perform antrotomies in the auditory canal, operate spermatic-cord torsions or even perform abdominal surgery without any anaesthesia.[59] The absence of cries was considered the main argument, and paradoxically so in view of the fact that the significance of crying had previously been rejected as proof of pain. In addition, attempts were made to elicit certain inhibitory mechanisms by providing a baby bottle, for instance, and by activating the suction reflex.

In terms of medical practices and education, the ultimate resultant attitudes rested on a certain number of beliefs which seemed corroborated by the state of scientific research at the beginning of the 20th century: these included the assumption that the projection and association paths functioned rigidly, that mental phenomena were limited to consciousness, that stimuli eliciting reflex actions were dissociated from emotional factors. The accepted practices continued despite new knowledge, creating a situation which clearly reflects the drawbacks and limitations of coming to conclusions on the sole basis of anatomical and histological data.

### Velocity and Conduction of Nervous Impulses

In order to illustrate the new approach adopted in the 20th century, a simple incident such as the pain experienced after a pinprick in the finger serves as a good example. For a long period, researchers sought to locate the specific pain receptor involved and wondered if a similar pinprick at another point would provoke the same sensation, or again, if a stronger pinprick in which the pin penetrated more deeply into the flesh would increase the pain. With Sherrington, and later with Adrian's work, the question focused on how quickly the stimulus generated by an electric current would take to elicit pain, how long the pain would last and, using the same example, what would happen if several pinpricks followed one another in quick succession? A close analysis of this problem would not have been possible without the help of a cathode-ray oscillograph which allowed the impulse triggered by the stimulation to be recorded by amplifying the wave. This research method, linking stimulus to sensation, made it possible to provide clearer explanations as Edgar Douglas Adrian shows in his fundamental work *The Basis of Sensation,* published in 1928: "Whatever our views about the relation of mind and body, we cannot escape the fact that there is an unsatisfactory gap between such events as the sticking of a pin into my finger and the appearance of a sensation of pain in my consciousness. Part of the gap is obviously made up of events in my sensory nerves and brain, and the psychological method by itself can tell us nothing at all about these events."[60] Adrian thus did what he could to bridge this gap in his research.

The conduction velocity of the nerve impulse as it was termed, its nature being unknown, had already been studied during the 19th century by Helmholtz, a pioneer in the electrophysiology field, and several laws concerning the direction of conduction had already been formulated. At the beginning of the 20th century, however, a number of physiologists sought to learn what, in terms of impulse propagation, was dependent upon changes in the nerve's excitability and what was dependent upon its conductivity. Studies on reactions to pain had shown the importance of duration, frequency, and the speed of stimuli, as much as the importance of intensity. The summation and inhibition concepts already mentioned in connection with Goldscheider and Sherrington were to be systematically examined by Keith Lucas and by Edgar Douglas Adrian. Two problems were posed by summation and inhibition: how was it possible to explain that even an intense stimulus was sometimes incapable of triggering a reaction, while two successive, but weaker, stimuli could bring one about on condition they were administered within a short lapse of time of one another? Moreover, it had been known since the Weber brothers' work that stimulating the vagus nerve could interrupt the heart beat, and Moritz Schiff had elaborated a theory concerning the "arresting nerves": how could this paradoxical inhibition which was elicited by a stimulation be explained? Under what conditions did nerve disturbances add up or annul one another? This question was all the more essential in that, aside from under research conditions, multiple stimuli affecting individuals were the norm. Elucidating such evasive mechanisms was fundamental to understanding the rhythmical nature of pain and its duration, as well as to discovering possible therapeutic solutions.

Lucas and Adrian's method consisted in measuring impulses in terms of the distance they could travel, rather than in terms of the intensity of the reaction they provoked, based on the underlying concept that there was no fundamental difference between conduction in peripheral nerves and conduction in the central nervous system. They had studied the attenuation of impulses travelling along portions of "narcotised" nerves using a morphine solution: "The fundamental fact is that the greater the distance already travelled by an impulse through a narcotised nerve, the lesser the distance the impulse is still able to travel before its total extinction,"[61] exactly like a traveller in the desert advancing without food or water, who is all the less likely to reach his destination the greater the distance already travelled has been. This decrease can actually be considered to exactly duplicate the normal conditions of impulse conduction because the junctions between neurons or between nerves and muscles may be considered to be "subsidence" zones. Keith Lucas showed that the impulse-conduction capacity fully returns when the impulse leaves the subsidence zone. He used the same metaphor as Willis to explain impulse conduction as a long fuse of gunpowder that can be set off by a small spark over a great distance. However, while Willis had emphasised the speed of propagation and the explosive nature of the phenomenon, Lucas was pointing out that at each point along the way, propagation of the impulse was ensured by the energy provided on the spot; just as a gunpowder fuse is sometimes impeded by a damp stretch, the impulse may also come upon a sufficiently long subsidence zone and come to a full halt or, if it manages to get through, to continue its trajectory intact. In short, the intensity of a nervous impulse depends above all on the immediate circumstances affecting its conduction and, if the subsidence zone is considerable, no matter how intense the stimulus applied may be, impulse conduction will be checked. This alternative, either a full stop or a return to normal, is another way of expressing the "All-or-None Law," or in other words that for any given stimulation, provided it is sufficient to elicit a reaction, the response will be as complete as is possible.[62] This concept can be illustrated using the analogy of a gun trigger which is either pulled back sufficiently to release the shot, or is not: there is no intermediate stage between getting the action underway and inaction.

What then could have caused the variations recorded in the sensory or motor nerves or, in other words, how could the effects of variations in the intensity of excitement of the nervous impulses be explained? The "All-or-None Law" prompted research not in the direction of a heightened intensity in the fibre, but in search of a greater number of fibres perhaps participating in the response, bearing in mind also the possibility that an intense stimulus might elicit several different impulses in a fibre. This law may be said to govern the fundamental conditions applicable to each differentiated fibre in a nerve trunk, particularly in terms of the concept that nervous impulses do not all travel at the same speed in all the fibres, a concept which was to prove decisive in identifying the "pain fibres."

At the same time, Keith Lucas and Adrian were using the same methods to study summation and inhibition phenomena, paying particular attention to the back-to-normal process subsequent to the refractory period during which the nerve was neither excitable nor conductible. Adrian thought he had detected a phase of exaggerated excitability, characterised by a "supranormal response" following the refractory period which might account for summation. He did not retain this hyperexcitability hypothesis in his 1928 work but he kept the principle whereby an initial impulse, though unable to provoke a response itself, "facilitated" the passage of a second. In a peripheral nerve, such as the motor nerve in a crayfish's pincer which had been studied by Richet[63]— as had been shown by research on stimuli transmitted to the brain— summation could not be interpreted as the result of the local addition of stimuli which multiplying their effects, but as a cumulation of factors which included an initial preparatory change in the passage to the centre and a subsequent change which succeeded, thanks to the first, in its own transmission. The central nervous system then appeared to be a vast network of conductors, subject to a multitude of stimuli, provoking different refractory phases, communicating back and forth and having to constantly deal with such problems as fatigue, inhibition, and increased stimulation.

This outlook resulted in redefining the stimulus concept by interpreting the response it elicited as a form of adaptation rather than as a reaction: "We will define a stimulus as any change in the environment of an excitable tissue which, if sufficiently intense, will excite the tissue, *i.e.* will cause it to display its characteristic activity."[64] The speed with which this change occurs, or the rhythm at which the intensity of the stimulation grows, proves to be a decisive factor in determining the response, and a reaction is not evoked if the progression is too gradual. For it to occur, not only must the intensity of the stimulus be sufficient, but the duration. On the basis, for one, of this theory that a tissue becomes adapted to a stimulus and, secondly, on the basis of observations that nerve fibres behave as distinct units where the velocity conduction is concerned, Adrian recorded the impulse trains arising from reactions to various stimuli such as pressure, touch, pain, etc. He showed, quite independently of any histological findings about receptor organs, that the frequency of impulses varied according to the intensity and the speed with which the sensation increased; but when pressure remained constant, for instance, the frequency of the impulses slowly diminished. This slow abating could be seen as a specific adaptation measure compared to other types of reactions, such as the quick speed of decline in the case of touch, for example.

Adrian proposed extending Sherrington's terms: "postural reflexes" and "phasic reflexes" to the organs themselves. The first would be characterised by their slow adaptation to a stimulus and would maintain their activity for a long time even when the environment remained static, while the "phasic reflexes" would describe rapid responses to abrupt changes; the skin's sense receptors would belong to this latter category. The next step was to determine whether the response to a painful stimulus presented different characteristics, either in terms of the nature of the impulses provoked or in terms of their frequency or their distribution. Indeed, Adrian knew intuitively that there must be a difference, but he could not find it either in the nature of the impulses, nor in their frequency. He hesitated in designating conduction velocity as a major criterion in pain transmission: "The recent work of Erlanger and Gasser has shown that the impulses set up in the different sensory fibres of a mixed nerve trunk do not all travel with the same velocity, and it is possible that the pain fibres may conduct at a slower or faster rate than those concerned with touch and/or pressure";[65] however, he could find nothing to back this hypothesis. Nonetheless, he recognised in the "massive" quality of the response to a painful stimulus, both in its duration and intensity, a distinct characteristic, although he did not dismiss the hypothesis that the fibres conducting tactile impulses were the same ones which conducted painful sensations. The duration and intensity of the disturbance produced by a painful stimulus would, through a summation process, allow the impulses to reach the central nervous system, a feat which brief stimuli could not achieve. Finally, the intensity might itself depend upon the size of the fibres involved which might itself influence the number of fibres participating in the reaction, according to a theory he had backed in his earlier work and which observed the "All-or-None Law."

Despite his doubts on the exact explanation of pain responses, Adrian's work identified all the major questions studied by later researchers. If his adaptation concept can be closely linked to Sherrington's views, and to those of the evolutionist trend to a lesser degree, he also contributed the outlook and experimental techniques of an electrophysiologist to the field of pain research and methodically explored the various aspects of the newly introduced temporal question in relation to conduction. By limiting the problem of the intensity of a stimulus to a question of distance travelled and number of fibres disturbed, he set up the necessary conditions for undertaking objective, quantifiable research on the relationships between stimulation, conduction, and sensation, while he also clearly demonstrated his ambition to establish the psycho-physiological nature of pain. Finally, he was no less remarkable in avoiding false leads than he was in providing coherent interpretations of the frequency and duration curves generated by nervous-impulse conduction.

### *"Pain Fibres":* 
### *Conduction Velocity and a Two-Phase Response*

One of the paradoxes in arriving at an understanding of the mechanisms at work in the painful sensation consisted first in studying the velocity of conduction in nervous fibres before returning to the apparently simple notion that there was a direct link between the size of the fibres and conduction velocity. Such a realisation was not as obvious as it might initially seem and instead required that a number of other plausible explanations be discarded, such as the link between the frequency of nervous impulses and the velocity of conduction or the role played by temperature. This type of research required that the single nerve fibre be studied as an individual working unit and that techniques to dissociate the fibres and to record the curves generated by their action potential at axon level be available. Research on the electrical changes arising in stimulated nerve fibres came about as a result of work done in the previous century which had shown that even at rest a nerve fibre has a polarised electrical potential because the outside membrane of the fibre is electrically positive while the inside is electrically negative. When a point is stimulated, it propagates a nerve impulse like an electronegative wave along the fibre away from the point of stimulation.

The classification of nerve fibres according to their conduction velocity was undertaken by Herbert S. Gasser and Joseph Erlanger in St. Louis, Missouri. Their research extended over a period of several years and it was only in 1927 that they established the role played by the actual diameter of the fibres in the velocity of nerve-impulse transmission, and published their findings in an article which appeared in the *American Journal of Physiology*.[66] Though the fact is seldom mentioned, Gasser and Erlanger clearly gave credit for part of their success to some of the work carried out by Lapicque who, along with Legendre, had been able to "establish experimentally the relation between the chronaxy of a nerve and its morphology";[67] they also mentioned the visit one of them had made to see Lapicque in his laboratory at the Sorbonne in Paris, recalling that he strongly encouraged them to continue in that direction. They published a joint paper, in fact, which appeared in the *Comptes rendus de la Société de biologie* in 1925, emphasising the different action potentials between the phrenic nerve and the saphenous nerves, and, as the latter were made up of fibres whose size varied considerably, they theorised that this must be the reason for variations in the speeds of conduction. The classification into three groups, alpha, beta, and gamma in order of decreasing velocity—initially with reference to the types of curves rather than to the fibres—was tested in several different nerves of a single species as well as in different species, first in the frog and then in mammals. In later research, they demonstrated that the velocities were 90, 30, and from 1 to 2 metres per second respectively; later still, studies were carried out to determine the difference in the myelinated versus nonmyelinated fibres. It would have been very tempting to conclude that pain was only associated with one type of fibre but, against all expectations, it seemed that pain transmission occurred just as easily through nerves of the alpha as of the gamma type.

The results obtained in electrophysiology were soon supplemented by the clinical findings that, after a painful stimulus, the response occurred in two stages, and that there were, so to speak, two types of painful sensations. This was the conclusion arrived at by Thomas Lewis who was in charge of the clinical research department of London's University College Hospital and whose book, soberly entitled *Pain* and published in 1942, has provided such an informative general synthesis on the subject. Once again, the association between experimental physiology and clinical research proved to be especially fruitful. The concept of a dissociated response to a stimulus was not entirely new as it should be remembered that Goldscheider had shown that a pinprick is initially felt as a pressure sensation and then, shortly thereafter, as a feeling of pain.

Other research, much of it undertaken from the very outset when anaesthetics first came into use, had been carried out to determine in which order sensations became abolished either by chilling the nerve, cocaine injections, or asphyxia. The order of suppression was not identical in all cases as cocaine first inhibited the feeling of pain, then the sense of touch while in the case of asphyxia, especially studied by Lewis, the contrary was true. These phenomena remained enigmas though their different conduction velocities might have suggested that all the fibres of a single nerve were not affected simultaneously. Where pain was concerned, however, the "echo" phenomenon indicated that something else was at work since the pain itself could be broken down into sensations of differing quality and duration depending upon their moment of occurrence. In order to demonstrate this dual response, Lewis generally used the pain provoked by the briefest contact (less than 3/10ths of a second) of a metal rod heated to between 60° and 65°C. He initially felt a smarting pain akin to a sting or a pinprick followed, soon afterwards, by a feeling more like an intense flash of pain which at times lasted longer than the first sensation. The idea that this echo response arose as a result of the reactions in two types of fibres with different conduction velocities seemed the obvious conclusion, particularly as this theory agreed with the results obtained by Gasser and Erlanger,[68] and corroboration could be provided. Lewis observed that the lapse of time before the occurrence of the second reaction increased in proportion to the distance from the nervous centre: it was thus longer from the foot than from the knee, and longer from the knee than from the thigh. In patients suffering from tabes, the slowness of reactions to stimuli had long been observed and might possibly find an answer in these new findings. Lewis concluded that there were two types of fibres responsible for transmitting pain sensations to the central nervous system, some reacting rapidly and involving reflex responses while others responded more slowly, as opposed to the view of such scientists as Ranson[69] who believed that pain borrowed slow channels. In order for his conclusions to be more widely accepted, Lewis still needed to demonstrate that this dual response was not limited to cutaneous nerves but applied also to the viscera and to proprioceptive sensibilities.

The body could thus be viewed as a vast communications network in which painful sensations were simultaneously transmitted along slow as well as rapid tracts. This temporal dimension was evident at tissue level, in the rhythmical functioning of the cells which paced the body's involuntary activities and thought processes, quite independently of any personal volition. Such fundamental changes in outlook made completely new therapeutic approaches possible as a result. Instead of severing bundles of nerve fibres in an attempt to relieve pain, these communication routes could be preserved by implementing new methods which either slowed, or blocked, or in some way modified pain transmission. Two pain-fighting strategies were thus possible: one affecting the transmission pathway itself, the other modifying pain's progression along the way. Even though the latter strategy was hardly used between the two world wars, it was ultimately to prove the most promising and we are now beginning to enjoy its benefits.

In his analysis of pain, Thomas Lewis was to demonstrate another no less baffling aspect of the pain mechanism, partly associated with diffuse pain felt well beyond affected zones, and partly associated with what are known as "subjective" pains or those arising from the body itself, rather than from the external environment; these pains continue despite the fact that the factors which caused them have gone away and any lesions have healed. Such pains are found in cases of causalgia, hyperesthesia, and the nerve pain of shingles whose viral origins had only recently been discovered, and all have the common trait of provoking persistent pain long after the cause has disappeared. The temporal and areal extension of this pain in relation to its initial cause was the departure point for Lewis' research.

He found, for instance, after carefully observing all the facts, that subsequent to the skin's exposure to ultraviolet rays, intense and prolonged pains persisted along with a hypersensibility to temperatures or to pressures which would not have had such an effect under ordinary circumstances. In another example, an electric current used to stimulate a carefully circumscribed area of the skin caused pain that became more intense over a period of several hours and gradually extended beyond the original area according to a well-defined pattern. Lewis developed the hypothesis that certain chemical agents were released locally and radiated outwardly from the affected point, thus reiterating a theory already formulated by von Frey. The difficulty arose in trying to prove that this extension was not due to a magnified reaction elicited once the central nervous system had been reached but to a purely local mechanism with its own precise delineation. To demonstrate that such was the case, all that was needed was to block transmission of the localised lesion towards the centre by injecting novocain subcutaneously just before applying the stimulus; while the local anaesthetic was still in effect, there was no zone of hyperalgesia, but it did reappear when the novocain wore off. As it would have been difficult under these circumstances, in other words several hours after the lesion, to imagine a delayed reaction being transmitted to the central system and back to irritate the zone, it was much simpler to conclude that a local phenomenon was occurring. However, what was its exact explanation?

By examining the actual design of the painful areas which resembled an arborisation and by comparing it to the shape of the neuron and its dendrites, Lewis became convinced that this particular type of pain sensibility could be explained by the axon-reflex theory which Langley had demonstrated. In contrast to the reflex arc, the axon reflex takes place entirely within a neuron, and the impulse, which can be either afferent or efferent, may travel in another direction upon coming to a new ramification and activate other receptors. "These experiments considered together indicate the kind of arrangement the relevant nerves possess. They indicate that the impulses from a small area of damaged skin are conveyed at first through nerves lying in the skin itself, and they indicate that these nerves are in the form of arborisations rather than a network. If we picture nerve axons forming finely branching, rich end plexuses lying mainly or entirely within the skin, and if we picture the parent axons lying deeper, and they themselves running into common subcutaneous stems, such a system would adequately explain the diffuseness of hyperalgesia."[70] In such a system, which Lewis termed a "nocifensor" system, the cutaneous nerve endings play a definite role, not in the sense that they are the source of pain, but in the sense that they associate the small area harmfully stimulated with portions of a more extensive zone.[71] If, anatomically speaking, these nerves seemed to belong to the same group as the nerves arising in the posterior roots, Lewis nonetheless considered that they were part of a separate system with other functions which were also associated with the vascular mechanisms. The highly diverse nature of the mechanisms involved in the production of painful sensations, already obvious in fairly simple examples such as the ones mentioned above, seemed even more complex when trying to understand visceral pain and the associated problem of "referred" pain or, in other words, the pain perceived in a different zone from the damaged area, even quite distant at times, and known in ancient times as "sympathetic" pains.

Before examining Leriche's contributions to the better understanding of pain and, more generally, his philosophy of pain, a closer look at the research being carried out in France between the two world wars in this field should be taken. Though not insignificant by any means, it may seem somewhat slim compared to the research being done in Great Britain and the United States, or compared to the work undertaken between 1850 and 1880. Among its highlights, we might mention the continued quality of the clinical-examination tradition established at the Salpêtrière Hospital which, without being as fruitful as the contributions made by Déjerine at the beginning of the century, remained highly productive; it earned considerable recognition for its research on "central pains" with the work of Ajuriaguerra,[72] Thurel,[73] André-Thomas and Alajouanine.[74] The other major accomplishment of note is the work undertaken by Leriche, which we will consider in some detail later as his professional career provided highly instructive lessons due in large part to his experiences during the First World War.

As a preliminary, however, we might examine such hazy areas as the reasons why so little truly original or advanced research was being carried out in France between the two wars on neurons, synaptical transmission, or such notions as adaptation to pain and integration; it seems to have been largely due to questions concerning the relevance of analytical approaches and may, in addition, have been linked to the way the cell theory was received in France and also, on a different level altogether, the way Darwin's theory was originally greeted here. Haven't the scientific areas in which France has distinguished itself during the 20th century been the ones in which these two theories have had no influence? The dominant research trends in a given country at any one time are undoubtedly the result of a number of institutional factors, the prestige of certain fields, as well as the impossibility of providing more than a few such advanced areas of research with the required logistical and financial backing. In this respect, the success of Pasteur's revolutionary findings certainly gained the public authorities' attention durably as well as the interest of the scientific community. But other reasons on quite another level also play a part, including a certain notion of national research traditions: the decision whether to approach research from the level of individual organs or tissues rather than on the cellular level, or determining how to tie these various levels together are all questions which are rarely examined in terms of scientific research's fundamental nature and goals in relation to man's aspirations. These traditions also tend to favour certain methods and disciplines above others based on past successes. Claude Bernard's physiological experimentation method was more readily useful for studying pains of vasomotor origin or visceral pains, for instance, than to study axon reflexes or synaptical transmission, a fact clearly borne out by Leriche's work.

In this limited overview of France's general situation, it nonetheless seems essential to include the "Lapicque case" which does not fit into any of the previous categories. His research on the excitability of tissues in relation to temporal factors virtually dominated biological research in France for nearly fifty years, right up through the two wars, and was widely taught. It was rather a lacklustre episode in terms of scientific results and also because it acted as an impediment to other potential research; however, the time has perhaps come to reevaluate his work in order to better understand the reasons for such an enduring failure.

Lapicque's general objective was to uncover the underlying mathematical principles governing biological phenomena as he could appreciate the benefits which this would provide for diagnostic purposes, especially in cases of paralysis, and to facilitate pharmacological dosing (by determining the exact physiological effects of various medications and poisonous substances). He was searching for a way of measuring the excitability of tissue but without having to contend with variations arising from the choice of stimulant or its intensity. "We will therefore undertake this experimental study using the electrical stimulation of motor nerves and muscles with the specific objective of better understanding excitability phenomena in general, in any living entity and for any given stimulus. We will pay particular attention to measuring the time factors involved in each case. This concept seems highly important to me; it seems curious that the physiological field has not conferred greater importance to this dimension in its systems though a number of physiologists, including the best among them, have mentioned its value for quite some time."[75] The two concepts which Lapicque developed to arrive at the mathematical constants expressing the time required to excite a living tissue, and specific to each type of tissue, were rheobase (the minimum strength of a direct current capable of exciting a neuron and having a long duration, while minimum duration determined the useful or threshold excitation) and chronaxy (the minimum duration of an electric current required to excite tissue with a current twice as strong as that used for rheobase). Chronaxy could also be used as a means of identifying and classifying every type of tissue and provided an electrophysiological calculation for determining morphological factors. We have seen the good use Erlanger and Gasser made of these properties, but Lapicque chose to follow a different direction and never explored the possibilities in terms of nervous fibres. Instead, he was more interested in using chronaxy to identify the specific reactions of individual tissues and, beyond that, in finding a nonchemical explanation for the nervous system's functioning.[76] He was assailed by lively criticism, not so much about the chronaxy concept but concerning his explanation of neuromuscular activity. He believed that the normal functioning of the neuromuscular junction was based on iso chronal considerations in both tissues and that delayed reactions or problems arose as a result of heterochronia. If these themes are only of indirect interest to the pain question, it remains nonetheless that Lapicque's influential position within French scientific circles, not only at the Sorbonne but also at the Muséum, the Académie des sciences, and the Société de biologie established a veritable school which managed to put French research "on hold" for quite some time.

A rebirth came about in France largely as a result of Alfred Fessard's work and that of his research team. Alfred Fessard began by working alongside Piéron, a specialist in the field of psychophysiology who concentrated on problems of vision and touch and who was also responsible for the *Année psychologique*. Though he was himself reasonably in favour of Lapicque's theories, the intellectual climate in his immediate surroundings was sufficiently open and tolerant to allow certain researchers to disregard the chronaxy issue. Fessard initially worked on muscular fatigues by making oscillographic recordings of the electric current generated in the flexor muscles by a voluntary contraction of the fingers. His first memorandum for the Académie des sciences was presented by d'Arsonval. His work gave him considerable expertise in handling the various electrophysiological techniques and he was probably the first in France to record an action potential, using an electromagnetic amplifier to start with and then a cathode-ray oscillograph whose potential Gasser had thoroughly demonstrated. A major portion of his research was devoted to the electrical activity of the nervous cell, a phenomenon he studied in various electric fish, and a field he became an expert in, along with his close collaborator D. Albe-Fessard. A visit to Cambridge in 1937 to work with Adrian confirmed his long-term interest in microphysiological synaptical transmission and especially in the sensory responses of muscular sensibility receptors which led to the concept of "synaptical potential." This aspect of his research was tied to the highly analytical approach prevailing in Cambridge, but Alfred Fessard generally had a broad intellectual curiosity and a non-specialised outlook on scientific research. It was as a result of this attitude that he participated in the original work on electroencephalograms in France, a technique developed by the German, Hans Berger, as well as continuing to pursue his interest in "psycho-physics" which he had initially studied with Piéron. It was only after the war that he was able to use CNRS (the national scientific research center) research facilities at the Institut Marey of the Collège de France, where he trained a whole new generation of scientists.

## Pain Surgery

The term "pain surgery," which became widespread as a result of Leriche's celebrated work, *La Chirurgie de la douleur*, originally published in 1937, covers two types of surgical operations: one performed on the sensibility tracts of the central nervous system and the other performed on the sympathetic nervous system, an area where Leriche was able to make his essential contributions. In order for this second type of surgery to become an efficient aid in fighting pain, however, a thorough knowledge of anatomy and an understanding of the physiology of the sympathetic system as well as its role in vascular and visceral pains was necessary and, though its "autonomy" was so often referred to, its interconnections with the higher activity centre governing all conscious interactions.

These various questions were still very much the subject of lively debate, and all the more so as the theory of visceral insensibility continued to be staunchly defended. Pain or "autonomic" surgery therefore initially remained limited to a series of interventions which sought to interrupt the pathways of cerebrospinal sensibility at different levels: a neurotomy to section a nerve branch, a radicotomy to section posterior spinal nerve roots and, as a result of research which brought to light the role of the spinal cord's spinothalamic tract in conducting thermal and painful sensations, the section of the antero-lateral bundle (cordotomy). Attempts were also made to section nerve tracts at higher levels and psychosurgical lobotomies were performed for a certain time, as well as selective operations of the frontal lobe to alter the subject's relationship to his pain.

The period between the two world wars witnessed the enormous development of these techniques which had required the resolution of a number of major problems concerning aseptic conditions, hemorrhage control, the maintenance of cerebrospinal fluid levels, etc. Pioneers in this field included Horsley in Great Britain, and Halsted in the United States who inaugurated gentler, less traumatic and more highly skilled surgical techniques which were soon adopted by Cushing.[77] The latter, who was particularly successful in performing surgical ablations of brain tumours and of the pituitary gland, had introduced a method of controlling blood pressure during operations, and also a technique for limiting hemorrhages by using hemostatic clamps made of silver and later, as of 1925, through electrocautery. René Leriche frequently reminisced about the extraordinarily positive impressions he had acquired in the course of his contacts with American surgeons and he also recalled, in *Souvenirs de ma vie morte* ("Memoirs of my past life"),[78] that a trip to the United States was not then a standard part of every young doctor's medical training. He made his first trip in 1913 at the urging of Alexis Carrel whom he had met in Lyons and who had been living in the United States for a number of years, working at the Rockefeller Institute. This voyage played a determinant role in his career orientation and, as early as 1913, he performed his first arterio-sympathectomy.

This emerging neurosurgical field exhibited fascinating audacity and seemed to proffer tempting solutions when every other available means—particularly the administration of morphine—was ineffectual or insufficient. It was also a time when morphinomania was increasingly coming to the attention of medical practitioners,[79] particularly as this phenomenon was largely of iatrogenic origin or, in other words, was caused by the medical treatments themselves. In comparison, pain surgery may have seemed to provide more radical solutions by attacking the source of pain directly.

It was for a case of trigeminal neuralgia that Walter Rose and Robert Abbe first attempted a resection of the fifth pair of cranial nerves; Frank Hartley and Fedor Krause excised the Gasserian ganglion during approximately that same period or around 1891–92, as did Harvey Cushing and Charles Frazier. In 1901, Spiller sectioned the posterior nerve root and it was only on the eve of the Second World War that the Gasserian ganglion was first injected with alcohol. These two techniques were widely used and attempts were made to section the posterior root for conditions other than trigeminal neuralgia.[80] These operations presented such access problems and risks of confusion between anterior and posterior nerve roots, particularly in the lumbar region, that practitioners preferred to perform them under local anaesthesia in order to be guided by patients' reactions. Spiller and Frazier are also credited with performing the first cordotomies. The ultimate results of these various surgical techniques were, at best, mixed: quite aside from their mutilating and irreversible nature, they rested on the rigid principle of a single pathway conducting pain, and the relief they provided was often questionable and short lived, not to mention the other various undesirable, if not highly disagreeable, secondary paresthetic effects experienced including partial anaesthesia, pins-and-needles sensations and numbness.

### Pain and the Sympathetic Nervous System

We know today that the sympathetic nervous system plays a major role in a whole series of pains of vascular origin such as migraine, or of visceral origin, the typical example being angina pectoris. It is also considered to play a role in "referred pain" of visceral origin but actually felt on the surface of the skin. This knowledge is fairly recent, however, and the complete relationship between the sympathetic nervous system and the central nervous system was not then entirely elucidated. We must therefore look back briefly to gain a better understanding of the obstacles which stood in the way of appreciating this system and its therapeutic consequences.

Though a clear distinction had been made in much earlier times, by Bichat in particular, between the nervous system which governs the individual's conscious interactions and the nervous system in charge of the individual's involuntary vegetative processes, research during the 19th century focused mainly on this or that function of the best known sympathetic nerves rather than on this system's general anatomical make-up or functions. Claude Bernard had brought to light the effects of sectioning the "major sympathetic" in the cervical region, especially the increased feeling of heat beyond the lesion, and showed its influence on circulation. The concept of vasomotor nerves thus became established, along with their dual constricting and dilating functions, and a number of writers described the system, including Vulpian. But until the work carried out by Dastre and Morat[81] around 1884, there had been no return to a unified concept of the sympathetic nervous system with its role in nutrition and other involuntary functions. They were able to demonstrate the need to uncover the global concept behind such basic phenomena as vasodilation and vasoconstriction, or activation and inhibition which were evident in all the nervous ganglia; they attributed a certain autonomy and considerable scope to this system which included nerves leading to the viscera (in the three main cavities: thoracic, abdominal, and pelvic), the vagus nerve, and the facial nerve, then known as the "lesser sympathetic."

In his article on the Nervus Sympathicus or "Major Sympathetic," written for the *Dictionnaire Dechambre* in 1884, François Franck proved to be a partisan of this unitary concept: "The major sympathetic nerve's physiology is but one chapter in the organic nervous system's physiology. In fact, not only are the innervations of organs assured by both the sympathetic and the vagus nerves, but a great number of the organic nerves are part of the cerebrospinal system itself throughout their length; they are only differentiated from it when they reach the periphery, more or less in the immediate vicinity of the organs, vessels, or glands which they subserve. Therefore, to limit research on the major sympathetic nerve to only the anatomical system bearing that name would be to stay confined within a single part of the organic nervous system."[82] He therefore went on to include this system's influence on circulation and heat regulation, secretion and excretion, as well as its effects on the digestive, urinary, and genital system's activities.

A further decisive step was made as a result of the work carried out in Cambridge by Gaskell,[83] who first showed the system's anatomical, histological, and physiological complexity and sought to locate the origins of visceral nerves arising in the spinal cord from its three segments in the cervico-cranial, thoracic or sacral regions. By studying the origins of spinal nerves in the cervical region, he was able to distinguish a "somatic nerve" (with its anterior and posterior roots) as well as a "splanchnic nerve" corresponding to the lateral root which includes the visceral ramus,[84] and to establish a close correlation between cranial nerves and spinal nerves. As a result, he presented a global view of the "involuntary nervous system," as opposed to the voluntary nervous system, establishing a distinction in it between the muscles which contract in the presence of adrenaline and whose nervous fibres all arise in the sympathetic nervous system, and those which contract in the presence of acetylcholine, whose fibres arise within the enteral system which Langley shortly thereafter named the "parasympathetic" system.

This work, which sought to shed light on the histological, morphological, and functional coherence of the sympathetic system, was exclusively concerned with efferent or motor nerves. However, one of the principal aspects of the polemic which follows concerns the existence of afferent fibres in the sympathetic nervous system and the problem of visceral sensibility. At about that time, Langley was experimenting with nicotine which selectively interrupted the nervous impulses being transmitted to the sympathetic ganglia in order to distinguish and determine the course of pre- and postganglionic fibres. He was able to draw up quite a precise anatomical and functional description of what he called "the autonomic nervous system"; the term was not used to express its independence with regard to the other system, but to emphasise its characteristics: "When I initially introduced the term 'autonomous,' there were two theories concerning the innervation of non-striated muscles and glands. According to one, these organs were subserved in part by fibres from the spinothalamic tract and in part by fibres from the sympathetic nerves. According to the other theory, upheld by Gaskell, these organs were considered to be innervated by a system which was separated into three anatomical parts through the structural development of the nerves leading to the limbs. Neither of these two theories seemed to correctly reflect the true picture. The fact that the sympathetic system is distributed throughout the body whilst the cranial and sacral nerves only subserve certain areas, and the fact that the functional effects of the sympathetic system generally oppose those produced by the other autonomic nerves, testify that the sympathetic is a distinct system."[85] Langley used the selective action of poisons to further reinforce this distinction between the sympathetic system, in which adrenaline has effects similar to those it normally produces though not absolutely identical, and the parasympathetic system which responded to the effects of pilocarpine, thereby reflecting an approach similar to Cannon's in this respect.

In relation to pain, however, the essential element was the preclusion of any afferent fibres in the autonomic nervous system. The question concerning the innervation of the viscera was reduced to the action of motor or inhibitory fibres, and Langley clearly felt there were no afferent fibres in the sympathetic system, nor was there any possibility of conceiving a reflex arc at ganglion level. Pain could not be directly conducted from the sympathetic system and therefore had to arise in the central sensory tracts. It was precisely in opposition to this theory and Mackenzie's interpretation of it that Leriche undertook his initial research. Nonetheless, Claude Bernard in his lectures on common traits in the animal and plant kingdoms had already endorsed "the existence of afferent fibres in the sympathetic system just as in the cerebrospinal system."[86] Both by his methodology, and by his philosophy of life in general, Leriche can be considered a disciple of Claude Bernard.

When Leriche began undertaking pain surgery as the logical consequence of the surgical experience he gained during the war in a specialised teaching unit set up in 1915, there were three conflicting views on the origins of visceral pain. One, defended by Lennander, maintained that the viscera's insensibility was absolute, based on observations that "solid" viscera could be cut, sewn, or burned without provoking pain, unlike hollow viscera whose parietes composed of smooth muscles were subject to pain when distended; in these circumstances, any pain which patients complained of arising in the viscera might result from some irritation to their serous envelopes or their parietal muscle fibres which, for their part, were innervated by spinal nerves, accounting for the term "parietal pain." Another view, defended by Ross, was that certain pains could arise directly from the viscera though he distinguished a direct "splanchnic" pain, and a "somatic" pain arising in the viscera and extending to the spinothalamic tract through the sensory roots. This opinion paved the way for James Mackenzie's theory[87] who interpreted this type of pain as the spinal cord's translation of painless nervous impulses arising in the viscera into painful stimuli which conducted their nervous impulses to the sensory roots; these might ultimately be mirrored by the peripheral nerves, giving the impression of referred pains. Unlike Ross, however, he excluded the idea that these pains arose directly from the viscera though his radical stance was never fully accepted.

Since Head's thesis, "On Disturbances of sensation with especial reference to the pain in visceral disease," published in *Brain* in 1893, many attempts had in fact been made to establish a clear correspondence between the viscera and the spinal cord's different segments by considering that such pains had a radicular origin, explaining why a lesion to an internal organ was often perceived as a cutaneous pain: "If impulses pass up from an organ which is diseased to the cord, they will set up a disturbance in the segment of the cord to which they are conducted. Now any second sensory impulse from another part conducted to the same segment of the cord will be profoundly altered. Under normal circumstances it would have set up its own proper disturbance in the grey matter and this disturbance would have been conducted to the brain. But in this case, the impulse does not reach an ordinary segment of the cord at rest, but a segment whose activity has already been excited. The resulting stimulus which is then conducted to the higher levels is therefore very different from ordinary impulse transmission occurring under normal circumstances. In many such cases, the second stimulus will be exaggerated, very much like light rays reaching the eye through a convex lens. As a result, if a segment of the cord is stimulated by a painful impulse arising from an internal organ, then a stimulus applied to the skin in a zone innervated by nervous roots connected to that segment would be exaggerated and the resulting stimulus—which would ordinarily only seem uncomfortable—would seem very painful."[88] In short, the messages from the diseased area and the messages from the cutaneous area would converge on a central level and thereby explain localisation errors. As is obvious, this concept was actually a great deal more flexible and also considerably more complex than Mackenzie's view—though their ideas have frequently been equated—nor did it specifically exclude the possibility of afferent visceral fibres. Finally, where "visceral pain" was concerned, and despite a number of varied views, opposition was rife between those who believed genuine visceral pain existed and those who did not, a stance which held considerable sway at the beginning of the century, particularly as a result of research carried out on the autonomic nervous system by the English physiological school.

Leriche, on the other hand, believed there definitely was a specific sympathetic sensibility. As proof, he referred to the pain felt when the sympathetic nerves are irritated and, above all, put forward the ideal example provided by angina pectoris, of cardiac origin and characterised by an intense pain behind the sternum in the chest region, the left arm, and the shoulder. The heart's innervation system for one, and the positive results he obtained through the surgical treatment of angina pectoris by operating on the inferior cervical ganglion, confirmed Leriche's view that visceral sensibility existed and led him to confer on the sympathetic ganglia the role of reflex centres.

Most examples were, in fact, quite difficult to interpret even once the presence of afferent fibres had been recognised in the viscera. Were these really specific fibres belonging to the sympathetic system, or actually somatic fibres following alongside the course of sympathetic and parasympathetic nerves, similar to ordinary sensory fibres and connecting up with the posterior root by the rami communicans which join the sympathetic system to the central nervous system? And although the idea of afferent fibres in the viscera might be accepted, did this mean the reality of referred pain should be denied? None of these questions were completely answered during the first half of the 20th century and the role of the sympathetic system in relation to pain was not yet clear.

Leriche, who trained in Lyons, undoubtedly had his attention drawn to the problems posed by the sympathetic system by the work of Jaboulay, a Lyons surgeon who performed surgery on the sympathetic system for thyroid conditions. In 1894, Jaboulay proceeded to section the cervical sympathetic nerve of an epileptic patient and observed, aside from the usual vasodilation phenomenon, that the eyeball became retracted. This prompted him to perform the same operation on patients suffering from Basedow's disease, caused by a thyroid condition and exhibiting exophthalmic goiter (protrusion of the eyeball), along with a number of other symptoms such as tremor and tachycardia, which the procedure suppressed. In a series of articles which largely appeared in *Lyon médical,* he explained his procedures and his results, along with his method for treating trigeminal neuralgia, and concluded that the sympathetic nervous system acted as a kind of "link between organs whose irregular behaviour reveals pathological conditions, and the nervous centres which control them."[89]

However, the most important work may have been carried out by François Franck who, in a long dissertation entitled "Physiological consequences of sectioning the sympathetic nerve in Basedow's disease, epilepsy, mental deficiency, and glaucoma," published in the *Bulletin de l'Académie de médecine* in 1899, attempted to go beyond a strict explanation of the consequences of sectioning the sympathetic trunk limited to just vasomotor effects. "In performing resections of the major sympathetic, surgeons were aiming at this nerve's efferent impulses; they sometimes believed they were affecting the thyroid gland by sectioning its vasodilator and secretory nerves . . .; or they believed that this sectioning interrupted cerebral vasomotor activities, both vaso-constriction and vasodilation; at times, finally, they were attempting to affect tachycardia by sectioning the cardioaccelerators. At no time, however, was there any question that the sympathetic nerves sectioned had any sensibility, an attitude which is hardly surprising as physiologists paid little attention to this nonetheless essential question."[90]

It was this specific aspect, however, which Franck felt was the most likely explanation for the various effects caused by sectioning the major sympathetic nerve. He, for one, had observed that stimulating the superior segment of the cervical segment of the sympathetic nerve in non-anaesthetised animals provoked painful reactions, as Claude Bernard had already shown; he also noted a set of general circulatory reactions which closely resembled the symptoms of Basedow's disease, and also demonstrated that the vertebral nerve and the superior thoracic rami communicans have an afferent function and connect the heart and aorta to the central nervous centres; finally, he concluded that sympathetic fibres have a sensibility function and explained their practical applications: "The entire sympathetic division, both at the thoracic and cervical levels, or the superficial as well as the deeper levels, is endowed with sensibility and conducts afferent nerve impulses arising principally in the heart and aorta to the bulb and the cord's cervico-dorsal segment. We believe it is logical to think that resection of the sympathetic nerve acts as much to suppress impulse conduction of abnormal cardioaortic excitation towards nervous centres as to suppress afferent impulses from the thyroid, or those of encephalic or cardiac origins. Aortal or cardiac irritations, as well as those affecting the sympathetic nerves, have the capacity of provoking a range of circulatory reactions reminiscent of the pathologies seen in Basedow's disease, including vasodilation in the thyroid: as a result, the effects of a complete resection of the major sympathetic could be explained because of the suppression of afferent conduction pathways for the reflexive form of reactions to stimuli of aortic origin. This new concept of aortal sensibility conducted through the cervico-thoracic sympathetic nerve may ultimately lead to the idea of performing resections for angina pectoris."[91]

These ideas, particularly the need to abandon the concept of a purely motor sympathetic nervous system and the assertion that it did not benefit from borrowed sensibility, were taken up by Leriche who, during the war as well as in Strasbourg in what may have been his richest career years, had had to treat a considerable number of causalgias and post-traumatic pains. He quickly realised that surgery of the major sympathetic, "at the confines of the physical and the mental," had other positive applications aside from simply relieving painful conditions, but preferred to explore the possibilities offered by this field in particular. Characterised by their progressive extension and distant repercussions due to the sympathetic system's "closed network construction,"[92] disorders of this nature could not be reduced to a straightforward pathological lesion. As a result, he saw surgery of the sympathetic system as "physiological surgery" which treated the effects and sought to reverse them, an approach which reveals a strong Bernardian influence. In an address before the 42nd Surgery Congress in 1933, he explained in what sense it was not possible to reduce the disorder to an anatomical lesion which often only constituted a single stage in a process in which physiological disturbances preceded it as well as caused it. "Surgery," he said, "has reached the end of its clinical and technical period":[93] by such a statement he meant that he intended to transform surgery into a knowledgeable science, a physiological science capable of actively intervening in "humoural and nervous pathologies," intuitively realising the future importance of neuro-endocrinology. "The knowledge we have gained over the past several years about the physiological consequences of certain neurovascular operations has encouraged us to treat these disorders through new surgical operations aimed at obtaining curative effects by setting off purely physiological processes. We are now able to perform certain neurological resections which permit us either to diminish or suppress glandular secretions and others, more precious still, which reinforce the normal activity of certain organs and enable improved tissue nutrition."[94] He saw examples of this in the pain suffered by arteritic patients in whom the adrenal gland which regulates the sympathetic system's tonicity may secrete excessive amounts of a vasoconstrictive hormone, causing a painful condition witnessed during menopause.

It is within this general framework that René Leriche's ideas about pain should be interpreted and particularly his definition of so-called "pain disorders" as opposed to a more limited category of "experimental" laboratory pain. By this he meant "certain little-understood conditions whose determinant factors remain unknown but which are frequent and in which pain is the entire, or almost the entire, disorder itself; this pain is so overwhelming that the rest of the symptoms are quite secondary; it is virtually continual or comes in recurrent closely-spaced paroxysms, or in terrifying bursts of sharp stabbing pains. It is to this type of pain, a disorder in itself rather than a symptom that I am referring. Often it has no specifically determined anatomical cause and it is frequently not transmitted by a lesion affecting an organ. The disorder and its expression are confined within the nervous system. Localised in appearance, it affects virtually the whole individual. Its origins and its apparent causes seem at times to be virtually intrinsic. In fact, everything about it is strictly internal."[95]

In this particular conception of a "pain-disorder," in contrast to established microbiological principles which were tending to encroach on the field of pathology, it is clear that there existed a deeper underlying desire to understand, as well as a genuine shared experience and close personal contact with those who were suffering; this aspect is perhaps best related by another war surgeon, Georges Duhamel, who describes the deeply moving intensity of such relationships in the pages of his *La Vie des martyrs* or *La Pesée des âmes*.[96] The patient confronts pain as a single entity, but Leriche interpreted individuality or temperament as but one more physiological characteristic. It was in this sense that he identified the "temperament" which improves men or makes them less able to deal with the pain responsible for endocrine disorders or calcification disorders and, in this fashion, he opened the way for research on nutrition-related pains.

In his book, *La Chirurgie de la douleur*, Leriche provides a minute description of the treatment for trigeminal neuralgia, emphasising his preference for performing a retrogasserian section (posterior to the Gasserian ganglion in the area from which the neuralgia arises) rather than using alcohol injections in the ganglion which provides a less radical cure. At the same time, he criticised French surgeons between the two wars who—with a certain degree of cruelty—operated using only "half-measure" therapies believing they could ultimately perform a whole range of possible successive interventions to this or that branch of the trifacial nerve, when in fact the retrogasserian operation being performed by American surgeons such as Cushing, or by the English, provided a radical cure. He thus played a role in disseminating techniques which were developed outside France. As early as 1913, however, he also personally contributed a highly useful new procedure, the periarterial sympathectomy, to relieve diffuse, post-traumatic pains, in cases of causalgia (starting in 1915), and in vascular disorders such as Raynaud's disease characterised by ischemia of the extremities. A little later, in 1920, he started to intervene on the rami communicans that ensure the connection between the sympathetic nervous system and the central nervous system, in cases of stump pain after amputations, for scleroderma (deep indurations of the skin), and for angina pectoris.

He considered that surgery itself was traumatic, even when only minimally so, and went so far as to develop the idea of post-operative "pain-disorder," a subject which has yet to be fully explored. More than just a simple procedure, whose precise limitations he carefully outlined, the periarterial sympathectomy required a new perception of pain as a self-perpetuating sensation: "The initial trauma has affected the tissue's nervous elements and conducted a stimulus to the centres which has become mirrored by the vasomotor system. The vasomotor disturbance is the very cause of the painful phenomenon. But the pain does not originate in the sympathetic system; it is not sympathetic pain, but its mechanisms are essentially those of the sympathetic system."[97] The fight against pain could simply focus on attempts to suppress its effects, which was a useful approach as this could interrupt the vicious circle of reciprocal phenomena which had been endlessly repeated. Surgery performed on the sympathetic system proved to be a powerful means of relieving pain because by acting on vasomotricity, it acted upon the internal environment which was responsible for many such disorders. In fact, in his final texts, Leriche increasingly suggested substituting chemical means for surgical means.

Based on a number of observations made following successful sur-

## ⤙ THE "DOLORIST" TREND BETWEEN THE TWO WARS ⤚

The term "dolorist"—which generally exalts the value of pain—first appeared in France in 1919 in a review about a book by Georges Duhamel, *La Possession du monde*, published in the newspaper *Le Temps*, though in this instance the term had definite pejorative connotations. But it quickly gained the momentum of a credo for the writer and journalist Julien Teppe, who published two books in rapid succession, first *Apologie pour l'anormal* or "Dolorist manifesto" in 1935, and *Dictature de la douleur* in 1937. During the same period, Julien Teppe launched the *Revue doloriste*, an episodic publication which struggled to survive after the war but managed to include such prestigious writers as Gide or Valéry, and other celebrated names such as Benda, Colette, Léautaud, and Daniel-Rops.

Julien Teppe's essays are foremost a reaction against the "tyranny of the fit" in which he underlines that pain is a means of self-discovery and a way to understand basic truth in relation to oneself: "I am suffering, therefore I am," said Teppe who saw in pain a type of catharsis, a means of purification from non-essentials, incidentals, and falsehoods: "Pain, of all the psychological states, is the one which takes over the entire being, both the flesh and the spirit, with the greatest urgency and force. It is a disposition which sweeps away, blots out, and annihilates all the rest. It does not allow for cheating or compromise. It is there and is enough to eliminate all the rest," which is why it exercises such tyranny over us. This physical pain which takes over the entire being and liberates the individual from any earthly ties should in consequence make him more compassionate, in the term's true sense, towards others and more lucid about himself. Foolishness, meanness and hostility, as well as war, all come about because men cannot truly imagine suffering: Teppe therefore suggested that visits to the hospitals and sanatoriums be included in the curricula of all educational programmes.

But the "Dolorism Manifesto" was not simply satisfied with finding the common thread, however tenuous, linking Socrates' *gnôthi seauton*, or "know thyself," and the Ecclesiastes' doctrines on the corresponding development between science and pain; it aspired to uncover the merits of pain and to draw up formulas through which to benefit from a practical method of personal enrichment, in a similar fashion to Rimbaud who practised a kind of controlled dissolution of all his senses in order to discover the unknown. What secrets will his *Journal de souffrances et de vérité*—a work written by one who set out to "reinstate literature" and establish a new common good through the endurance of chronic pain— lay bare when it is finally disclosed to the public? "I consider extreme anguish, particularly that of somatic origin, as the perfect incitement for developing pure idealism, created anew in each individual," proclaimed Teppe in his campaign in favour of dolorism. Leriche's vigorous reaction is easily understood as a result! That so many intellectuals should have contributed to the *Revue doloriste* and debated about such issues in the years which preceded the Second World War adds further evidence confirming the climate of the times, with its ambivalent compromises between a complacent attitude to pain and other momentous issues—openly fascist—on restoration and regeneration; it also bears witness to the equivocal nature, whether God was involved or not, of the utility or the value of pain.

In 1941, Desclée de Brouwer published what amounted to a debate about pain entitled *Qu'est-ce que la douleur?* in its "Science et Charité" collection, between a highly reputed physiologist, Jean Lhermitte, and a theologian, Father Morineau, who both contributed their views in this pre-agreed format. The physician recalled the Hippocratic precepts that "The first duty of medicine is to heal when it can, to relieve often, and to comfort always" and that "As a result, a physician's role is not only to relieve moral suffering often but also, insofar as is possible, to alleviate pain and its repercussions on the different organs." For his part, the theologian recalled that for the Christian, pain was caused by sin but that "Less than the actual road travelled, even if it is hard, strenuous or cruel, it is the final destination which counts. And the final destination is always Divine life," in other words, pain for a Christian could only be understood in terms of the final redemption and resurrection.

If we leave such principles aside for a moment in an attempt to understand the behaviour they have incited on an entirely different level, then the picture darkens considerably and sinister overtones can be heard. Lhermitte, who used the example of interrupted sensory fibres in cases of syringomyelia which prevents patients from feeling they are burning themselves, for instance, was not content with only underlining the physiological importance of pain, but added: "From this we may conclude that pain is a necessary evil." In diametric opposition to Leriche's views about the value of pain, but also in the explanation of visceral pain, Lhermitte principally saw in morphine an effective means of avoiding potential suicides among sufferers of causalgia, cancer, or neuralgia. He preferred opium, however, as it caused less dependence, and he went on to add: "I insist strongly on this particular point as it seems so essential to me: some physicians give out morphine too freely; morphine should only be injected if one is quite certain a patient is suffering a great deal. Morphine should not be given out lightly." How much suffering has been endured by patients in the name of this therapeutic conservatism, in the name of this ambiguous reticence to alleviate the suffering of others? How many deaf ears have been turned to patients' complaints of pain, how much misunderstanding and suspicion has confronted those who say "I am in pain"? And the very manner in which he distanced himself from the notion of "dolorism" clearly revealed his views on the practical attitudes it might lead to when confronting pain: "I believe this is a highly dangerous trend and, naturally enough, a trend repudiated by Catholics who are willing to suffer when faced with pain, but who do not appreciate pain for pain's sake though only for the moral value it has to offer." Does a "good pain" exist? Behind the parsimonious distribution of analgesics still sometimes practised today, behind purely medical reasons which might prevent relieving post-operative pains, are we quite certain that obscure and reprehensible arguments like some we have just examined don't still exist?

In this somewhat dated analysis which most likely does not reflect the unanimous point of view of the Church any more than it does that of the medical corps, the theologian did not stay confined to recalling the meaning of pain as revealed by Christ's sacrifice, according to Christian doctrines; neither did he restrict himself to simply drawing lessons therein as a means of detachment from worldly pleasures, nor as a sign from Providence and a link between all living beings, but sought to go beyond as well: "Suffering, and particularly patiently endured suffering, has always been considered one of the Church's major riches; Christian patience and suffering have served as proof that the untold wealth arising from Christ's Passion has not been wasted." To teach patients endurance, and to push back the limits of resistance thus became a practical injunction for all those caring for the sick, especially for hospital sisters and visiting nurses: "Happy are those who endure. Be as teachers among your patients, teachers who instruct how to interpret and how to overcome; in this way we will be working for the glory of God and the good of our brethren." These comments, which clearly show that Bourneville's battles at the close of the 19th century were yet to be fully won, go beyond specific doctrinal teachings and demonstrate that fundamental outlooks need "educating." To understand and overcome pain, on a different level, is an essential moral obligation which entails studying it carefully, explaining its mechanisms, and mobilising all the rational means and intelligence available to human beings in order to overcome it.

gical procedures to relieve pain on the sympathetic system, and on hearing the complaints of patients operated on under local anaesthesia for a gastrectomy, for instance, Leriche once again posed the more general question concerning visceral pain and contested the whole theory of referred or reflex pain. In order to reaffirm the genuine sensibility of the viscera in opposition to Lennander and Mackenzie's viewpoints, he provided the following argument, among others: to suppress visceral pain, it was sufficient to anaesthetise the splanchnic nerve and the rami communicans between the sympathetic nerves and the spinal nerves, or the rami level with the dorsal nerves for example (D6, D7, D8 for the pyloris and the duodenum, D8, D9, D10 for the gallbladder). Nonetheless, two convincing arguments had been put forward by the partisans of referred pain which Leriche spoke of as "the myth of referred pain"; first, the absence of overlapping between the actual seat of the pain and the topography of the diseased or injured organ and, secondly, "parietal signs" in abdominal pains which seemed to indicate that the pain was arising from the organ's walls. There was undoubtedly a very tempting hope of establishing a definite correlation between deeply set organs and dermatomal areas which corresponded to their area of predilection, a hope which, probably unknowingly, matched those of medical theories on the significance of "sympathies" during the Age of Enlightenment. To these two arguments were added the therapeutic results obtained by A. Lemaire in 1928 when he used superficial local anaesthetic to fight referred pains. Leriche responded not only with the counter-examples mentioned above including the pains of angina pectoris, the pains of renal colic, the painful reactions to stimulation of the sympathetic trunks, but also by attempting to find explanations for the signs which gave the idea they might be parietal pains. Though the argument does not seem very convincing, he began by incriminating the lack of a proper educational outlook concerning internal pains which hindered any attempts to correctly analyse the location of the pain, and transforming cenesthesia into an obscure sensibility that required an apprenticeship we are unaccustomed to since "this sensibility is normally unconscious" and that "health is life enjoyed in the silence of one's organs."[98] Above all, he referred to a reflex phenomenon arising from the sympathetic ganglia: "In their normal physiological state, they mirror many of the impulses they receive out to the periphery and produce vascular reflexes in various directions. In the pathological state, even when the impulses travel to the centres and give rise to painful sensations, they continue to receive the impulses. Thus parietal vascular reflexes are set up which provoke diffuse pain," and when oedema or congestion is present in addition, "then the central nervous system also becomes involved. The peripheral nerve endings quickly become excited. Parietal pain occurs, with hyperesthesia and contracture juxtaposed onto the visceral pain."[99]

Leriche's view, which did not completely exclude the participation of ordinary sensibility in a secondary phase, emphasised the reality of visceral sensibility and the therapeutic possibilities opened up for relieving a whole range of intractable pains, those without anatomical lesions, or gastric, renal, and feminine pelvic pains. This insistence was undoubtedly backed by all his experience as a surgeon, and his reasoning was based on frequently repeated observations and the belief that surgery was not a technical act but a learning experience. These concepts evidently earned him his appointment to the Collège de France and, immediately following the Second World War, his nomination to the Académie de médecine and the Académie des sciences. This recognition was at least as much due to his philosophical outlook based, it should be noted, on a vitalist ideology following in the footsteps of Bichat's beliefs. There is also reason to believe that this outlook was characteristic of the one prevalent in the biomedical community in Strasbourg before the war. As an illustration of this perspective, Leriche wrote: "Sensibility is not something isolated within us. It is not a privilege reserved to the skin. It is not a purely peripheral sensory phenomenon. It exists wheresoever there is life—a life that cannot exist without it."[100] If all parts of a living being are alive and if all living things are thus sensitive, there is then no justification for excluding such a property in the viscera which, by slipping from a normal into a pathological state, would—to use Bichat's terms—graduate from an unconscious and obscure sensibility to a conscious awareness and, in this case, an awareness of pain.

Perhaps because he was so often confronted with pain in his daily practice, and also because early in his career he had tried to relieve the pain of a young man suffering from neuralgia who had attempted suicide several times, Leriche was more than a great "pain surgeon," he was also a militant humanist, fighting every kind of "dolorist" tendency attempting to exalt the value of pain—a palpable trend between the two wars that was encouraged by highly varied social groups. He was violently opposed to the notion of beneficial or useful pain, no matter what it represented, because pain had little value in his eyes, neither from a diagnostic standpoint nor a prognostic one. "It reveals only a minute proportion of illnesses and often, when it is one of their accompaniments, it is misleading. On the other hand, in certain chronic cases it seems to be the entire disorder which, without it, would not exist. Despite this harsh reality, doctors readily state that pain is a defence mechanism, a welcome warning which cautions us about the danger of impending illness, that it is useful, and, I was about to add, necessary. . . . My belief is entirely different. . . . Defence mechanism? Welcome warning? In actual fact, most illnesses, even the most severe, take hold without giving any warning. Sickness is generally a drama which unfolds in two acts: the first one appears stealthily and plays its part in the dark, in the drab silence of our tissues. . . . When the pain does arrive, it is already too late. The outcome has already been decided; it is imminent. The pain has only made the whole battle, lost early on in the game, sadder and more unbearable. . . . Pain is always a sinister bequest which diminishes the individual, making him sicker than he would have been without it, and a doctor's most pressing duty is to do his utmost to suppress it if he can—always."[101] Pain surgery presumes an ethical stance with regard to pain, and represents a vibrant plea in favour of living a full life under the best possible circumstances. It is also a resolute indictment against the practices Leriche witnessed around him, against the hypocrisy of all those who speak of "handling" pain when they have never experienced it, against those who are not driven by the impatience to fight, to answer the challenge presented by pain, that unspeakable experience above all.

# ⇥ *Conclusion* ⇤

I t may doubtlessly be easier to fully assess the value of past research when a scientific theory has managed to encompass its clinical and experimental findings and presents a coherent system bearing out its various hypotheses. However, and according to the specialists themselves, such does not seem to be the case where the study of pain is concerned. There still remains a great deal of latitude for further exploration and the unresolved questions may be more numerous than those which have been answered; this is true despite the fact that over the past thirty years certain researchers have developed a new theory which attempts to incorporate disparate or contradictory explanations such as the theory of specificity or that of summation. This may be the appropriate moment to recall Thucydides' ironic words about the men who are quite convinced that the war they have fought, the one which has just taken place, surpasses all earlier wars in magnitude and importance. It is not the historian's role to judge the truth or value of a given model which is still being debated by modern scientists; nonetheless, his task is to situate it in relation to earlier historical developments by providing a useful perspective that discloses what it owes to its precursors and what it contributes that is entirely new. The historian's work, in this sense, consists in undertaking a critical re creation of a given subject and it is in this modest fashion that he can contribute to the construction of this *ktèma es aei* (κτῆμα ἐς ἀεί), or essential endeavour for succeeding generations which, from our standpoint, the transmission of scientific knowledge represents.

When attempting, within this broad arena, to define the major ori-
entations of the various theories about pain, it does seem that the
specificity theory holds a paradoxical place because it has never ceased
to maintain its heuristic value since it was first formulated, or to prompt
a search for specific elements of a highly diverse nature such as receptors, fibres or centres, though at the same time it has never ceased to
be considered inadequate. How can its longevity be explained? Could
it perhaps stem not so much from a theoretical level, as such, but rather
from a solidly anchored belief dear to scientists and in keeping with a
division of labour concept and a law of economy—satisfactory in the
vast majority of cases—that holds that for any given function there is
one, and only one, corresponding structure? Is it possible as well that
invalidating it definitively would pose a certain number of basic problems? Broadly speaking, two sorts of criticisms have been levelled at the
specificity theory. The results obtained by von Frey, for instance, have
not been duplicated by other researchers and his mapping of specific
pain points in the skin cannot be matched. It might prove useful at this
point to recall what Claude Bernard had to say about scientific facts:
when findings which were originally established with the utmost care
cannot be repeated when an experiment is tried subsequently, it must
not necessarily be assumed that the results were imaginary, but only
that the experimental conditions themselves have changed. The second
type of criticism concerns the gap between the experimental conditions
in which a painful stimulus is inflicted in the laboratory versus the actual
"pain-disorder" itself which is always a much more complex phenomenon involving a number of stimuli along with the individual's previous
experiences and an affective dimension that plays an important role as
well. Such criticism is undoubtedly justified as is that which emphasises
the distance between the animal and the human example. It should also
be stressed, however, that we only have these "methodologically reduc-
tionist" methods available in order to make pain more intelligible. Here
again, as with the first criticism, we are touching upon the possible
conditions for any experimentation. One solution might be to behave
"as if" the specificity model worked, while keeping in mind the necessity of not confusing a hypothesis' methodological utility with its psychological validity. Specificity would seem, then, to be destined for
criticism and ultimately to be surpassed.

On the other hand, there are the theories which favour the space-
time ordering of responses to volleys of stimuli which may be amplified,

inhibited, or modified in some way by the different relays through which they are conducted. These obviously offer a far more dynamic and inter-active concept of the organism's relationship to its environment or its inner workings. These theories do not really exclude specificity but they are more concerned with the conduction itself or the actual transmission of information; they are not totally satisfactory in accounting for the interplay between the ascending and descending systems either; the very notion of a descending control system only appeared in the late 20th century, towards the end of the 1960s, in point of fact. Why, in the history of theory evaluation, did the human mind settle on a binary solution which opposes two theories covering different territories and which are not mutually exclusive? The increasing elucidation of pain-producing mechanisms might have led to maintaining that the skin's free nerve endings were involved in the sensation of pain, for instance, or that fine non-myelinated fibres played a role; it did not permit the exclusion of encapsulated nerve endings or that of thicker fibres. A great deal of contention has arisen as a result of this rather inadmissible shift from a strictly scientific approach to one which transforms a single observation into the sole and exclusive factor involved in producing a whole set of consequences; as if there were a fundamental reticence in admitting that a single organ may be pluri-functional, or in recognising that equivalence, overlapping and compensation exist between different parts.

If useful indications may be drawn from an historical approach, they are of a triple nature in this instance. First, they underline the necessity of associating all the basic disciplines together (experimental physiology, electro-physiology, and bio-chemistry), along with the clinical method which remains irreplaceable. Clinical examination provides not only the means with which to confirm the hypotheses formulated in the laboratory, but also yields a rich store of unanswered problems and suggestions. Such is the case where visceral pain is concerned, for instance, as much still needs to be explained; this is also true of the vast sector covering pains of deafferentation which occur even when conduction is interrupted, or of the banal and ancient trigeminal neuralgia that may be set off by the lightest breeze or slightest touch, a phenomenon situated at the transition between the pressure sensation and the pain sensation. This conclusion clearly poses the problem of providing a formal arena where the two approaches can interrelate, though this in no way implies interrupting specialised research.

Secondly, the pain historian cannot help but being struck by the important inroads already made in gathering knowledge about the mechanisms of pain. Even if the field is still seen as being in its "exploratory phase" at times, even if comprehensive theoretical discussions are still underway, and even if many factors still remain unexplained such as the placebo effect or the mechanisms of hypnosis, to mention but two of the most striking examples, we cannot deny the definite sense of progress. This is all the more obvious if one considers the problems that had to be overcome to prove that crossed impulse conduction occurred in the cord and to demonstrate the role played by the antero-lateral tract in pain transmission. Identifying the "limbic system," which seems so familiar today, only dates back to 1953 and the discovery of endorphines is much more recent. Why, then, are individuals still suffering and why are we so powerless when it comes to certain intractable pains? Doesn't the time lapse between technical progress and the proposed treatments seem striking? The history of the relationship between pain-related knowledge and pain-relieving practices reveals that it has often been governed by differing rhythms and logic. Generally speaking, there was no decisive progress between the methods available in Graeco-Roman times and those in use at the beginning of the 19th century, though there may actually have been some regression. If the opening of the 19th century enjoyed a major leap forward with the discovery of morphine and its close with the advent of aspirin, our own century has yet to have produced anything of such revolutionary moment. The rhythm of new discoveries has therefore proved to be either spread over vast periods of time or has occasionally been tremendously accelerated. Sometimes, too, certain substances have provided relief though we could not explain the basic mechanisms at work. This discontinuity in the rate of discoveries leads to questions about pharmacology's role within the biomedical field as a whole, and the association between pharmaceutical laboratories, the chemical industry and universities. Such questions were posed in the second half of the 19th century and are part of yet another story.

Finally, as a result of the many texts examined as well as the reactions and battles fought by physicians to relieve pain, a certain number of problems inherent in the practice of medicine come to light that may at times give the impression that pain is not sufficiently taken into account. Rather than incriminating a medical corps that has taken an oath to relieve pain and which, in the final analysis, is subject to the same political, religious, and ideological currents as the rest of society, we must question the conditions under which medicine is actually practised. Such an analysis includes everything from the dilution of responsibilities witnessed in hospitals, to the problem of nurses' status and training because it is they who are ultimately the ones in closest contact with suffering patients, to questions related to the way in which work is organised, as well as all the many sociological factors involved which shed considerable light on medicine. The pain problem touches on something linked to the very nature of the medical act or, in other words, to the doctor/patient relationship, itself determined by the degree of certainty ascribed to medicine. Since approximately the middle of the 19th century, every actual—or potential—patient has held the basic conviction that medicine is a science rather than an art, and the success obtained in the course of the 19th century with anaesthetics and in areas other than in pain relief has meant that patients and physicians alike have shared this faith in the powers of medicine which, it is said, increases the likelihood of recovery. Only those unfamiliar with all the progress achieved might consider they were sharing an illusion. It now seems that the painful experience is the stumbling block which undermines both this conviction and these certainties, however. Pain, because it is such an extreme experience, seems to call into question the traditional roles allocated to the physician and the patient; it may only be surmounted, in each case, by reappropriating—or by overcoming—one's status as an experimental subject, at the risk of one's life.

# ⇒ *A Modern View* ⇐
## *Pain Today*

*J. Cambier*

This "history of pain," which begins with a chronicle of mankind, comes to a close at a moment in history when the past, the focus of this work, gives way to the modern day. A study carried out right up until the present would certainly have produced quite a different story, as the past thirty years have modified our relationship to pain more than all the previous centuries combined.

An inventory such as this, terminating during the 1950s, provides a clear overview of all the research accomplished until that time. However, the fruits of any "cross-pollination" had not yet come to light, though these ultimately led to a re-evaluation of pain's position in our society. The tribute to René Leriche has a symbolic value, in fact, as the pain surgeon's actual scientific contribution proved limited; he may nonetheless be viewed as a pioneer and prophet of the new era because he unfailingly refused to accept pain as a necessary evil.

Up until the nineteen-fifties, research on the subject was very widely dispersed: physicians described pain and concentrated on its semiology, philosophers queried its original status among the various sensations, moralists saw it as the price to be paid for the human condition, the physiologists identified nociceptive stimuli and studied their conduction paths and mechanisms, while therapists made the most of various empirical discoveries. Any communication among these different disciplines was generally nonexistent due to the lack of a common objective and language. Explanations remained piecemeal at best. No theory took the reality of man's pain into account as a psychological experience which depends as much on the subject's personal proclivities as on the physical aggression sustained. A change has now occurred as a result of a pragmatic and multidisciplinary approach: pragmatic because it is based on the prevention and treatment of pain in the specific individual (pain consultations), and pluridisciplinary because it makes a concerted effort to marshal every competent branch in order to explore specific themes in unison through periodic national and international gatherings.

This concertation has influenced such facets as pain semantics, neurophysiology, psychology, and therapeutics by: (*i*) the establishment of a common vocabulary, including a definition of pain and a pain lexicon; (*ii*) the adoption of a physiological concept subordinating painful stimuli to the interplay of pro- and anti-nociceptive mechanisms; (*iii*) the undertaking of a study of pain's manifestations as a behavioural mode, while paying particular attention to language as it is man's specific way of expressing pain; (*iv*) the development of a therapeutic strategy based on pharmacological remedies and the physiopathology of pain, adjusting each treatment to the individual patient's personality.

Among specialists, this cultural evolution has gradually become the accepted norm, having established a consensus which remains compatible with the multiple hypotheses and opposing theories. Bearing in mind the possible risks of oversimplification, we will provide a broad outline of these concepts for the benefit of non-specialists whose interest, as a result of this history of pain, has undoubtedly been kindled by a topic so revealing of man's relationship to his own body, and the role granted the individual in each civilisation. May this account of present-day research lead to a more optimistic view of the future when it comes to pain.

## A Definition of Pain

"Pain is a disagreeable experience which we originally associate with a bodily lesion, or describe in terms of tissue damage, or both simultaneously" (Merskey, 1976).

This definition, which does not underestimate the nature of the painful stimulus, attributes an equal significance to the appreciation of pain on the basis of its affective connotations and perceptual interpretation.

### The Controlled Integration of the Painful Message

Inspired by H. Head's work, the gate-control theory of pain (Wall and Melzack) explains the pain perceived in a bodily region which is deprived of its innervation by an abnormal state of excitation in the neurons of the spinal cord's dorsal horn. This hyperactivity results from a failure of the normal inhibition mechanism exercised by the collateral fibres of the large myelinated fibres relaying impulses to the posterior funiculi on the fine fibres conducting impulses in the dorsal horn. The theory has been enhanced and diversified over the past quarter century. This input modulation is not limited to a simple competition among afferent stimuli, as a multitude of descending influences from the brain incite neurons in the dorsal horn to modulate nociceptive impulses. The pro-nociceptive mechanisms vie with the anti-nociceptive mechanisms, and the perception of pain results from the outcome of their interchanges.

*A dual processing of nociceptive information* accounts for the dual nature of the painful experience which is both a brief and immediately localised physical sensation as well as a diffuse noxious experience, delayed and prolonged over time. The initial function depends upon the fine myelinated fibres and specific nociceptive receptors, while the second is ensured by nonmyelinated fibres and polymodal receptors. When the stimuli are relayed in the dorsal horn of the spinal cord, the autonomy of the message's two components is preserved. The discriminatory content is conducted to the anterior nuclei of the thalamus, while the noxious message is directed to the medial nuclei of the thalamus, and to the limbic and frontocingular cortex regions. The characteristics of the ascending information are determined by the integration which occurs in the dorsal horn. The neurons in the superficial layers respond exclusively to nociceptive stimulation and they are termed "nociceptive specific." The neurons in the base layers are activated both by nociceptive stimuli as well as by various non-nociceptive stimuli (light-touch sensations) and are known as "non-specific nociceptive neurons" or as "convergent neurons"; these underlie the affective dimension of pain.

*The physiopathology of convergent neurons* has been studied in several experimental models of chronic pain. Their hyperactivity is associated with the adoption of "pained behaviour" and exaggerated responses to non-nociceptive stimuli. In man, the hyperactivity of convergent neurons is thought to play a part in chronic pain. It explains the pains projected when visceral afferents converge with cutaneous afferents relayed in the same neurons. In the same fashion, the exquisite sensibility of skin in an area overlying joint inflammation is attributed to the convergence of nociceptive information arising from the lesions and non-nociceptive information stemming from light-touch sensations in the skin. Pain anaesthesia in polyneuropathology is based on a similar explanation. It should be added that the convergent neurons control segmentary reflex activity in the spinal cord. Their hyperactivity is responsible for the tonic motor response underlying abdominal contracture or muscle retraction in the neighbourhood of a fracture. Even more important, though, is their influence on the autonomous nervous system: the neurons of the intermedio-lateral tract account for the vasomotricity and tonicity of hollow viscera. In addition, they are capable of sensitising the mechanical receptors which give rise to the self-perpetuating algogenic processes which are the basis of causalgia.

*The sustained stimulation of convergent neurons has multiple causes:* the primary cause of somatic pain is due to an excess of afferent nociceptive impulses which are amplified by the sensibility of receptors as a result of tissue damage. Neuropathological pains are caused by the production of ectopic impulses along the damaged fibres and the consequences of deafferentation. In both cases, pain is a reflection of a failure or of an overflow of the gate controls. Such a shortcoming may be the *primum movens* of psychogenetic pains.

*The suprasegmentary control* calls into play the descending tracts whose inhibitory influence on the dorsal horn's neurons has been demonstrated by stimulating various regions in the brain stem. These tracts are serotoninergic, adrenergic, even dopaminergic. They carry out their functions, at least in part, through the intermediary of an encephalinergic neuron. They provide the mechanisms necessary for the reticular tissue in the brain stem, the limbic system, and the cerebral cortex to be able to modulate the flux of afferent nociceptive impulses, on the basis of the messages' acquired significance at the various stages of integration and in terms of awareness, conditioning, and rational interpretation, respectively. In addition to this modulatory function in relation to the subject's particular disposition, the descending system plays a part in differentiating precise nociceptive information as opposed to the incessant background of afferent messages.

## The Language of Pain

Pain cannot be measured concretely. It is possible, however, to define a pain threshold by using increasingly strong electric stimuli and assigning this threshold to the defense or flexion reflex; nonetheless, this threshold varies from one individual to another and even within a single individual, depending upon circumstances. We may only gain an idea of the subjective experience of pain through its outward reflection or, in other words, through retraction motions, facial contortions, or involuntary organic reactions, and we may only perceive the pain suffered by others through their descriptions of it. As a result, the practice of trying to extrapolate the pain felt by human beings from the supposed pain suffered by test animals is certainly questionable.

Pain has a specific vocabulary which has been carefully inventoried. Words pertinent to pain have been classified according to whether they fall into the affective category or the descriptive category; the intensity of pain is included under affective criteria while the individual characteristics of each pathology are descriptive traits. The terms used to describe a pain are constant from one examination to another, which confirms the role played by language in the *recollection* of pain.

Language, which is used to express pain, is also the means of communicating with others. It also determines the relationship the patient develops with his physician, his family, and society as a whole. In these relationships, the expression of pain is a part of the individual's operational mode of behaviour. As a cry for comfort, it is also a way of influencing the reactions of one's environment. In such group dynamics, pain has a language of its own and this pain language contributes to amplifying and perpetuating a pained behavioural pattern. Finally, language is also the symbolic representation of pain; the patient uses it to understand his present condition in the light of his past experience and what he knows about the pain suffered by others, of his cultural background, and of the social conventions of his day. The overwhelming influence of this cognitive dimension in reacting to nociceptive stimuli is characteristic of man's pain.

## A Diversified Therapeutic Strategy

Most methods used to control pain, whether through medication or physical means, have been known of for a long time. A thorough knowledge of their mode of action has made it possible to design a therapeutic strategy based on the physiopathology of pain.

The use of anti-inflammatory analgesics (corticosteroids or non-steroids) reduces the activation of nociceptors and only acts indirectly on the central mechanisms. They have a slight influence on chronic pain in which central hyperexcitation is constant. The antispasmodics and the sympathetic blocks are able, if necessary, to arrest self-perpetuating algogenic processes.

Opiate analgesics have an influence on nociceptors and a central action. The latter is both spinal and supraspinal. The opium derivatives influence receptors whose natural ligands are the endorphines. Their pharmacological properties have been fully determined; the choice of medication and their mode of administration (oral, parenteral, even subarachnoidal) depends upon the particular objective. Opiates have little effect on neuropathic pains and carry the risk of perpetuating psychogenic pains. On the other hand, they constitute the best treatment for somatic pains, particularly those of cancer for which their use has unfortunately been limited by restrictive rules regulating their prescription. As a result of the pressure exercised by physicians providing palliative therapy, a trend is gaining momentum to ensure that suffering patients cease being innocent victims of the struggle to control narcotics.

Certain antiepileptic medications (carbamazepine) have a selective action in relieving shooting pains. Tricyclic antidepressants influence, at least in part, the serotoninergic mechanisms. Their analgesic effectiveness, particularly in cases of pain due to deafferentation, has been shown, but it is difficult to measure because of the frequent relationship between chronic pain and depression.

The role attributed to inhibitory mechanisms has renewed the interest in stimulation therapies, including stimulation of the peripheral nerves, of the cord's posterior funiculi, or even of the thalamus. The effectiveness of this stimulation has been ascertained in cases of pain due to deafferentation, though it has not been established for other types of pain.

Whatever therapy is adopted, one must take into account the non-specific effects of any medical intervention in choosing it. The especially marked placebo effect of analgesic therapies confirms that psychological components play an important part in the physician/patient relationship. In extreme cases, certain chronic pains have been controlled through behavioural methods which called on the punishment/reward incentive.

The renewed interest in studying pain has been promoted by a group of specialists and it has changed the behaviour of physicians as well as the demands of the public. If the importance allotted to the study of pain in the first cycles of medical school is still limited, a qualification is available in the third cycle in many universities, and post-graduate courses about pain are increasingly becoming of interest to practicing physicians. The public has become much more aware of this issue due to media coverage and is, quite justifiably, more demanding.

Today, physicians no longer feel they have a right to allow patients to suffer "for their own good" when undergoing an exploratory procedure or operation. They know that acute pain is never beneficial and that an urgent response is called for in times of painful crisis. They consider that chronic pain must be closely analysed in order to determine its underlying somatic origins (excessive nociceptive afferents), and its neurophysiological (deafferentation, causalgia) and psychological causes. For them, the fight to control pain is no longer limited to acting on the symptoms alone. Whether it is a question of preparing a patient for an examination, or for an operation, of relieving a medical crisis, or of interpreting the significance of chronic pain, the physician is henceforth fully convinced of the importance of his attention, the repercussions of his words, and the consequences of his actions. In today's terms, he is aware of his duty to intervene and his only plea is that he not be deprived of the appropriate means to do so.

# ⇌ *Notes* ⇌

## Chapter 1

1. See for example, R. Melzack, *The Puzzle of Pain,* Penguin Books (London, 1973), Chap. II, "The Psychology of Pain," pp. 20–7.
2. M. D. Grmek, *Les Maladies à l'aube de la civilisation occidentale,* Payot (Paris, 1983), 527 p. See in particular the introduction and examples of the problems posed by the "toux de Périnth" ("Périnth's cough"), Chap. XII.
3. F. Mawett, "Recherches sur les oppositions fonctionnelles dans le vocabulaire homérique de la douleur (autour de πῆμα et ἄλυος)" ("Research into functional opposites in Homeric vocabulary on pain, based around πῆμα and ἄλγος"), *Mémoires de la classe des Lettres,* Académie royale de Belgique, 2nd series, fasc. 4, 1979; 442 p.
4. H. G. Liddel, R. Scott, *A Greek-English Lexicon,* new ed. revised by Henry S. Jones, Clarendon Press (Oxford, 1966).
5. Homer, *Iliad,* text compiled and translated by P. Mazon, Les Belles Lettres (Paris, 1972), book XI, 268–72.
6. *Iliad,* book V, 336: ὀξεῖ δουρὶ, and 395: ὠκύν.
7. *Iliad,* book XI, 268–72: in this passage pains are compared with the pains of childbirth: ὀξεῖαι δ'ὀδύναι δῦνον μένος Ατρείδαο, ὡς δ'ὅτ'ἄν ὀδινουσαν ἔχῃ βέλος ὀξύ γυναῖκα δριμύ τό τε προϊεῖσι μογοστοκοι Εἰλεἴθυιαι Ἥρης θυγατέρες πικράς ὠδῖνας ἔχουσαι.
8. *Iliad,* book V, 658: "The point plunged itself in through and through, painfully": αἰχμή δέ διαμπερές ἦλθ' αλεγεινή *(diampérès).*
9. A feeling of weight is invoked by the expression ἄχθομαι ὀδύνῃαι: "I am overcome with pain" (*Iliad,* book V, 354), and λίην ἄχθομαι: "I suffer too much" (*ibid.,* 361), and also ὄδυναι βαρεῖαι κατηπιόωντο (*ibid.,* 417).
10. F. Mawett, *op. cit.,* p. 162.

11. *Iliad,* book V, 886: πήματ᾽ ἔπασχον, "I had suffered a thousand pains."
12. P. Vidal-Naquet, *Mythe et tragédie en Grèce ancienne,* La Découverte (Paris, 1981). For a precise study, see M. Martinez-Hernandez, *La esfera semantico-conceptual del dolor en Sofocles,* Universidad Complutense (Madrid, 1981), 2 vols. This work cites 17 uses of *algos* in Sophocles' work, 2 uses of *algèma* and 9 uses of *odunè* (3 in *Philoctetes* and 1 in *Ajax*).
13. Sophocles, *Philoctetes,* text compiled by A. Dain and translated into French by P. Mazon, Les Belles Lettres (Paris, 1967), 7.
14. *Philoctetes,* 313, "I am dying in sustaining such a devouring pain": ἀδηφάγον νόσον. Jacques Jouanna showed the permanence of these metaphors, analogous in the Greek tragedy and the *Hippocratic Collection* although the tragedy might resort to a more archaic concept than that of illness: "Disease as an aggression in the *Hippocratic Collection* and Greek tragedy: savage and devouring illness," in P. Pottier, G. Maloney, J. Desautels (eds.), *La Maladie et les maladies dans la Collection hippocratique,* Actes du VI^e Colloque international hippocratique, Editions du Sphynx (Québec, 1989), pp. 39–60. The same image can be found in the *Trachiniœ* (985–6 and 1054) where one finds the verb βρύκει *(brukei)* used for the horse who gnaws at his bit.
15. τρυσάνωρ *(trusanor)* made up of τρύω and ἀνηρ; τρύω *(truo)* must without doubt be from the same root as *teiro* and *tribo.*
16. *Philoctetes,* 787, which in a minute way evokes the build-up of the paroxysm, and 1026: θρῴσκει.
17. *Trachiniœ,* 786: Εσπᾶτο γάρ πέδονδε καί μετάρσιος βοῶν ἰύζων.
18. Formed from ποτί (poti), the Dorian form of πρός *(pros:* towards) and βαίνειν *(bainein:* to walk, to go).
19. Sophocles, *Trachiniœ,* text compiled by A. Dain, translated into French by P. Mazon, ed. revised by J. Irigouin, Les Belles Lettres (Paris, 1981), see 777: διώδυνος/σπαραγμος *(diodunos/sparagmos);* and *Philoctetes,* 267: ἀγρίῳ χαράγματι (from χαρασσω, *charasso:* to sharpen, to make a cut).
20. *Philoctetes,* 760: διὰ πόνων πάντων φαίνεις; the image of the spectrum is not found in the Greek text, but serves to give the value of *\*dia.*
21. In Sophocles, *algos* is still not really rivalled by *algèma* (2 uses) and *algèsis.* One can note that *algèma* is used in a very general sense, for example, *Philoctetes,* 340: "Your pains are enough for you," as is *algos* (734); but in 791, *algèsis* indicates a physical pain. But one cannot draw general conclusions from this.
22. *Philoctetes,* 649–50: "I have here a plant which, better than anything else, calms my wound in such a way that it soothes it totally" (cf. *Iliad,* XVI, 574). The verb used by Sophocles, κοσμά-ω *(kosmao),* means "to stretch out on a bed, to be made to rest or to be made to sleep"; one also finds πραύνειν *(praunein):* to calm, to soften.
23. The *Trachiniœ* evoke the powerlessness of being able to forget pain

($\lambda\alpha\theta\acute{\iota}\pi o\nu o\nu$, 1021–2); cf. *Iliad,* V, 394: "an incurable pain" ($\mathring{\alpha}\nu\eta\kappa\epsilon\sigma\tau o\nu$ $\mathring{\alpha}\lambda\gamma o\varsigma$).

24. A. Souques, "La douleur dans les livres hippocratiques" ("Pain in the Hippocratic Books"), *Bulletin de la Société française d'histoire de la médecine* (1942), 166 p. The second part of the article, dedicated to the analysis of pain as a function of different systems and to the dangerous exercise of retrospective diagnoses, is highly debatable in that the notion of a system is totally anachronous in relation to the categories of thought in the *Collection,* and is only based on a philological study of the names of organs.

25. For a complete representation of these studies, see: R. Joly, "Hippocrates of Cos," in C. Scribner (ed.): *Dictionary of Scientific Biography* (London, 1979); and by the same author: *Le Niveau de la science hippocratique,* Les Belles Lettres (Paris, 1966); M. P. Duminil, "La recherche hippocratique aujourd'hui" ("Hippocratic research today"), *History and Philosophy of the Life Sciences* (1979), I-1: 153–81; J. Jouanna, *Pour une archéologie de l'École de Cnide,* Les Belles Lettres (Paris, 1974); as well as the Introductions to different volumes of Hippocratic texts in the Collection des universités de France; A. Thivel, *Cnide et Cos? Essai sur les doctrines médicales dans la Collection hippocratique,* Les Belles Lettres (Paris, 1981), p. 435; and for a more recent appraisal: "Médecine hippocratique et pensée ionienne, réponse aux objections et essai de synthèse" ("Hippocratic medicine and Ionian thought, reply to the objections and attempted synthesis"), *Formes de pensée dans la Collection hippocratique,* Actes du IV$^e$ Colloque international hippocratique, Lausanne, 1981, Droz (Genève, 1983); for a very detailed synthesis, see J. Jouanna, *Hippocrate,* Fayard (Paris, 1992), 648 p.

26. The aetiology is humoural, but there is no agreement in the *Collection* on the number of humours, nor on the exact role that they play in the morbid process; some attribute it to their super-abundance, others to the absence of crasis; others still hold the pneuma responsible. See R. Joly, "Le système cnidien des humeurs" ("The Cnidian system of humours"), *La Collection hippocratique,* Colloque de Strasbourg, 1972, Brill (Leiden, 1975).

27. See Jacques Jouanna, *Pour une archéologie de l'École de Cnide, op. cit.,* and the Introduction to *Maladies II,* Les Belles Lettres (Paris, 1983).

28. *Concordantia in Corpus Hippocraticum,* ed. by Giles Maloney and Winnie Frohn, Olms-Weidmann (1986). We use this correspondence to make our deductions. See also Joseph-Hans Kuhn and Ulrich Fleisher: *Index Hippocraticus,* Cui elaborendo interfuerunt sodales Thesauri Linguae Graecae Hamburgensis, Gottingae, Vandenhoeck & Ruprecht (1989).

29. *De l'Art,* III, 2, text established and translated into French by J. Jouanna, Les Belles Lettres (Paris, 1988) (Littré VI, 5) and herein into

English by J. A. & S. W. Cadden: "I said that the object in general is to avert the suffering of the sick and to lessen the violence of the illnesses, by abstaining from touching those where ailment is the strongest; a case placed, as we know it, below the resources of art." (We are mentioning the edition of *Œuvres complètes* by Emile Littré [Baillière, Paris, 1839–61, 10 Vols.], in the form "Littré," followed by the number of the tome and the page.)

30. *De l'ancienne médecine*, text compiled and translated into French by J. Jouanna, Les Belles Lettres (Paris, 1990), II, 3 (Littré I, 574).

31. *Ibid.*

32. See: L. Bourgey, "La relation du médecin au malade dans les écrits de l'École de Cos" ("The relationship of the doctor to the patient in the writings of the School of Cos"), *La Collection hippocratique et son rôle dans l'histoire de la médecine,* Brill (Leiden, 1975), pp. 228–9; D. Gourevitch, *Le Triangle hippocratique dans le monde gréco-romain,* BEFAR 251, École française de Rome, Palais Farnèse (1984), 569 p.

33. *Pronostic,* Littré II.

34. M. D. Grmek, "Pronostic, diagnostic et conceptualisation des maladies chez Hippocrate" ("Prognosis, diagnosis and conceptualisation of illnesses in Hippocrates"), *op. cit.,* Chap. XI, pp. 409–36.

35. *Aphorismes,* Littré IV, 482.

36. *Aphorismes,* Littré IV, 512.

37. *Maladies I,* Littré VI, 167.

38. *Ibid.,* 171.

39. *Aphorismes,* Littré IV, 514.

40. This is explained by the weakness of anatomical knowledge.

41. *Des lieux dans l'homme (Of Parts in Man),* text compiled and translated into French by R. Joly, *Hippocrate,* Les Belles Lettres (Paris, 1978), T. XIII, XLII, pp. 1–4; Littré VI, 334.

42. *Des affectations,* Littré VI, 211.

43. *Maladies III,* Littré VII, 119.

44. *Des affections,* Littré VI, 244.

45. *Aphorismes II,* 46; Littré IV, 482.

46. L. Bourgey, *Observation et expérience chez les médecins de la Collection hippocratique,* Vrin (Paris, 1953), 304 p.

47. M. Moisan, "Les plantes narcotiques dans le Corpus hippocratique" ("Narcotic plants in the *Hippocratic Collection*"), in *La Maladie et les maladies dans la Collection hippocratique* (Actes du VI$^e$ colloque international hippocratique), *op. cit.,* pp. 381–92.

48. This text appeared in a collective work, *Douleurs, société, personne et expressions,* under the direction of B. Claverie, D. Le Bars *et alii,* Eschel (1992), pp. 179–96.

49. Heinrich von Staden, *Herophilus. The Art of Medicine in Early Alexandria,* Edition, Translation, and Essays, Cambridge University Press

(1989), 666 p. We have referred to this amazing work throughout this section. The author has compiled all the available accounts on Herophilus, as his own original texts are lost, and clearly presents all the opinions on controversial points and gives his own views as well.

50. F. Solmsen, *Greek Philosophy and the Discovery of the Nerves,* Museum Helveticum (1961), 18, 150 p.

51. Rufus of Ephesus, *De Anatomia Partium Hominis,* 71–5, English translation by M. von Staden, *op. cit.,* p. 201.

νεῦρόν ἐστιν ἁπλοῦν σῶμα καὶ πεπυκνωμένον, προαιρετικῆς κινήσεως αἴτιον, δυσσαίσθητον κατὰ τὴν διαίρεσιν. κατὰ μὲν οὖν τὸν Ἐρασίστρατον καὶ Ἡρόφιλον αἰσθητικὰ νεῦρα ἔστιν κατὰδέ Ἀσκληπιάδην οὐδὲ ὅλως. κατὰ μὲν οὖν τὸν Ἐρασίστρατον δισσῶν ὄντων τῶν νεύρων αἰσθητικῶν καὶ κινητικῶν, τῶν μέν αἰσθητικῶν ἅ κεκοίλανται ἀρχὰς ἂν ἐν μήνιγξι, τῶν δὲ κινητικῶν ἐν ἐγκεφάλῳ καὶ παρεγκεφαλίδι. κατὰ δὲ τὸν Ἡρόφιλον ἃ μέν ἐστι προαιρετικά, ἃ καὶ ἔχει τὴν ἔκφυσιν ἀπὸ τοῦ ἐγκεφάλου καὶ νωτιαίου μυελοῦ, καὶ ἃ μέν ἀπὸ ὀστοῦ εἰς ὀστοῦν ἐμφύεται, ἃ δὲ ἀπὸ μυὸς εἰς μῦν, ἃ καὶ συνδεῖ τὰ ἄρθρα . . .

52. Aulus Cornelius Celsus, *De Re Medicina, Proemium,* 23–4; there is only one translation into French of Celsus from the 19th century in the Collection des Auteurs latins, published under the direction of M. Nisard, Firmin Didot (Paris, 1876), p. 4 (the translation is often inaccurate); the *Proemium* was recently translated by Philippe Mudry, *Preface du De re medicina de Celse,* edition, translation and annotations, Bibliotheca Helvetica-Romana (1982); the authoritative edition is *Celsus, De Re Medicina,* translated by W. G. Spencer, Loeb Classical Library (1971), 3 Vols.

53. See *Les Écoles médicales à Rome,* "Actes" from the 2nd International Colloquium on ancient Latin medical texts (September 1986, Lausanne), ed. prepared by Philippe Mudry and Jackie Pigeaud, Droz (Geneva, 1991), 317 p.

54. Celsus, *Proemium,* 13, p. 9: "abditarum et morbos continentium causarum notitiam, deinde evidentium; post haec etiam naturalium actionum, novissime partium interiorum."

55. See Danielle Gourevitch, "La pratique méthodique: définition de la maladie, indication, traitement," in *Les Écoles médicales, op. cit.,* pp. 57–82.

56. See M. D. Grmek, *Les Maladies à l'aube de la civilisation, op. cit.*

57. Celsus, IV, 13, 1–3, p. 405: "Remedium [vero] est magni et recentis doloris sanguis missus; at sive levior sive vetustior casus est, vel vacuum vel serum id auxilii est; confugiendum ad curcubitas est, ante summa cute incisa. Recte etiam sinapi ex aceto super pectus imponitur, donec ulcera pusulasque excitet, et tum medicamentum quod umorem illuc citet. Praeter haec circumdare primum oportet latus abso lanae sulpuratae; deinde, cum paululum inflammatio se remisit, siccis et calidis, fomentis uti. Ab his transitus ad malagmata est. Si vetustior dolor remanet, novissime resina imposita discutitur. Utendum cibis potionibusque

calidis, vitandum frigus. Inter haec tamen non alienum extremas partes oleo et sulpure perfricare; si levata tussis est, leni lectione uti, jamque et acres cibos et vinum meracius absumere."

58. Celsus, I, 9, p. 77, and II, 7, p. 119.
59. Aretaeus of Cappadocia, *Traité des signes, des causes et de la cure des maladies aiguës et chroniques (On the causes and indications of acute and chronic diseases)*, translated from the Greek by M. L. Renaud, F. Lagny (Paris, 1984). This translation is not reliable and it is preferable to refer to the Greek text of the *Corpus Medicorum Graecorum, Aretaeus*, ed. Carolus Hude, Berolini in Aedibus Academiae Scientarum (1958); we lack clues as to Aretaeus' exact dates but for Dioscorides' account and the fact that he mentions an antidote said to have been "invented" by the physician Andromaque in Nero's time.
60. *Ibid.*, II, 2, pp. 75–6.
61. *Ibid.*, II, 2, p. 77.
62. *Ibid.*, pp. 358–69.
63. *Ibid.*, pp. 195–6. However, in the same passage about sciatica, Aretaeus mentions the pain radiating to all the nerves.
64. *Ibid.*, II, 12, pp. 197–8.
65. Galen, *De l'utilité des parties du corps humain (On the Uses of the Parts of the Body of Man)*, VIII, 5–6, *Œuvres anatomiques et physiologiques et médicales*, French translation by Charles Daremberg, Baillière (Paris, 1854–6), 2 Vols., I, pp. 539–40.
66. *Ibid.*, VIII, 6, p. 544.
67. *Galeni Opera Omnia*, ed. Kuhn, Lipsiae (1821–30), 20 Vols., VII, p. 34 (my translation, and then into English by L. Wallace).
68. See Rosa Maria Moreno Rodriguez and Luis Garcia Ballester on this whole question, "El dolor en la teoria y la pratica medicas de Galeno," *Dynamis* (1982), 2, pp. 3–24. See also Rudolph E. Siegel, *Galen, On the Sense of Perception. His Doctrines, Observations and Experimentations on Vision, Hearing, Smell, Taste, Touch and Pain, and their Historical Sources*, S. Karger (Basle, 1970), particularly Chap. V, pp. 174–93; a thoroughly researched work which tends, however, to translate Galenic concepts in modern equivalents (specific receptors, referred pain). This remark is even more applicable to A. Souques' work, *Les Étapes de la neurologie dans l'Antiquité*, Masson (Paris, 1921), which therefore makes it virtually unusable for our purposes.
69. Galen, *De la différence des symptômes* (Kuhn, VII, p. 739).
70. Galen, *De l'utilité des parties*, V, 9, ed. Daremberg, I, p. 361 (Kuhn, IV).
71. Galen, *De Locis Affectis*, II, *De la douleur comme moyen de diagnostic*, in Daremberg, *op. cit.*, 2, p. 507.
72. Hippocrates, *Aphorismes*, V, 25.

73. Galen, *De Locis Affectis*, II, 3, p. 510.
74. See M. D. Grmek, *op. cit.*, pp. 1–34.
75. Galen, *De Locis Affectis*, II, 2, p. 508.
76. *Ibid.*, II, 3, p. 511.
77. *Ibid.*, II, 5, p. 512.
78. *Ibid.*
79. Celsus, *De Re Medicina, Proemium*, 6–8.
80. Cicero, *De Finibus (Des termes extrêmes des biens et des maux)*, text established and translated by J. Martha, Les Belles Lettres (Paris, 1961), I, X, 33, p. 25. See Jackie Rigeaud, *La Maladie de l'âme*, Les Belles Lettres (Paris, 1989), particularly Chaps. II and III, pp. 139–372.
81. Plato, *Phaedo*, 59e.
82. Cicero, *De Finibus*, I, XI, 37, p. 27.
83. Lucretius, *De Natura Rerum*, I, v. 62–79.
84. Cicero, *De Finibus*, III, VI, 20 (2), p. 18
85. *Ibid.*, III, VIII, 29, p. 23.
86. *Histoire naturelle*, book XXX, lxxvi, 199–200, ed. and translated by J. André, Les Belles Lettres, pp 97–8.

## Chapter 2

1. Owsei Temkin, *Galenism. The Rise and Fall of a Medical Doctrine*, Cornell University Press (Ithaca and London, 1973).
2. Guy de Chauliac, *La Grande Chirurgie*, first published by E. Nicaise (Paris, 1363), Germer Baillière (Paris, 1890), p. 18.
3. *Ibid.*, p. 620.
4. Vivian Nutton, *From Democedes to Harvey*, "From Galen to Alexander, Aspects of Medicine and Medical Practice in late Antiquity," Galliard Ltd, Variorum reprints (London, 1988), X, 1–14.
5. See Danielle Jacquart and Françoise Michaud, *La Médecine arabe et l'Occident médiéval*, Maisonneuve et Larose (Paris, 1990), pp. 32–44; see also Gotthard Strohmaier, "Galen in Arabic: Prospects and Projects," *in* Vivian Nutton, ed., *Galen: Problems and Prospects*, The Wellcome Institute for the History of Medicine (1981), pp. 187 & fol.
6. Avicenna, *Poem of Medicine*, written in Arab, translated into Latin during the 13th century, French translation by Henri Jahier and Abdelkader Noureddine, Les Belles Lettres (Paris, 1956), p. 53.
7. *Ibid.*, p. 86.
8. *Ibid.*, p. 84.
9. Georges Duby, "Réflexion sur la douleur physique au Moyen Age," in *La Douleur*, under the direction of Geneviève Lévy, Édition des Archives contemporaines (Paris, 1992).

## Chapter 3

1. Jean Delumeau, *Les Malheurs du temps,* Larousse (Paris, 1987), p. 268.
2. Jean-Noël Biraben, *Les Hommes et la Peste,* Mouton (Paris, La Haye, 1976), 2 Vols.
3. Michel de Montaigne, *Essais* (Ed. Villey), III, 12, p. 1048.
4. *Ibid.*
5. Ambroise Paré, *Œuvres complètes,* Malgaigne ed., J. B. Baillière (Paris, 1840–1), II, p. 356.
6. See the discussion between J. Garrison and Natalie Z. Davis on this point.
7. Pierre Ronsard, "Discours à la Reine," v. 43–4, *Œuvres complètes,* XI, Soc. des Textes français modernes (Paris, 1973), p. 21.
8. Agrippa d'Aubigné, *Les Tragiques* (1st ed. 1616), I, v. 97–8, *Œuvres complètes,* Gallimard, "La Pléiade" (Paris, 1987).
9. *Ibid.,* v. 89–94.
10. Jean de La Ceppède (1548–1623), *Les Théorèmes spirituels* (1st ed. 1613–21), in *Anthologie de la poésie baroque française,* Armand Colin (Paris, 1968), 2 Vols., II, p. 123.
11. D'Aubigné, *Les Tragiques,* VII, "Le Jugement dernier."
12. *Le Corps à La Renaissance,* Actes du XXX$^e$ Colloque de Tours (1987), under the direction of Jean Céard, Marie-Madeleine Fontaine and Jean-Claude Margolin, Aux amateurs de livres (Paris, 1990), 502 p.
13. Jean Calvin, *Le Catéchisme de Genève,* in *Œuvres,* I, Ed. Je sers (Paris, 1934–1st ed. 1542). For an overall view, see E. G. Léonard: *Histoire générale du protestantisme,* P.U.F. (Paris, 1961–4), 3 Vols.
14. See Philippe Denis, "La Réforme et l'usage des cinq sens spirituels," in *Le Corps à la Renaissance, op. cit.,* pp. 187–98.
15. Ignatius Loyola, *Exercices spirituels,* preceded by the *Testament,* Arléa (Paris, 1991), "First annotation," pp. 119–20.
16. *Ibid.,* "Additions pour mieux faire les exercices (Additional things in order to perform the exercises better), 10th Addition," p. 159.
17. Montaigne, *Essais,* III, 10.
18. *Ibid.,* III, 12.
19. Paré (Malgaigne, ed.), II (*De la peste* [Of the Plague], 1568), p. 354.
20. *Ibid.,* p. 356.
21. Andreas Vesalius, *De Fabrica Corporis Humani,* Basiliae, per Oporinum, 1543, 663 p.
22. *La Fabrique,* Préface, Editions Actes Sud/INSERM (1987), p. 23.
23. *Ibid.,* p. 29.
24. *Ibid.,* p. 37.
25. Juan Valverde de Hamusco, *Historia de la composicion del cuerpo humano* (1556), Biblioteca de Clasicos de la Medicina española, Fundacion de la

ciencias de la salud (Madrid, 1991), p. II (translated into French by R. Rey and thence into English by J. A. and S. W. Cadden).

26. Marie-Paule Dumaître, *Ambroise Paré, chirurgien de quatre rois de France,* Perrin & Fondation Singer-Polignac (Paris, 1986).

27. Jean Fernel, *Universa Medicina,* Francofurti, apud, Andream Wechelum, 1581, 670 p.: "Hunc proinde in Physiologia demonstravi a temperatis suique similibus nihil affici, nec ea percipere, a dissimilibus vero et contrariis varie perpeti, eaque non nisi offensione et laesione sentire," Liber II, *De Symptomatis, Dolorem symptomata tactus esse et quae illius causa,* Cap. VI, in *Pathologiae Lib. VII,* p. 400 (translated into French by R. Rey and thence into English by J. A. and S. W. Cadden).

28. *Ibid.*

29. Paré, Épître, in *Dix Livres de la chirurgie, avec le magasin des instruments nécessaires à icelle,* Jean Le Royer (Paris, 1564), facsimile.

30. Paré, Malgaigne, ed., II, p. 443.

31. Paré, "Discours de la Mumie et de la licorne," 1582, in *Œuvres,* III, p. 587.

32. Paré, *ibid., Vingt cinquième Livre des médicaments,* Chap. XIX, "Des médicaments anodyns," p. 547.

33. *Ibid.,* p. 547.

34. *Ibid.,* p. 549.

35. *Ibid.*

36. Valerius Cordus, *De Artificiosis Extractionibus Liber,* in Valerii Cordi Simefusii, *Annotationes in Pedacii Dioscoridis Anazarbei De Medica Materia Libros V,* . . . omnia summo studio atque industria doctiss. atque excellentiss. viri Conradi Gesneri medici Tigurini collecta, 1561, pp. 225–9: "Oleum vitrioli quod a quibusdam oleum vitae appellatur, aqut melancholia artificialis, nihil aliud est, quam aluminosa qualitas et substantia ex vitriolo per artem extracta, modicoque sulphuri mixta. . . . Est autem duplex hoc oleum nempe austerum & dulce. Austerum duplici constat mixtura, scilicet alumine multo, et modico sulphure. Dulce vero simpliciter sulphure constat," p. 226*b* (translated into French by R. Rey and thence into English by J. A. and S. W. Cadden).

37. "Hoc oleum quod valeat potenter ad omnes in corpore putrefactiones et praecipue ad pestem, ad educendum a pulmone in pleuritide et perpneumonia et difficili tussi, pus et crassos viscososque humores: tuto enim absque omni periculo intracorpus sumitur. Calculum neque in venibus neque vesica coalescere finit, exculceratam vesicam sanat," *ibid.,* p. 229*a,* Vires ejus quod segretatum est, Cap. XII (translated into French by R. Rey and thence into English by J. A. and S. W. Cadden).

38. Paracelsus, *Operum medica-chemicorum sive Paradoxorum tomus genuinus primus undecimus,* Francofurti, a collegio Musarum Paltheniananum, 1603–5, 11 tomes in 4 vols.

39. Paré, *Dix Livres de chirurgie,* pp. 152–8.

40. *Ibid.,* p. 172.
41. *Ibid.,* p. 19.
42. Montaigne, *Essais,* I, 14, p. 59.
43. *Ibid.,* I, 14, p. 58.
44. *Ibid.,* I, 14, p. 56.
45. *Ibid.,* II, 12, "Apologie de Raymond Sebond," p. 492.
46. *Ibid.,* II, 37, "De la ressemblance des enfants aux pères," p. 761.


*Chapter 4*

1. Mirko, D. Grmek, *La Première Révolution biologique,* Payot (Paris, 1990), 358 p.
2. Robert G. Franck, *Harvey and the Oxford Physiologists: Scientific Ideas and Social Interactions,* University of California Press (Berkeley, 1980), XVIII–368 p.
3. René Descartes, *Dioptrique* (1st ed., 1637), Discours Quatrième, in *Œuvres et Lettres,* Gallimard, "La Pléiade" (1953), p. 203 (all our references to Descartes are taken from this edition).
4. *Ibid.,* see also *Les Passions de l'âme,* article 12, "Comment les objets du dehors agissent contre les organes des sens" ("How external objects react against sensory organs"), p. 701.
5. *Principes de la philosophie,* Part IV, § 196, p. 659; for Ambroise Paré, see *Œuvres,* ed. Malgaigne, T. II, pp. 221–31.
6. *Ibid.,* § 191, p. 657.
7. *Les Passions de l'âme,* Part I, article 30, p. 710.
8. *Traité de l'homme,* p. 823.
9. *Ibid.,* pp. 865–6.
10. *Les Passions de l'âme,* article 36, where Descartes analyses the reactions of an individual to a strange and worrying form in relation to his understanding and previous reactions to the same phenomenon "as this makes the brain so disposed in some men that the spirits reflected from the image so formed on the gland go from there and run partly into the nerves which are used to turn the back and to move the legs in order to run away, and partly into those which thus expand or contract the orifices of the heart . . . that [the blood] sends spirits to the brain which are the very ones to maintain and intensify the passion of fear" (p. 713). For G. Canguilhem, see note 25, below.
11. *Les Passions de l'âme,* Part I, article 94, p. 741.
12. For a complete presentation of this problem which has been studied in depth, see *Corpus,* No. 16/17 (1991); *Histoire et Nature,* No. 7 (1985) and No. 9 (1987); F. Dagognet's introduction to the *Traité des animaux* by Condillac, Vrin (Paris, 1989).
13. *Les Passions de l'âme,* Part I, article 16, "Comment tous les membres peuvent être mus par les objets des sens et par les esprits sans l'aide de

l'âme" ("How all the limbs can be moved by sensory objects and by the spirits without the aid of the soul"), p. 704.

14. *Ibid.*

15. Here, one could not completely enumerate the passages where Descartes deals with this problem; let us mention the *Discours de la méthode,* Part V, pp. 164–7, and the *Quatrièmes Réponses* to the objections made by Arnaud, pp. 447–8, as well as the *Sixièmes Réponses,* pp. 530–1, the *Lettre à Plempius* dated 30th October 1637, and the *Lettre au marquis de Newcastle* dated 23rd November 1646, pp. 1254–7.

16. *Sixièmes Réponses,* p. 531.

17. On this problem, see in particular the controversy between Stahl and Leibniz; and for a general presentation, Georges Canguilhem, *La Connaissance de la vie,* Vrin (Paris, 1975), pp. 101–27.

18. Pierre Bayle, *Dictionnaire historique et critique,* Boehm (Rotterdam, 1715, 1st ed., 1696, 2nd ed., 1702), "Rorarius" article

19. *Ibid.,* note C.

20. Leibniz, *Commentatio de Anima Brutorum* (1710), in *G. G. Leibnitii opera philosophica omnia,* Erdmann (Berlin, 1840), p. 465. "Le Système Nouveau de la Nature," published in 1795 in the *Journal des savants,* explains the link between dynamics and the problem of the soul of animals: "It was thus necessary to rehabilitate substantial forms . . . but in a way which made them intelligible. . . . I therefore found that their nature consisted of strength, and that from that came something similar to feeling and appetite, and that it was thus necessary to conceive them as the imitation of the notion that we have of souls. . . . I saw that these forms and souls had to be indivisible just as was our own spirit as, in fact, I recalled that this was St. Thomas' feeling regarding the soul of animals." Bayle puts forward Leibniz' theses in notes H and L (in the 1702 edition).

21. *Dictionary of Scientific Biography,* Charles Scribners & Sons, article titled "Thomas Willis" by R. G. Franck, p. 406. See also *op. cit.,* note 2; and Yvette Conry, "Thomas Willis ou le premier discours rationaliste en pathologie mentale," *Revue d'histoire des sciences* (1978).

22. Thomas Willis, *The Anatomy of the Brain and Nerves, Five Treatises,* printed for T. Dring, C. Harper, J. Leigh and J. Martin (London, 1681), 2 Vols.; Facsimile, William Feindel ed. Foreword by Wilder Penfield, McGill University Press (Montreal, 1965), 261 p. + 32 p.: Vol. II, p. 111.

23. *Ibid.,* Vol. II, p. 111.

24. *Ibid.,* Chap. XVI, p. 119.

25. Georges Canguilhem, *La Formation du concept de réflexe aux XVII$^e$ et XVIII$^e$ siècles,* Vrin (Paris, 1977), p. 66.

26. This commentary summarises what Willis said, p. 120; see also p. 299, and the *De Motu Musculari,* in *Opera Omnia,* Huguetan (Lyons, 1676), p. 673.

27. Thomas Sydenham, *The Works of Thomas Sydenham,* translated from the Latin edition of Dr. Greenhill, with a life of the author by R. G. Latham, in two volumes, London, printed for the Sydenham society, Vol. 1, 1848, Vol. 2, 1850. (The first edition of the *Opera Omnia* was published in London in 1683.)

28. Philippe Hecquet wrote an anonymous pamphlet in favour of opium; see Laureen Brockliss, "Philippe Hecquet," in Roger French, *Biology in the Seventeenth Century* (Cambridge U.P., 1989).

29. Thomas Sydenham, *op. cit.,* p. 173.

30. *The Works of (id.),* Vol. II, *A Treatise on Gout and Dropsy,* p. 124.

31. *Id.,* Vol. II, Epistolary to Dr. Cole, p. 91.

32. See Georges Duby & Michelle Perrot (eds.), *Histoire des femmes,* T. III: *XVIᵉ–XVIIIᵉ Siècles,* under the direction of N. Zemon Davis & A. Farge, Plon (Paris, 1991), 554 p.

33. Pascal, *Prière pour demander à Dieu le bon usage des maladies,* in *Œuvres complètes,* Le Seuil (Paris, 1963), p. 362.

34. *Ibid.,* p. 364.

35. *Ibid.*

## Chapter 5

1. Denis Diderot, *Lettre sur les aveugles à l'usage de ceux qui voient,* ed. Vernière (1749), Garnier (Paris, 1964), p. 122.

2. Jean-Noël Pluche, *Le Spectacle de la nature,* Vve Estienne (Paris, 1732–50), 8 tomes in 9 Vols.

3. Marc-Antoine Petit, *Discours sur la douleur,* given at the opening of the lectures on anatomy and surgery at the Hospice général des malades de Lyon, 28 brumaire, year VII, Reymann & Co. (Lyons, year VII), p. 90. For M. A. Petit, see Roselyne Rey: "Le corps et la douleur au temps de la Révolution. Le point de vue des médecins et chirurgiens," in *La Douleur,* L'Harmattan (Paris, 1992), pp. 47–65. Petit's *Discours* has just been published in Jean-Pierre Peter, *Observations sur les attitudes de la médecine prémoderne envers la douleur,* Quai Voltaire (Paris, 1993), pp. 73–124.

4. *Ibid.,* p. 91.

5. François-Joseph Double, "Fragment de sémiotique et considérations pratiques sur la douleur" ("A memorandum on semiotics and practical considerations concerning pain"), in *Journal général de médecine* (1805), p. 359.

6. *Encyclopédie, ou Dictionnaire raisonné des sciences, des arts et des métiers,* by an association of scholars, collated and published by M. Diderot; and the mathematical part by M. d'Alembert, Paris, Briasson, David, Le Breton, then Neufchâtel, S. Faulche, 1751–65, 17 Vols.: article on Pain, T. V, p. 85a and b.

7. Ignace Vincent Voulonne, *Mémoires de la médecine agissante et expectante*, the shortened title for the *Mémoire* which won the main prize from the Académie de Dijon, 18 August 1776, on the question posed in the following terms: "On determining which are the illnesses in which active medicine is preferable to expectant medicine, or the latter to the former, and which signs the physician must recognise in order to proceed or remain inactive awaiting the best moment to institute remedies?", J. J. Niel (Avignon, 1776), 248 p. (re-ed. Paris, Croullebois, year VII).

8. See the testimonies and descriptions provided by Michel Foucault in *Surveiller et punir, naissance de la prison*, Gallimard (Paris, 1975), 318 p.

9. *Encyclopédie*, T. V, article on "Douleur" ("Pain"), p. 85a.

10. *Dictionnaire des sciences médicales*, Adelon, Alard & Alibert, eds., Panckoucke (Paris, 1812): article on "Douleur" (Pain), T. X (1814), p. 185.

11. François-Xavier Boissier de Sauvages, *Nosologie méthodique*, trans. into French from the last Latin edition by Dr. Gouvion, Bruyset (Lyons, 1770–2), 10 Vols., T. VI, p. 8; see also p. 46.

12. Augustin-Jacob Landré-Beauvais (pupil of Pinel, professor of clinical medicine, doctor at the Hospice de la Salpêtrière), *Séméiotique ou Traité des signes des maladies*, Brosson (Paris, 1813), p. 315.

13. *Ibid.*, p. 323.

14. *Encyclopédie*, article on "Céphalalgie" ("Cephalalgia"), signed N. (Vandenesse), T. II, p. 831a.

15. *Encyclopédie*, article "Sciatique," signed Ménuret, T. XIV, p. 782a.

16. Bichat, *Anatomie générale*, Gabon et Brosson (Paris, 1801), T. I, p. 164.

17. Hippolyte Bilon, *Dissertation sur la douleur*, thesis presented before the École de médecine de Paris, 6 floréal, year XI, Feugueray (Paris, 1803), 154 p.

18. *Dictionnaire de médecine*, Adelon, Béclard & *al.*, eds., Béchet Jeune (Paris, 1823), T. VI, article entitled "Diagnostic" ("Diagnosis"), signed by Chomel, p. 547; see also the *Dictionnaire des sciences médicales, op. cit.*, the articles by Mérat on "Anamnestiques," T. II (1812), "Cri," T. VII (1813), pp. 354–63, "Diagnostic," T. IX (1814), and "Interrogations (des malades)," T. XXV (1818), pp. 523–8.

19. F. J. Double, "Fragment de sémiotique et considérations pratiques sur la douleur," *op. cit.*, p. 368.

20. Landré-Beauvais, *op. cit.*, p. 316.

21. *Ibid.*, p. 315.

22. *Encyclopédie*, "Douleur" (Pain), signed d. (d'Aumont), T. V, p. 86b.

23. Hermann Boerhaave, *Aphorismes de chirurgie*, with comments by M. Van Swieten, translated from Latin into French, Vve Cavelier & Fils (Paris, 1753), 6 Vols., T. I: Aphorisms 76 to 85, pp. 415–72. For Boerhaave, see Gerrit A. Lindeboom: *Hermann Boerhaave. The Man and His Work*, Methuen (London, 1968).

24. Hermann Boerhaave, *op. cit.*, p. 419.

25. *Ibid.*, pp. 421–2. See, on this point, François Azouvi, "Quelques jalons dans la préhistoire des sensations internes," *Revue de synthèse,* 3rd S., T. CXV, no. 113–114 (1984), pp. 113–33.

26. Friedrich Hoffmann, *La Médecine raisonnée,* translated into French by J. Bruhier, Briasson (Paris, 1739–43), 9 Vols., T. I, p. 58 (Preface). See F. Duchesneau, *La Physiologie des Lumières. Empirisme, modèles et théories,* Nishoff (The Hague, 1892), 611 p.

27. Hoffmann, *op. cit.,* T. I, p. 107.

28. Boissier de Sauvages, *op. cit,* T. VI, p. 8.

29. *Ibid.,* p. 31.

30. *Ibid.,* T. II, p. 591.

31. M. D. Grmek, "La notion de fibre vivante chez les médecins de l'école iatro-physique" ("The concept of the living fibre amongst physicians of the iatro-physical school"), *La Première Révolution biologique,* Payot (Paris, 1990), pp. 159–88; and R. Rey, *Naissance et développement du vitalisme en France,* thèse d'État, Paris-I, 1987, 3 Vols.

32. Georgio Baglivi, *De Fibra motrice et morbosa* (Perusia, 1700).

33. Albrecht von Haller, *Dissertation sur les parties irritables et sensibles des animaux,* translated into French by M. Tissot, M. M. Bousquet (Lausanne, 1755), pp. 5–6.

34. *Ibid.,* p. 9.

35. Robert Whytt, "Observations sur la sensibilité et l'irritabilité, à l'occasion du Mémoire de M. Haller" ("Observations on the Sensibility and Irritability of the Parts of Men and Other animals: Occasioned by M. de Haller's Late Treatise on These Subjects"), in *Essais physiologiques,* translated into French from the English by M. Thébault, Les Frères Estienne (Paris, 1759), II–296 p.

36. Viewpoints which revealed the political and general histories of mentalities were dealt with by Daniel Arasse, *La Guillotine et l'Imaginaire de la Terreur,* Flammarion (Paris, 1988), 224 p. For a history of scientific ideas, see R. Rey, "Le corps et la douleur au temps de la Révolution: le point de vue des médecins" ("The body and pain at the time of the Revolution: the physicians' points of view"), in *La Douleur, approches pluridisciplinaires,* essays collected by A. Lafay, L'Harmattan (Paris, 1991), pp. 47–65.

37. "Lettre de M. Oelsner aux rédacteurs du *Magasin encyclopédique* sur le Supplice de la guillotine par M. le professeur Soemmering" ("Letter from M. Oelsner to the authors of the *Magasin encyclopédique,* concerning the Punishment of the guillotine, by Professor Soemmering"), in *Magasin encyclopédique,* T. III (1795), pp. 463–77; and "Opinion de J. J. Sue sur la douleur qui survit à décollation" ("Opinion of J. J. Sue on the pain which follows beheading"), *Magasin encyclopédique,* T. IV (1795), pp. 51–76.

38. M. A. Petit, *Discours sur la douleur,* p. 28 and 29.

39. J. Joseph Sue, *Recherches physiologiques et expérimentales sur la vitalité*, read to the Institut national de France, 11 Messidor, year V, Impr. du Magasin encyclopédique (year VI), 76 p.

40. Fouquet, *Encyclopédie*, article on "Sensibilité" ("Sensibility"), T. XV, p. 51a and b; in this article, Fouquet returns specifically to Whytt's criticisms: "A sting which causes intense pain suppresses hiccups and there is no need to be surprised, said Mr. Whytt, by the fact that, after the section of the most sensitive parts, the animals which M. de Haller experimented on gave no sign of pain when he injured the parts which were the least sensitive" (p. 51b). For the Scottish physician, R. Whytt, see R. K. French, *Robert Whytt, the Soul and Medicine*, Wellcome Institute (London, 1969).

41. Nicolas A. Rupke (ed.), *Vivisection. Historical Perspective*, The Wellcome Institute Series in the History of Medicine, Routledge (London & New York, 1990), 373 p.

42. Haller, *Mémoire sur la nature sensible et irritable des parties du corps animal*, translated into French by Tissot, M. M. Bosquet (Lausanne, 1756–60), 4 Vols., T. II, p. 109.

43. *Ibid.*, p. 108.

44. Bordeu, *Recherches sur la position des glandes et leur action*, in *Œuvres complètes*, Caille et Ravier (Paris, 1806; 1st ed., 1751). In order to explain the secretion caused by the arousal of a gland which has a specific sensibility, Haller upheld the concept of a quasi-total absence of nerves in glands (*Dissertation*, p. 37).

45. Ménuret de Chambaud, *Encyclopédie*, T. XI (1765), article on "Oeconomie animale" (Animal Economy), p. 361a.

46. Fouquet, *Encyclopédie*, article on "Sensibilité" ("Sensibility"), T. XV, p. 38.

47. *Ibid.*, p. 42.

48. Pierre-Jean Georges Cabanis, *Rapports du Physique et du Moral de l'Homme*, Dr. Cerise, ed., Fortin et Masson (Paris, 1843; 1st ed., 1798), p. 117. See Martin S. Staum: *Cabanis, Enlightenment and Medical Philosophy in the French Revolution*, Princeton University Press (1980), 430 p.

49. Cabanis, *op. cit.*, p. 157.

50. *Ibid.*, p. 135. See on this point F. Azouvi's works, in particular "Quelques jalons dans l'histoire des sensations internes," art. cit.

51. *Ibid.*, p. 119.

52. See Catherine-Laurence Maire, *Les Convulsionnaires de Saint-Médard*, Gallimard & Julliard (Paris, 1985). Cabanis compared these so-called miracle makers, the French "convulsionaries" and German "visionaries," to "charlatans from all walks of life and every land" (*ibid.*, p. 119).

53. F. A. Mesmer, *Essai sur le magnétisme animal* (1784). See also *Le Magnétisme animal*, Payot (Paris, 1971).

54. Cabanis, *op. cit.*, p. 118.

55. Bichat, *Recherches physiologiques sur la vie et la mort*, Gérard & Co.

(Verviers, 1973; 1st ed., 1800), p. 242. For a bibliography of Bichat's works, see R. Rey, *op. cit.*, T. III, pp. 1–103.

56. François Pourfour du Petit, "Mémoire dans lequel il est démontré que les nerfs intercostaux fournissent des rameaux qui portent des esprits dans les yeux" (1727) ("Memorandum which demonstrates that the intercostal nerves provide the ramifications which transmit spirits to the eyes"), *Mémoires de l'Académie royale des sciences* (Paris, 1739), pp. 1–19.

57. Bichat, *Anatomie générale*, Gabon & Brosson (Paris, 1801), T. I, p. 115.

58. *Ibid.*, p. 229.

59. Bichat, *Recherches physiologiques*, p. 241.

60. Bichat, *Anatomie générale*, T. I, p. 213.

61. Bichat, *Recherches physiologiques*, p. 65.

62. *Ibid.*, p. 66.

63. *Ibid.*, p. 75.

64. Haller, *Dissertation sur les parties irritables et sensibles*, p. 34.

65. Benedetto Mojon, *Sull'utilità del dolore, De l'utilité de la douleur*, translated into French by the Baron M. de Trétaigne, Dentu (Paris, 1843; 1st ed., 1811), 139 p.: a work which was severely criticised in the *Dictionnaire des sciences médicales* by Adelon, Béclard, etc.

66. Théophile Bordeu, *Recherches sur le tissu muqueux*, Didot jeune (Paris, 1767), and *Recherches sur les maladies chroniques*, Ruault (Paris, 1775), 592 p.

67. Paul-Joseph Barthez, *Nouveaux Éléments de la science de l'homme*, Goujon and Brunot (Paris, 1806; 1st ed., 1778), 2 Vols., T. I, p. 33.

68. *Idem*, ed. of 1778 (1 Vol.), p. 147.

69. *Ibid.*, p. 142.

70. *Idem*, ed. of 1806, T. II, p. 19. This idea was developed in particular by J. B. Van Helmont.

71. *Ibid.*, Chap. X, "Sympathies from the forces of the vital principle in similar organs which are linked into individual systems in the blood vessels and nerves," pp. 52–99.

72. Jacques-Louis Moreau de la Sarthe, *Encyclopédie méthodique*, article on "Opium," T. XI (1824), p. 153a–61a.

73. Mérat, *Dictionnaire des sciences médicales*, article on "Opium," T. XXXVII (1819), pp. 465–507.

74. John Brown, *The Elements of Medicine*, a translation of the *Elementa Medicinae Brunonsis*, with large notes, illustrations and comments by the author of the original work, J. Johnson (London, 1788), Vol. I, note pp. 248–9.

75. Incitability is the property of living bodies to be affected either by external objects (warmth, air, food, etc.), or by their own actions (muscle contractions, action of the brain), which constitute inciting forces; incitation is the effect produced by the stimulus.

76. R. Rey and J. L. Fischer, "Électricité et médecine à la fin du XVIII$^e$

siècle" (Electricity and Medicine at the end of the XVIIIth century), in *Transferts du vocabulaire scientifique,* under the direction of Jacques Roger and Pierre Louis, C.N.R.S. (Paris, 1989); see also A. Perra, *The Ambiguous Frog,* Cambridge U.P. (1991) (*La Rana ambigua,* 1987).

77. Bilon, *op. cit.,* p. 75.
78. Fouquet, *Encyclopédie,* article "Vésicatoire" ("Vesicatory"), T. XVIII (1765), p. 193a.
79. See for example the articles on "Moxa" in the *Encyclopédie* (T. X, 1765) and the *Encyclopédie méthodique.*
80. *Ibid.,* p. 203a and b. Acupuncture was forgotten in the *Encyclopédie,* and does not appear to have been practised very much. The information given by Fouquet, providing two illustrations of the needles used, is all second hand and is a matter of curiosity regarding exotic practices; this was not the case for moxa.
81. M. A. Petit, *op. cit.,* p. 60.

## *Chapter 6*

1. Auguste Comte, *Cours de philosophie positive,* Hermann (Paris, 1975), "Première Leçon. Loi historique des trois états (théologique, métaphysique et scientifique ou positif)" (First lesson, Historical law of the three estates [theological, metaphysical and scientific or positive]), pp. 20–41.
2. Amédée Dechambre (under the direction of), *Dictionnaire encyclopédique des sciences médicales,* 1st series, T. XXX (1885), article on "Douleur" (Pain), p. 466.
3. *Ibid.,* p. 467. (The stress is ours.)
4. *Dictionnaire des sciences médicales,* under the direction of Adelon *et alii,* Panchoucke (Paris, 1812–22), T. X, article on "Douleur" (Pain).
5. François Magendie, *Leçons sur les fonctions et les maladies du système nerveux,* Baillière (Paris, 1839), T. I, p. 37. For experimental physiology, see John Lesch, *Science and Medicine in France. The Emergence of Experimental Physiology,* Harvard University Press (Cambridge, Mass., & London, 1984).
6. Georg Weisz, "The Medical Elite in France in the Early Nineteenth Century," *Minerva,* 25 (1987), pp. 150–70. See also John Lesch: "The Paris Academy of Medicine and Experimental Science, 1820–48," in *The Investigate Enterprise. Experimental Physiology in Nineteenth-Century Medicine,* William Coleman & Frederic L. Holmes, eds., University of California Press (Berkeley & Los Angeles, 1988), pp. 100–38.
7. For Larrey, see the biography by Dr. Antoine Soubiran, *Le Baron Larrey, chirurgien de Napoléon,* Fayard (Paris, 1966), 521 p.
8. Dominique Jean Larrey, *Mémoires de chirurgie militaire et campagnes,* J. Smith and F. Buisson (Paris, 1812–17), 4 Vols.; see T. I (1812), p. 307.

9.  Larrey, *op. cit.*, T. II (1812), p. 452.
10. *Ibid.*, p. 485.
11. *Ibid.*, p. 493.
12. Larrey, *op. cit.*, T. III (1812), p. 42.
13. The account of this meeting at the Académie de médecine on the 23rd September 1828 appeared in the *Archives générales de médecine*, T. XVIII (1818), under the title "Moyen de faire les opérations sans douleur" ("Means of performing operations without pain"): "M. Gérardin recalls a letter written to his Majesty Charles X by Mr. Hickmann, a surgeon in London, in which this surgeon announced a way of carrying out the most delicate and most dangerous operation without causing pain in the individuals who underwent them. This method consisted of suspending the faculty of feeling by the systematical introduction of certain gases into the lung. Mr. Hickmann proved this by experimenting on many living animals, and wishes the co-operation of the great Parisian doctors and surgeons in carrying out some experiments on man. This letter will be communicated to the whole Académie" (p. 453). Hickmann's attempt would not be taken any more into account in France than it had been in Great Britain four years previously but, in 1847, the defenders of etherisation would remember this.
14. Jean-Baptiste Sarlandière, *Mémoires sur l'électropuncture, considérée comme moyen nouveau de traiter efficacement la goutte, les rhumatismes et les affections nerveuses, et sur l'emploi du moxa japonais en France, suivis d'un traité de l'acupuncture et du moxa, principaux moyens curatifs chez les peuples de la Chine, de la Corée et du Japon*, Paris, in the works of the author, rue de Richelieu, 1825, p. 64. He practised medicine in an annex of the Hôpital militaire de Paris, at Montaigu, which specialised in chronic illnesses, and it was he who provided Percy and Laurent with the material for the article on Moxa in the *Dictionnaire encyclopédique des sciences médicales* although this publication preceded that of his own work; one could have reproached him for having kept silent for too long on his medical innovations, in particular concerning the application of galvanism to points on the body and the use of needles in acupuncture. Through his preoccupation with the way of administering electricity and his interest in the iconographical representation of human movements, he was able to attract the attention of Duchenne de Boulogne. In his preface, he explained the circumstances which had led him to obtain manuscripts and in particular to dispose of the Chinese figure called "tsoe-bosi," a sort of cardboard doll showing the exact locations of different acupuncture points. The illustration which he gave of it was to be used later in the *Dictionnaire des sciences médicales*.
15. *Ibid.*, p. 65.
16. *Leçons orales de clinique chirurgicale faites à l'hôpital de la Charité par M. le professeur Velpeau*, Germer Baillière (Paris, 1840), 3 Vols., T. I, p. 66.

17. Gabriel Andral, *Clinique médicale,* Deville et Cavellin (Paris, 1834: 3rd edition; 1st ed., 1823–7), 5 Vols.

18. For Henry Hill Hickmann (1800–30) (cf. note 13, above), and his experiments on animals which allowed him to obtain a "suspended animation" in organisms, see Victor Robinson, *Victory Over Pain. A History of Anaesthesia,* Henry Schuman (New York, 1946), 338 p.; in particular pp. 56–66.

19. In 1784 James Moore made an appliance for the compression of the nerves which allowed him to obtain insensibility in that part of the body in about half an hour, and Benjamin Bell considered this process in his *System of Surgery.*

20. Velpeau, *Leçons,* p. 59.

21. *Ibid.,* p. 59.

22. *Ibid.,* p. 56.

23. *Ibid.,* p. 13.

24. Humphrey Davy, *Researches, Chemical and Philosophical, Concerning Nitrous Oxide or Dephlogisticated Nitrous Air and its Respiration,* J. Johnson (London, 1800), XVI–582 p.: p. XI.

25. *Ibid.,* p. 457.

26. *Ibid.,* p. 464–5.

27. "Essai et dissertation sur un moyen à employer avant quelques opérations pour en diminuer la douleur" ("Essay and dissertation on a technique for employing before some operations to reduce pain in them"), by M. Sassard, *Observations sur la physique* (or *Journal de physique* by Abbé Rozier), October 1780, T. XVI, p. 259. (The stress on "analogy" is ours.)

28. *Ibid.,* p. 482.

29. *Ibid.,* p. 556.

30. *Éléments de philosophie chimique, by Sir Humphrey Davy, translated into French with additions to the text by J. B. Van Mons,* J. E. G. du Four (Paris, 1813), 2 Vols.; see T. I, p. 447.

31. *Encyclopédie méthodique,* T. XI (1824), article on Opium, p. 153b; see also *Dictionnaire des sciences médicales,* T. XXXVII (1819), article by Mérat on opium, pp. 465–507.

32. See for example C. Benedetti and L. Premuda, "The History of Opium and its Derivatives," in *Advances in Pain Research and Therapy,* Raven Press (New York, 1990), Vol. 14.

33. "Mémoire sur l'opium by Citizen Derosne, a pharmacist in Paris, read to the Société de pharmacie," *Annales de chimie, ou Recueil de Mémoires concernant la chimie et les arts qui en dépendent, et spécialement la pharmacie,* T. XXXXV, year XI (30 nivôse), pp. 257–85.

34. Leipzig's *Journal of pharmacy,* Vol. 14 (1806), pp. 47–93. Due to the young age of the author, and to difficulties in monitoring the results, this work did not receive the acclaim it deserved.

35. "Analyse de l'opium. De la morphine et de l'acide méconique, con-
    sidérées comme parties essentielles de l'opium" ("Analysis of opium.
    Concerning morphine and meconic acid, which are considered to be
    essential ingredients of opium") by H. Sertuerner, a pharmacist at Eim-
    beck, in the kingdom of Hanover, translated from *Gilbert's Annalen der
    Physik*, neue folge, Vol. XXIV, p. 56, by H. Rose, a pharmacist in Berlin,
    *Annales de chimie et de physique*, T. V (1817), pp. 21–42. In the same
    issue of the *Annales*, Robiquet inserted "Observations sur le Mémoire
    de M. Sertuerner relatif à l'analyse de l'opium" ("Observations on the
    Memoir by Mr. Sertuerner concerning the analysis of opium"), in which,
    contrary to Derosne's opinion, he confirmed that Sertuerner's morphine
    was really a radically different substance from narceine (an alkaloid of
    opium), but contested the results concerning meconic acid (pp. 275–88).
36. Derosne, *art. cit.*, p. 260.
37. Vauquelin, "Examen de l'opium indigène et réclamation en faveur de
    M. Seguin, de la découverte de la morphine et de l'acide méconique"
    ("Examination of indigenous opium and claim in favour of M. Seguin
    for the discovery of morphine and meconic acid"), *Annales de chimie et
    de physique*, T. IX (1818), pp. 282–6. Vauquelin stated: "I was very
    curious to re-read a Memoir which M. Seguin communicated to the
    Institut on 24th December 1804 and which was printed in the *Annales
    de chimie* only in December 1814. This Memoir contained everything
    which has been said, recently on morphine and meconic acid . . ."
    (p. 282). Seguin's text is not to be found either in the *Annales de chimie*
    of 1814 or later. As Vauquelin gave a lengthy extract from it, it is possible
    that it could have been published elsewhere or that the memoir, which
    was submitted to Vauquelin for publication, was not published perhaps
    due to the events which happened at the end of the Empire and during
    the Restoration.
38. This is the title of Orfila's work, *Éléments de chimie appliqués à la
    médecine et aux arts*, which was reprinted several times.
39. François Magendie, article on morphine, *Encyclopédie méthodique*, T. X
    (1821), p. 296a.
40. For this "malheureuse célébrité" ("unfortunate celebrity"), due to an
    episode of poisoning in 1823, see J. L. Alibert, *Nouveaux Éléments de
    thérapeutique*, Béchet jeune (Paris, 1826), T. II, pp. 167, 281–2. The
    term "narcotisme" was used to describe the effects of this poisoning,
    which can be fatal.
41. "Observation sur les effets thérapeutiques de la morphine ou narcéine"
    ("Observations on the therapeutic effects of morphine or narceine"),
    *Mémoires de l'Académie de médecine*, T. I (1828), pp. 99–180; see p. 118.
42. *Archives générales de médecine*, T. XVI (1818), p. 293–4.
43. Émile Aron, *Histoire de l'anesthésie*, Expansion scientifique française
    (Paris, 1954); Georges Arnulf, *L'Histoire tragique et merveilleuse de*

*l'anesthésie,* Lavauzelle (Paris, 1989), 239 p.; Thomas E. Key, *The History of Surgical Anaesthesia,* Dover Pub. (1963), 200 p. One can also find an excellent description in Claude Bernard's work, *Leçons sur les anesthésiques et sur l'asphyxie,* J. B. Baillière & fils (Paris, 1875), 536 p., in particular pp. 34–46.

44. Victor Robinson, *Victory Over Pain, op. cit.,* p. 84.
45. Jean-Louis Alibert, *Nouveaux Eléments de thérapeutique et de matière médicale, suivis d'un Essai français et latin sur l'art de formuler,* Caille et Ravier (Paris, 1817), 2 Vols.: T. II, p. 143.
46. Orfila, *Traité des poisons ou de toxicologie générale,* Crochard (Paris, 1814–5: 3rd ed.).
47. *Bulletin de l'Académie royale de médecine,* T. XII (1846–7), p. 395.
48. *Bull. de l'Acad. roy méd.,* T. XII (1847), p. 282.
49. *Comptes rendus des séances hebdomadaires de l'Académie des sciences (C. R. Acad. scien. hereafter),* T. XXIV (1847), p. 92.
50. *Ibid.,* p. 93.
51. *Bull. de l'Acad. roy. méd.,* T. XII (1847), meeting of 9th March 1847, p. 356.
52. *Ibid.,* meeting of 2nd February 1847, p. 318.
53. For an account of the events, see in the *Revue médicale,* T. III (1848), pp. 576–97, extract from *The Monthly Journal of Medical Sciences,* October 1848, p. 209.
54. *Bull. Acad. roy. méd.,* meeting of 23rd February, pp. 400–9.
55. *Ibid.,* p. 405.
56. Serres, "De l'action de l'éther liquide sur le tissu nerveux" ("Concerning the action of liquid ether on nervous tissue"), *C. R. Acad. scien.,* T. XXIV, p. 362.
57. Achille Longet, "Actions des vapeurs d'éther" ("Actions of ether vapours"), *Bull. Acad. roy. méd.,* T. XII (1846–7), meeting of 9th February 1847, pp. 361–71; see in particular p. 365; see also P. Flourens, "Note touchant l'action de l'éther sur les centres nerveux," *C. R. Acad. scien.,* T. XXIV (1847), pp. 340–4. For Flourens, see G. Legée: *Marie Jean Pierre Flourens (1794–1867), physiologiste et historien des sciences,* F. Paillard (Abbeville, 1992), 2 Vols.
58. *C. R. Acad. scien.,* T. XXIV, p. 137.
59. See, for example, Jean-Jacques Yvorel, "La transformation du rapport à la douleur au XIX$^e$ siècle. Le débat sur l'anesthésie" ("The transformation of the relationship to pain in the 19th century. The debate on anaesthesia"), *La Douleur. Approches pluridisciplinaires, op. cit.,* pp. 67–74.
60. *C. R. Acad. scien.,* T. XXIV (1847), p. 134.
61. See for example news items on the debate in the *Revue des Deux Mondes* or *La Presse.*
62. *C. R. Acad. scien.,* T. XXIV (1847), p. 134.
63. *Ibid.,* p. 172.

64. *C. R. Acad. scien.*, T. XXIV, pp. 253–8 and 340–4; during the meeting of 8 March 1847, Flourens stated: "One recalls that ethyl chloride gave me the same results as diethyl ether. Ethyl chloride led me to trying the new substance known as chloroform. After a few minutes, indeed a very few minutes (after 6 in the first experiment, and 4 in the second and third), the animal being subjected to inhalation of chloroform was completely etherised. Then the spinal cord was exposed; the posterior region, the posterior roots were insensible; for five anterior roots successively tested, only two retained their motility; the three others had lost it" (p. 342).

65. E. Soubeiran, "Recherches sur quelques combinaisons du chlore" ("Research on some compounds of chlorine"), *Annales de chimie et de physique*, T. XLVIII (1831), pp. 113–57.

66. Justus Liebig, "Sur les combinaisons produites par l'action du chlore sur l'alcool, l'éther, le gaz oléifiant et l'esprit acétique" ("On the compounds produced by the action of chlorine on alcohol, ether, ethylene and 'acetic spirit'"), *Annales de chimie et de physique*, T. XLIX (1832), pp. 146–203; "I will try to show that in the complete decomposition of alcohol, chlorine separates from hydrogen and replaces it. It forms a compound of chlorine, carbon and oxygen which I shall call, for lack of a more suitable name, chloral: the composition of this word is, as one can see, copied from that of the word ethal" (p. 156).

67. J. B. Dumas, "Recherches de chimie organique" ("Organic Chemistry Research"), *Annales de chimie et de physique*, T. LVI (1834), pp. 113–54; Dumas, who gave the chemical formula for chloroforms, added: "The substance which I have just studied possesses all the characteristics of an acid and its constitution reminds one of the ingenious opinions uttered by M. Dulong on the nature of hydrated acids and on the essential role played by water in determining their acidic reaction. This is what encourages me to describe it under the name of chloroform" (p. 120).

68. *Bull. Acad. roy. méd.*, T. XIII (1847–8), meeting of 25th July 1848, p. 1209.

69. The "Rapport sur le chloroforme" ("Report on chloroform") was read during the meeting of 31 October 1848, under the presidence of Royer-Collard, by Malgaigne in the name of a commission consisting of Roux, Velpeau, Begin, Jules Cloquet, Amussat, Jobert, Honoré, Poiseuille, Bussy, Renaud, Gibert, Guibourt and Malgaigne. It was published in the *Bull. Acad. roy. méd.*, T. XIV (1848–9), pp. 203–48, and then submitted for discussion.

70. Ducros, "Sur l'action thérapeutique de l'extrait de belladone et de l'éther sulfurique" ("On the therapeutic action of belladona extract and diethyl ether"), *C. R. Acad. scien.*, T. XXIV (1847), p. 74.

71. *Bull. Acad. roy. méd.*, T. XIV (1848–9), p. 222.

72. *Ibid.*, p. 255.

73. Ventricular fibrillation results in rapid circulation failure due to the disorganized spread of impulses throughout the ventricles because of irregular contractions of the muscle fibres. It is responsible for most cases of heart failure due to chloroform anaesthesia.

74. The details of this case and an excerpt of the legal decision can be consulted in the *Archives générales de médecine*, 1853 (1), 5th series, pp. 756–7.

75. John Snow (1813–1858), one of the first anaesthesia "specialists," wrote *On Chloroform and other Anaesthetics*, published after his death in 1858.

76. See Luc Lecron, *La Douleur maîtrisée*, Arnette (Paris, 1992), p. 50, who is quoting from *Essays on the First Hundred Years of Anaesthesia*, published by Churchill Livingstone.

77. Charles Lasègue and Jules Regnauld, *La Thérapeutique jugée par les chiffres*, Asselin (Paris, 1876), 53 p. (this study was reprinted in the *Archives générales de médecine* the following year); see p. 11. The organisation of the central pharmacy was the work of Soubeiran. Hospital historians, such as Olivier Faure for Lyons, pay special attention to the breakdown of expenditures.

78. *Ibid.*, p. 15.

79. Claude Bernard, *Leçons sur les anesthésiques*, Baillière (Paris, 1875), pp. 108–9.

80. The procedure consists of making an opening in the cranium which allows the state of the brain's vascularisation to be seen.

81. *Ibid.*, p. 141.

82. *Ibid.*, p. 150.

83. The chronology of events is not clear, perhaps because Claude Bernard wrote from memory: he noted that the same week in 1864 when he discovered this action in his laboratory, Nusbaüm made the same discovery in his surgical practice. However, Nusbaüm's note, "Prolongation de l'anesthésie chloroformique pendant plusieurs heures" ("Prolongation of chloroform anaesthesia for several hours"), appeared in 1863 in the 10th October issue of *Aertzl Intelligenzblatt* and also in the *Gazette médicale de Strasbourg*. The experiments were repeated on animals and man by the Société de médecine of Versailles and were published in the *Bulletin général de thérapeutique*, T. LXVI (1864), p. 233.

84. *Leçons sur les anesthésiques, op. cit.*, pp. 225–6.

85. *Ibid.*, p. 234.

86. *Ibid.*, p. 301.

87. Arnott, *Medical Times and Gazette*, 135 (1857), p. 7.

88. Nicolai Ivanovich Pirogoff, *Recherches pratiques et physiologiques sur l'éthérisation* (St. Petersburg, 1847).

89. Guérard, *L'Union médicale* (1854), p. 313; the technique was perfected by A. Richet, "Anesthésie localisée" ("Local anaesthesia"), *Gazette des hôpitaux de Paris*, 27 (1854), pp. 251–2, 263–4, and 267–8.

90. Sigmund Freud, "Über Coca," *Centralblatt für ges. Therapie,* 2 (July, 1884), pp. 289–314; this article as well as others which Freud had written on the subject and some extracts of his correspondence with Koller were translated into French in Sigmund Freud, *De la cocaïne,* edition annotated by Anna Freud, texts assembled and presented by Robert Byck, Ed. Complexe (Brussels, 1976), 350 p.

91. Paolo Mantegazza, *Sulle virtù igieniche e medicinali della coca* (Milan, 1859).

92. Albert Niemann, "Uber eine organische Base in der Coca," *Annalen Chemie,* 114 (1860), p. 213.

93. Al. Bennett, "An Experimental Inquiry into the Physiological Action of Theine, etc.," *Edinburgh Medical and Surgical Journal* (1874); Von Anrep, "Ueber die physiologische Wirkung des Cocaïns," *Pflügers Archiv,* XXI (1880).

94. Freud, *op. cit.,* p. 98. This power of local anaesthesia had been the subject of publications in France, notably in *L'Union médicale,* in 1877.

95. Due to a lack of money, Koller himself was not able to go to the meeting in Heidelberg. We have a personal account of Koller's views on these experiments: "Personal Reminiscences of the First Use of Cocaine as a Local Anaesthesic in Eye Surgery," read to the International Anaesthesia Society, *Bulletin of the International Anaesthesia Society,* VII, No. 1 (1928); his first publication on the subject was "Ueber die Verwendunig des Cocaine zur Anästhesierung am Auge," *Wiener Medizinische Wochenschrift* (1884), XXXIV, No. 43 and 44. The article was translated into English in *The Lancet,* 2 (1884), p. 990. Numerous reports in the French medical press followed immediately.

96. William S. Halsted, "Water as a local Anaesthesic," *New York Medical Journal,* 42 (1885), 327; and "Practical Comments on the Use and the Abuse of Cocaine, suggested by its invariably successful employment in more than a thousand minor surgical operations," *ibid.,* p. 294.

97. John Wood, "New Method of Treating Neuralgia by the Direct Application of Opiates to the Painful Points," *Edinburgh Medical and Surgical Journal,* 82 (1855), p. 265.

98. C. G. Pravaz, "Sur un nouveau moyen d'opérer la coagulation du sang dans les artères applicable à la guérison des anévrismes" ("On a new means of producing coagulation of blood in the arteries applicable to the curing of aneurisms"), *C. R. Acad. scien.,* T. XXXIV (1855), p. 88.

99. On this point, see Georges Lanteri-Laura, *Histoire de la phrénologie,* P.U.F. (Paris, 1970), 263 p.

100. François-Joseph Gall and G. Spurzheim, *Anatomie et physiologie du système nerveux en général et du cerveau en particulier,* F. Schoeel (Paris, 1810), 4 Vols.

101. Claude Bernard, *De la physiologie générale,* Hachette (Paris, 1872),

340 p. See in particular pp. 14–21 and 216–22. The question was studied equally by Edme Vulpian and by J. M. D. Olmsted, *François Magendie. Pioneer in Experimental Physiology and Scientific Medicine in XIXth-century France,* Schuman's (New York, 1944), 290 p. (see in particular pp. 93–122).

102. Charles Bell (1774–1842), *An Idea of a New Anatomy of the Brain* (1811). This passage was reproduced by Magendie himself in his *Journal* (1822, 2, p. 370) and is often quoted, inexplicably, against Magendie. It proves only that Bell had had the idea of the function of the anterior roots.

103. Charles Bell, "Recherches anatomiques et physiologiques sur le système nerveux" ("Anatomical and physiological research on the nervous system"), *Journal de physiologie expérimentale et pathologique,* 1 (1821), pp. 384–90.

104. François Magendie, "Expériences sur les nerfs rachidiens" ("Experiments on the spinal nerves"), *Journal de physiologie expérimentale et pathologique,* 2 (1822), pp. 276–9 (p. 277).

105. Nicolaas A. Rupke (ed.), *Vivisection in Historical Perspective,* Routledge, The Wellcome Institute Series on the History of Medicine (London and New York, 1987), 373 p.

106. François Magendie, *op. cit.,* pp. 366–71.

107. François Magendie, *Leçons sur les fonctions et les maladies du système nerveux, professées au Collège de France,* collected and edited by Constantin James, Ebrard (Paris, 1839), 2 Vols., T. 2, p. 62; "Quelques nouvelles expériences sur les fonctions du système nerveux" ("Some new experiments on the functions of the nervous system"), *C. R. Acad. scien.,* 8 (1839), p. 865 and "Note sur la sensibilité récurrente" ("Communication on recurrent sensibility"), *ibid.,* 24 (1847), p. 1130–835.

108. On this question, see F. A. Longet, "Fait physiologique relatif aux racines des nerfs rachidiens" ("Physiological finding concerning the roots of spinal nerves"), *C. R. Acad. scien.,* 8 (1839), p. 881, and "Influence des nerfs de la sensibilité sur les nerfs du mouvement" ("Influence of the sensory nerves on the nerves for movement"), *ibid.,* p. 919. Longet returned to this question in his *Anatomie et physiologie du système nerveux de l'homme et des animaux vertébrés,* Fortin, Masson (Paris, 1842), 2 Vols.

109. Robert Remak, *Observationes Anatomicae et microscopicae de systematis nervosi structura,* Berolini, sumpt. Remarianis (1838), 42 p.

110. Johannes Müller, *Manuel de physiologie,* translated into French from the fourth edition of the German (1844), with annotations by A. J. L. Jourdan, J. B. Baillière (Paris, 1845), 2 Vols.; T. 1, p. 516 (translated from French into English by J. A. & S. W. Cadden).

111. *Ibid.,* p. 598: "When the limb in which the nerve trunk is distributed has been removed by amputation, this trunk, provided it includes all the

shortened original fibres, can have the same sensations as if the amputated limb still existed, and this state persists for the whole [of the patient's] life." (Translation.)

112. *Ibid.*, p. 602; this schematic outline based on electric conduction is linked to the discussions on the analogy between electrical and nerve fluids, on which Müller seemed to hesitate; for Reil, see *Archiv für Physiologie*, VII.

113. E. H. Weber, *Annotationes Anatomicae et physiologiae*, pp. 44–81; a chart of the degrees of acuity in the sense of touch had been drawn up by Valentin in his *Lehrburch der Physiologie*, T. II, p. 565.

114. Müller, *op. cit.*, II, p. 256.

115. *Ibid.*, II, pp. 251–74.

116. *Ibid.*, II, p. 478.

117. *Ibid.*, II, p. 482.

118. F. Pacini, *Nuovi organi scoperti nel corpo umano* (Pistoia, 1840); these corpuscles had already been envisaged by Vater.

119. Rudolf Wagner, "Sur l'appareil propre du sens du tact (en commun avec M. Meissner)" ("On the specific apparatus for the feeling of touch [in collaboration with M. Meissner]"), in *C. R. Acad. scien.*, T. XXXIV (1852), pp. 771–2; first mention *ibid.*, p. 336.

120. Pierre Nicolas Gerdy, *C. R. Acad. scien.* (26th January 1847), pp. 303–4.

121. Joseph H. S. Beau, "Recherches cliniques sur l'anesthésie, suivies de quelques considérations physiologiques sur la sensibilité" ("Clinical research into anaesthesia, followed by some physiological considerations on sensibility"), *Archives générales de médecine*, 4th series, T. XVI (1848), pp. 5–24; see in particular p. 7.

122. The insensibility of hysterical people was known of, but until then it had been interpreted as a pretence or the result of the exclusive concentration of the attention on an object to the indifference of all else. This purely psychological interpretation had also been put forward to explain the insensibility of convulsionaries. However, in the work of Gendrin, Henriot and Calmeil, prior to Charcot, the physical reality of insensibility was noted, no matter what the aetiology was.

123. Beau, *op. cit.*, p. 20.

124. *Ibid.*, p. 24.

125. Octave Landry, "Recherches physiologiques et pathologiques sur les sensations tactiles" ("Physiological and pathological research into tactile sensations"), *Archives générales de médecine*, 29 (1852), p. 55.

126. *Ibid.*, 30 (1852), p. 55.

127. Augustus Waller, "Nouvelle méthode pour l'étude du système nerveux, applicable à l'investigation de la distribution anatomique des cordons nerveux et au diagnostic des maladies du système nerveux pendant la vie et après la mort" ("New method for the study of the nervous system, applicable to the investigation during life and after death of the anatomi-

cal distribution of nerve fibres and the diagnosis of illnesses of the nervous system"), *C. R. Acad. scien.,* T. XXXIII (1851), pp. 606–11.

128. A. Waller, "Recherches expérimentales sur la structure et la fonction des ganglions" ("Experimental research into the structure and function of ganglia"), *C. R. Acad. scien.,* T. XXXIV (1852), pp. 524–7.

129. Mirko D. Grenk, *Raisonnement expérimental et recherches toxicologiques chez Claude Bernard,* Droz (Paris and Geneva, 1973), 474 p.

130. Claude Bernard, *Leçons sur les effets des substances toxiques et médicamenteuses,* J. B. Baillière et fils (Paris, 1857), 488 p. (see in particular p. 311).

131. *Ibid.,* p. 314. Claude Bernard then made the experiment more complex by reproducing the two situations in the same frog by means of a ligature of the vessels in one of its limbs, which prevented the introduction of curare into the blood below the ligature, and represented the frog as decapitated.

132. *Ibid.,* p. 331.

133. *Ibid.,* p. 333. The question had already been put in these terms during the debates on anaesthesia and the transient nature of the action of curare made Claude Bernard think along these lines; but the proposed experiments could not take account of the problem of consciousness and memory.

134. Marshall Hall (1790–1857), "On a Particular Function of the Nervous System," *Proceedings of Zoological Society* (1832), and "On the Reflex Function of the Medulla Oblongata and the Medulla Spinalis," *Philosophical Transactions of the Royal Society* (1833), for Johannes Müller (1801–58), see below. For an historical appraisal of the whole question, see Georges Canguilhem, *La Formation du concept de réflexe aux XVII^e et XVIII^e siècles,* Vrin (Paris, 1977: 2nd ed.).

135. Claude Bernard, *op. cit.,* pp. 340–5.

136. A. E. Vulpian, *Leçons sur la physiologie générale et comparée du système nerveux,* Germer Baillière (Paris, 1866), p. 206.

137. *Ibid.,* p. 211.

138. Gabriel Andral, *Clinique médicale,* Crochard, Fortin et Masson (Paris, 1839–40: 4th ed.), 5 Vols, T. V, p. 658.

139. Charles Robin, a *professeur agrégé* (*i.e.* someone who has passed the *agrégation* examination) at the Faculté de médecine, the author with Émile Littré of a famous *Dictionnaire de médecine,* reprinted many times. In his *Cours de philosophie positive,* he explained, following the principles of Auguste Comte's classification of the sciences, the place occupied by biology in the hierarchy of the sciences on the basis of the historical order of their development and their degree of complexity.

140. Charles Robin, "Sur la direction que se sont proposée en se réunissant les membres fondateurs de la Société de biologie" ("On the direction the founding members of the Société de biologie in joining together

proposed to take"), read on 7th June 1849, in *Comptes rendus et mémoires de la Société de biologie,* T. I (1850), p. IX.

141. *Ibid.,* p. VIII.

142. Charles-Édouard Brown-Séquard, "De la transmission croisée des impressions sensitives par la moelle épinière" ("On the crossed transmission of sensory impressions by the spinal cord"), *Comptes rendus et mémoires de la Société de biologie,* first year, 1849, T. I (1850), pp. 192–3; for C. E. Brown-Séquard (1817–94), see M. D. Grmek, *Dictionary of Scientific Biography,* T. II, pp. 524–6; and J. M. D. Olmsted, *Charles-Edouard Brown-Séquard. A Nineteenth Century Neurologist and Endocrinologist,* Johns Hopkins University Press (Baltimore, 1946), 253 p.

143. See for example the very detailed rebuttal of it by Brown-Séquard in his *Recherches expérimentales sur les voies de transmission des impressions sensitives et sur des phénomènes singuliers qui succèdent à la section des racines des nerfs spinaux,* communicated to the Société de biologie in July, August and September 1855, Thunot (Paris, 1856), 6–25.

144. "Sur le siège de la sensibilité et sur la valeur des cris comme preuve de perception de la douleur" ("On the site of sensibility and the value of cries as proof of perception of pain"), *C. R. Acad. scien.,* T. XXIX (1849), p. 672.

145. *Ibid.,* pp. 25–6.

146. Claude Bernard, *Leçons de pathologie expérimentale,* 5th lesson, Baillière (Paris, 1880) (Lessons given at the College de France 1859–60), p. 201.

147. Report by Broca to the Société de biologie, *Comptes rendus et mémoires de la Société de biologie* (1855), pp. 23–55.

148. Moritz Schiff, *C. R. Acad. scien.,* T. XXXVIII (1854), p. 929.

149. Charles-Édouard Brown-Séquard, "Recherches sur la transmission des impressions de tact, de chatouillement, de douleur, de température et de contraction (sens musculaire) dans la moelle épinière" ("Research into the transmission of impressions of touch, tingling, pain, temperature and contraction [in the muscular sense] in the spinal cord"), *Journal de la physiologie de l'homme et des animaux,* T. 6 (1863), pp. 124–43 and 232–48, 581–6; see in particular pp. 131–2.

150. *Ibid.,* p. 125.

151. *Ibid.,* p. 610.

152. Alfred Edme Vulpian, article entitled "Moelle épinière (Physiologie)" ("Spinal cord [Physiology]"), *Dictionnaire des sciences médicales,* under the direction of Amédée Dechambre, 1877, p. 383. The article represents an expanded version of the *Leçons* of 1866.

153. *Ibid.,* p. 380.

154. *Ibid.,* p. 392.

155. *Ibid.,* p. 415.

156. W. R. Gowers, *A Manual of Diseases of the Nervous System,* J. A. Churchill (London, 1886), Vol. 1, p. 131.

157. Moritz Schiff, *Recueil des mémoires physiologiques,* B. Benda (Lausanne, 1894–6), 4 Vols.; see in particular, in volume 1, the article on cenaesthesia and for the research into localisation within the cord, see the memoir presented to the Académie des sciences in 1854 *(op. cit.),* in which he asserts that the white substance of the posterior funiculi was sensory and transmitted sensory impressions, while the antero-lateral funiculi could be excised without sensibility being modified. Moritz Schiff dealt with this question again in his *Lehrbuch der Physiologie des Menschen,* Lahr (1858–9).

158. See *Clinical Society's Transactions,* Vol. XI (1878), p. 24.

159. W. R. Gowers, *A Manual of Diseases of the Nervous System,* p. 135. The presentation of the problems is particularly clear and well-balanced (see pp. 130–5).

160. A Goldscheider, *Die Lehre von den Specifischen Energieen der Sinnesorgane,* L. Schumacher (Berlin, 1881), 40 p. Thereafter Goldscheider was opposed, as we shall see, to the interpretation of the points proposed by von Frey.

161. Max von Frey published a first article on the subject in 1895; the clearest explanation can be found in *Untersuchungen über die Sinnesfunctionen der Menschlichen Haut, Erste Abhandlung: Druckempfindung und Schmerz,* Des XXIII. Bandes der Abhandlungen der mathematisch-physischen Classe der Königl. Sächsischen Gesellschaft der Wissenschaften, No. III, S. Hirzel (Leipzig, 1896), taken up again in *Deutschsprachige Klassiker der Schmersforschung,* H. O. Handwerker ed., Physiologische Institut (Heidelberg, 1987), p. 91.

162. *Ibid.*

163. *Ibid.,* p. 95.

164. *Ibid.,* p. 127: "Die organe der Druckempfindung sind an der behaarten Körperstellen die Nervenkränze der Haare, an der unbehaarten Stellen die Meissner'schen Körperchen. . . . Die Schmerzempfindung der Haut stammt, soweit es sich um oberflächliche Schmerzempfindung handelt, von den frein intraepithelialen Nervenendigungen."

165. A. Goldscheider, *Physiologie der Hautsinnesnerven,* J. A. Barth (Leipzig, 1898), taken up in *Deutschsprachige Klassiker der Schmersforschung, op. cit.,* p. 53.

166. *Ibid.,* p. 145.

167. A. Goldscheider, *Das Schmerz Problem,* Springer (Berlin, 1920), pp. 6–17 and 77–91.

168. Domenico Cotugno, *De ischiade nervosa Commentarius,* Venetii, Bartolomaei Occhi (1782; 1st ed., 1764).

169. François Chaussier, *Recueil de tables synoptiques d'anatomie et de physiologie, suivant la méthode adoptée au cours de l'École de médecine de Paris,* Barrois (Paris, year XI).

170. Montfalcon, article on "Névralgie" ("Neuralgia"), *Dictionnaire des sciences médicales,* T. XXXV (1819), p. 502.

171. Ollivier d'Angers, article on "Névralgie" ("Neuralgia"), *Dictionnaire des sciences médicales* (2nd ed., 1840).

172. *Ibid.* (1st ed.), p. 512.

173. Valleix, *Traité des névralgies ou affections douloureuses des nerfs,* Baillière (Paris, 1841), p. 652.

174. Alfred Trousseau, *Clinique médicale de l'Hôtel-Dieu,* Baillière (Paris, 1864). The first mention of "Trousseau's illness" is found in *L'Union médicale,* the 13th and 18th February issues, 1864, pp. 274–82 and 307–29.

175. Alexandre Axenfeld, "Traité des névroses" ("Treatise on Neuroses"), in Requin, *Éléments de pathologie médicale,* T. III (1863).

176. Arloing and Tripier, "De la sensibilité récurrente" ("Concerning recurrent sensibility"), *Gazette hebdomadaire de médecine et de chirurgie* (1874). Tripier had been interested previously in reflex pains.

177. L. Lereboullet, article on "Névralgie" ("Neuralgia"), *Dictionnaire encyclopédique des sciences médicales,* T. XII, 2nd series (1877), pp. 615–98; see in particular p. 669.

178. Alfred Edme Vulpian, *Leçons sur les nerfs vaso-moteurs* (Paris, 1875), T. II, p. 625.

179. S. Weir Mitchell, *Des lésions des nerfs et de leurs conséquences,* translated from English into French and annotated by A. Dastre; prefaced by Professor Vulpian, Masson (Paris, 1874), p. 70 (translated back into English by J. A. & S. W. Cadden).

180. *Ibid.,* p. 70.

181. *Ibid.,* p. 162.

182. *Ibid.,* p. 299.

183. *Ibid.,* p. 233.

184. The bibliography concerning Charcot is considerable; the main elements are found in Jean Thuillier: *M. Charcot à la Salpêtrière,* Laffont (Paris, 1993), 310 p.

185. Émile Zola, *Le Docteur Pascal* (Paris, 1887).

186. Jean-Baptiste Charcot, *Discours prononcé aux funérailles de Vulpian,* 21st May 1887, p. 7.

187. Duchenne de Boulogne, "De l'ataxie locomotrice progressive" ("Concerning progressive locomotor ataxia"), *Archives générales de médecine,* 5th series (1858), 12, pp. 641–52, and 13, pp. 36–62, 158–81, 417–51.

188. See Alphonse Daudet, *La Doulou,* in *Œuvres complètes,* Club du Livre (1967), T. XI. The writing of *La Doulou* was spread over an eight-year period and was interrupted three years prior to the writer's death. This notebook was published only some thirty years later.

189. Jean-Baptiste Charcot, *Leçons sur les maladies du système nerveux, Œuvres complètes,* Delahaye and Lecrosnier (Paris, 1885), T. II, First lesson, p. 14. The description of ataxia and its differentiation from multiple sclerosis takes up the first two lessons.

190. Article on "Goutte" ("Gout"), *Dictionnaire encyclopédique des sciences médicales,* T. X (1884), p. 190.

191. Armand Trousseau, *Clinique médicale de l'Hôtel-Dieu, op. cit.*

192. Article on "Goutte" ("Gout"), *op. cit.,* p. 9.

193. Garrod was translated into French and quoted by Charcot.

194. Article on "Goutte" ("Gout"), *op. cit.,* p. 226–7.

195. Sée, Communications presented to the Académie de médecine, 25th June 1877 and 23rd August 1887; see Georges Dieulafoy, *Manuel de pathologie interne,* Masson (Paris, 1904), T. IV, pp. 609–13.

196. Article on "Goutte" ("Gout"), *op. cit.,* p. 229.

197. On the polyneuritis of alcoholic origin in the working classes, see Duménil, *Archives générales de médecine* (1864) and the histories of alcoholism: in particular, Jean-Charles Sournia, *Histoire de l'alcoolisme,* Flammarion (Paris, 1986), 323 p.; and Didier Nourrisson, *Le Buveur du XIX^e siècle,* Albin Michel (Paris, 1990), 383 p.

198. W. Gowers, *op. cit.,* p. 52 and p. 92 ff.

199. See, for example, R. Rey and J. L. Fischer, "L'électricité médicinale au XVIII^e siècle" ("Medicinal electricity in the 18th century"), in *Transferts de vocabulaire dans les sciences,* Éd. du C.N.R.S. (Paris, 1988), pp. 49–66.

200. Luigi Galvani, *De viribus electritatis in motu musculari Commentarius, cum Joannis Aldini Dissertatione et notis,* Ap. Soc. Typ. (Mutinae [Modena], 1792), XXVII, 79 p.

201. For the controversy between Galvani and Volta, see Walter Bernardi, *L'Électricité animale: les savants italiens et leurs relations avec les milieux français à la fin du XVIII^e siècle. Échanges d'influences scientifiques et techniques entre pays européens de 1780 à 1830,* C.T.H.S. (Paris, 1990), pp. 161–70. For an overall view of these problems, see Marcello Pera, *The Ambiguous Frog, The Galvani-Volta Controversy on Animal Electricity,* translated by Jonathan Mandelbaum, Princeton University Press (Princeton, 1992; 1st ed., 1986).

202. Margaret Rowbottom, Charles Susskind, *Electricity and Medicine. History of their Interaction,* San Francisco Press (San Francisco, 1984), 303 p.

203. Duchenne reported the accident which happened to one of his patients who had originally been treated for ocular paralysis by induced electricity, and finding that the results were not obtained rapidly enough, asked him to try galvanic electricity. It felt like a ball of fire and caused the loss of sight on one side (see *De l'électrisation localisée, op. cit.,* note 188, pp. 19–22, and "Recherches sur les propriétés physiologiques et thépeutiques *[sic]* de l'électricité de frottement, de l'électricité de contact et de l'électricité d'induction" ["Research on the physiological and 'thepeutic' properties of electricity by rubbing, contact and induction"]), *Archives générales de médecine* (1851), pp. 70–2 and pp. 82–3.

204. Iwan Rhys Morus, "Marketing the Machine: the Construction of Elec-

trotherapeutics as Viable Medicine in Early Victorian England," *Medical History*, 36 (1992), pp. 34–52.

205. François Magendie, "Note sur le traitement de certaines affections nerveuses par l'électropuncture des nerfs" ("Note on the treatment of certain nervous afflictions by electroacupuncture of the nerves"), *C. R. Acad. scien.*, T. V (1837), pp. 855–6: "M. Magendie makes . . . known the happy results which he obtains every day by using electric currents for illnesses of the senses, and particularly for that affliction which is so awful due to the acuteness and the persistance of the pains which accompany it, called neuralgia. A single application was enough, in certain cases, to provide immediate and definitive relief of the pain" (translation). Magendie employed galvanism and Clarke's volta-faradic apparatus interchangeably.

206. Guillaume Duchenne de Boulogne, "Exposition d'une nouvelle méthode de galvanisation, dite galvanisation localisée" ("Description of a new method of galvanisation called localised galvanisation"), excerpt from the *Archives générales de médecine* (1850), p. 4. It was in these terms that he presented his first work.

207. Duchenne de Boulogne, *De l'électrisation localisée et de son application*, Baillière (Paris, 1855), 926 p.; the work was reprinted several times. For the therapeutic aspects, see in particular pp. 625–912. It is evident that the method was used principally in the treatment of motor paralysis.

208. Émile Du Bois-Reymond, *Untersuchungen über thierische Elektricität* (Berlin, 1848); a good explanation of his work can be found in Auguste Tripier, *Manuel d'électrothérapie*, J. B. Baillière (Paris, 1861).

209. Robert Remak, *Galvanothérapie ou de l'application du courant galvanique constant au traitement des maladies nerveuses et musculaires*, translated from German into French by Dr. A. Morpain, Paris, J. B. Baillière et fils (1860; 1st ed., 1858), p. 239 ff. (and hence into English by J. A. & S. W. Cadden).

210. *Ibid.*, p. 299. The articles on "Électricité (Physiologie)" ("Electricity [Physiology]") and "Électricité (Physique médicale)" ("Electricity [Medical Physics]"), from Jaccoud's *Dictionnaire de médecine et de chirurgie*, T. XII (1871), gave an excellent presentation of the different apparatus and the work achieved, notably by Du Bois-Reymond, see pp. 452–560.

211. Julius Althaus (1833–1900), *A Treatise on Medical Electricity, Theoretical and Practical* (London, 1859); this work was reprinted several times.

212. Auguste Tripier, *op. cit.*, pp. 207–9.

213. Among these manuals, apart from the one by Tripier, it is worth noting those of Doctors Onimus and Legros, of Bergonié and of Bordier, at the very beginning of the 20th century.

214. Jacques Arsène d'Arsonval (1751–1840), "Ondes électriques. Charatéristiques d'excitation" ("Electrical waves. Characteristics of exci-

tation"), *C. R. et mémoires de la Société de biologie,* 1st April 1882, reprinted in *Titres et travaux scientifiques de d'Arsonval* (1894), pp. 37–8. For d'Arsonval, see the entry in the *D.S.B.,* and Léon Delhoume, *De Claude Bernard à d'Arsonval,* J. B. Baillière et fils (Paris, 1939).

215. A. d'Arsonval, "Sur les effets physiologiques de l'état variable en général, et des courants alternatifs en particulier" ("On the physiological effects of the variable state in general and of alternating currents in particular"), *Bulletin de la Société internationale des électriciens,* T. IX (1892), p. 6.

216. A. d'Arsonval, "Procédé pratique pour doser les courants d'induction et changer la forme de l'excitation électrique, de manière à agir plus spécialement soit sur le nerf, soit sur le muscle" ("Practical procedure for dosing induction currents and changing the form of electrical excitation, so as to have a more selective effect be it on a nerve or a muscle"), *Archives de physiologie,* 3 (July 1891), p. 586.

217. *Ibid.,* pp. 589–90.

218. A. d'Arsonval, *Titres et travaux, op. cit.,* pp. 48–50.

219. Jean-Jacques Ménuret de Chambaud, *Essai sur l'histoire médicotopographique de Paris, ou Lettres à M. d'Aumont,* Serpente (Paris, 1786; 2nd ed. Méquignon l'aîné, 1804).

220. *Bulletin de l'Académie de médecine,* T. I (1836), p. 346.

221. *Ibid.,* pp. 957–73.

222. Paul Richer and Gilles de la Tourette, article on "Hypnotisme" ("Hypnotism"), *Dictionnaire encyclopédique des sciences médicales,* 4th series, T. XV (1889), p. 67.

223. C. E. Brown-Séquard, *Recherches expérimentales et cliniques sur l'inhibition et la dynamogénie* (Paris, 1882); taken up again in the preface of Braid's *Neurypnologie* (see following note), p. IX.

224. James Braid, *Neurypnologie. Traité du sommeil nerveux ou hypnotisme,* translated from English into French by Dr. Jules Simon, with a preface by C. E. Brown-Séquard, A. Delahaye and E. Lecrosnier (Paris, 1883), pp. 32–3; (translated back into English by J. A. & S. W. Cadden).

225. *Ibid.,* p. 27.

226. *Ibid.,* p. 212; on the case of curing *tic douloureux,* see pp. 172 ff.; for cephalgia, p. 207 ff.

227. Jean-Martin Charcot, *Leçons sur les maladies du système nerveux, Œuvres complètes,* Delahaye et Lecrosnier (Paris, 1885–90), T. III; see in particular the 21st and 22nd Lessons, pp. 333–85.

228. In a note presented to the Académie des sciences in 1883, entitled "Essai d'une distinction nosographique des divers états nerveux compris sous le nom d'hypnotisme" ("Essay on the nosological distinction of diverse nervous states, included under the name of hypnotism"), Charcot had described the three states of the "great hypnotism"—lethargy, catalepsy and somnambulism. He would be challenged on the existence of these three states.

229. Charcot, *op. cit.,* p. 361.

230. B. C. Brodie, *Lectures illustrative of Certain Local Nervous Affections* (London, 1837), Lecture II (translated into French by Dr. Aigre, 1880; thence back into English by J. A. & S. W. Cadden).

231. Jean Babinski, "Ma conception de l'hystérie et de l'hypnotisme (pithiastisme)" ("My concept of hysteria and hypnotism [pithiastism]"), Lecture given to the Société de l'internat des hôpitaux de Paris on 28th June 1906, in *Mélanges,* p. 469.

232. *Ibid.,* p. 470.

233. *Ibid.,* p. 472.

234. Jean Babinski, "De l'hypnotisme en thérapeutique et en médecine légale" ("Concerning hypnotism in Therapeutics and Forensic Medicine"), published in *La Semaine médicale,* 27th July 1910, in *Mélanges, op. cit.,* p. 512.

235. Article on "Hypnotisme," Dechambre, p. 97.

### Chapter 7

1. S. Ramon y Cajal, "Sur la structure de l'écorce cérébrale de quelques mammifères" (On the structure of the cerebral cortex of certain mammals), *La Cellule,* 7 (1891), pp. 125–76; "New Conception of the histology of nervous centres," *Re. Cienc. Med.,* 18 (Barcelona, 1892), pp. 361–76; see also *Histologie du système nerveux de l'homme et des vertébrés* (French translation), Maloine (Paris, 1909–11); Golgi's work is earlier: *Sulla fina anatomia degli organi centrali del sistema nervoso,* Hoepli (Milan, 1886).

2. G. Fritsch and E. Hitzig, "On the Electric Stimulation of the Brain," *Arch. Anat. Physio. wiss. Med.,* 37 (1870), pp. 300–32; see also E. Hitzig, *Brain Research* (Berlin, 1874).

3. Paul Emil Flechsig (1847–1929), *Die Lietungsbahnen im Gehirn und Rückenmark des Menschen auf Grund entwicklungsgeschichtelicher Untersuchungen* (Leipzig, 1876); see also "Zur Anatomie und Entwicklungsgeschichte der Leitungsbahnen im Grosshirn des Menschen," *Archiv für Anatomie und Physiologie* (Anatomische Abteilung, 1881), pp. 12–75.

4. Jules Déjerine and Gustave Roussy, "Le syndrome thalamique," *Revue neurologique* (1906), pp. 521–32. Unfortunately, no good scientific biography of Déjerine exists, nor does he have entry in the *D.S.B.*

5. *Ibid.,* p. 532.

6. Jean-Martin Charcot, *Leçons sur les localisations dans les maladies du système nerveux,* collected and published by Bourneville, Bureau du Progrès médical and A. Delahaye (Paris, 1876), 425 p.; see especially 3rd lesson, pp. 105–7. Therein, Charcot criticises the concept that the

optic thalamus was the common sensorium, the convergent point of all sensory pathways.

7. Jules Déjerine, *Clinique des maladies du système nerveux,* Leçon inaugurale, Masson (Paris, 1911), p. 1.

8. Jules Déjerine, *Sémiologie des affections du système nerveux,* Masson (Paris, 1914), see in particular pp. 722 940. In his Leçon inaugurale, Déjerine mentions that research on radicular lesions had been under taken only as of a few years earlier. With regard to staining techniques employed to study the nervous system at the start of the 20th century, see W. V. Bechterew, *Les Voies de conduction du cerveau et de la moelle* (Conduction Tracts in the Brain and Spinal Cord), translation by C. Bonne, A. Storck (Paris, 1900), pp. 2–7.

9. Syringomyelia, perhaps caused by a malformation, was described by Laehr in 1896; a classical description was provided by Pierre Marie; see also Déjerine, *Sémiologie,* p. 904, and the case presented by F. Raymond and Henri Français in the *Revue neurologique* (1906).

10. Leçon inaugurale, p. 14.

11. Gustave Roussy, *La Couche optique (étude anatomique, physiologique et clinique). Le syndrome thalamique,* Steinheil (Paris, 1907), p. 353.

12. Henry Cushing, "Note on Electric Stimulation of the Postcentral Gyrus in Conscious Patients Using a Faradiser," *Brain,* 32 (1909), pp. 44–53. The method was later developed further by W. Penfield. In order to selectively stimulate the cortex, strychnine was also used.

13. Paul Broca, "Remarques sur le siège de la faculté du langage articulé, suivies d'une observation d'aphémie (loss of speech)," *Bulletins de la Société anatomique,* XXXVI (1861), pp. 330–57; "Nouvelle observation d'aphémie produite par une lésion de la moitié postérieure des deuxième et troisième circonvolutions frontales," *ibid.,* 2nd series, VI, pp. 398–407. On Broca, see Francis Schiller, *Paul Broca, Founder of French Anthropology,* University of California Press (Berkeley and Los Angeles, 1979), 350 p. (French trans., Ed. O. Jacob, 1990, 426 p.).

14. Pierre Marie, "Révision de la question de l'aphasie. La troisième circonvolution frontale gauche ne joue aucun rôle spécial dans la fonction du language" (The left 3rd frontal convolution plays no special role in speech), *Revue neurologique* (1907), pp. 12–15.

15. David Ferrier, *Brain Functions,* trans. *Les Fonctions du cerveau* by Henri de Varigny, Baillière (Paris, 1878), p. 2. The terms afferent and efferent are not only broader, as Ferrier points out; they refer not to a nerve's function, but to the direction of their activity, *i.e.* from the periphery to the centre, or from the centre to the periphery; for details on Ferrier, see Robert M. Young, "The Functions of the Brain: Gall to Ferrier (1808–1886)," *Isis,* 59 (1968), pp. 251–68.

16. *Ibid.,* p. 474: the word thus coined refers to *noesis* or conscious activity

and to motor activity (from *kinesis,* movement, with the preposition *kata* designating the path which descends from one to the other).

17. *Ibid.,* p. 282.
18. John Hughlings Jackson (1835–1911), "Psychology and the Nervous System," *Medical Press and Circular* (1879), II, pp. 199–201; 239–41; 283–5; 409–11; 429–30; see also "Remarks on Evolution and Dissolution of the Nervous System," in *Selected Writings,* ed. by J. Taylor, Hodder and Stoughton (London), Vol. 2, pp. 92–118.
19. Bechterew, *op. cit.,* p. 675.
20. Thomas Alajouanine and L. Cornil, "Localisations cérébrales corticales," in the *Traité de physiologie normale et pathologique,* supervised by G. H. Roger and L. Binet (Roger & Binet hereafter), T. IX, *Physiologie du système nerveux,* Masson (Paris, 1933), p. 264.
21. Theodor Meynert, *Psychiatrie. Clinique des maladies du cerveau antérieur basée sur sa structure, ses fonctions et sa nutrition,* trans. by Dr. Georges Cousot, Manceaux (1st ed., Bruxelles, 1888), VII, 294 p.
22. Alfred W. Campbell, *Histological Studies on the Localisation of Cerebral Functions,* Cambridge University Press (1905), XIX, 367 p.; Brodmann, *Vergleichende Lokalizationslehre der Grosshirnrinde,* Barth (Leipzig, 1909).
23. John Farquhar Fulton, *Physiology of the Nervous System* (Eng. ed. 1938), *Physiologie du système nerveux,* preface to Fr. trans. by L. Binet, Vigot (1947).
24. Alajouanine et Cornil, *op. cit.,* p. 354.
25. *Ibid.,* p. 355.
26. Henry Head, "Sensory Disturbances from Cerebral Lesions," *Brain,* 34 (1911), pp. 102–254.
27. *Ibid.,* p. 181.
28. *Ibid.,* p. 191.
29. Henry Head presented his thesis at Cambridge in 1892, "On Disturbances of Sensation with Especial Reference to the Pain of Visceral Disease," *Brain,* 16 (1893), pp. 1–132.
30. Henry Head, "The Afferent Nervous System from a New Aspect," *Brain,* 68 (1905), pp. 99–115.
31. *Ibid.,* p. 107.
32. W. Trotter and H. M. Davies, "Experimental Studies in the Innervation of the Skin," *Journal of Psychology,* 38 (1909), pp. 134–246.
33. G. Weddell, "Clinical Significance of the Pattern of Cutaneous Innervation," *Proceedings of the Royal Society of Medicine,* 34 (1941), pp. 776–8.
34. See Roger & Binet, *op. cit.,* T. IX, p. 397.
35. See Gerald Geison, *Michael Foster and the Cambridge School of Physiology. The Scientific Enterprise in Late Victorian Society,* Princeton University Press (Princeton, 1978), 401 p.
36. The term first appears in M. Foster and C. S. Sherrington, *Textbook of Physiology,* III (London, 1897). Sherrington uses it again in 1908 (see the following note), pointing up that, in view of the importance of this

type of connection between neurons, a specific term was essential, and that "synapse" would be used (p. 18). Sherrington conceived this area as being more physically elaborate (partitioning membrane with some form of osmosis) rather than in terms of a chemical reaction (p. 17).

37. Charles Sherrington, *The Integrative Actions of the Nervous System,* Archibald Constable (London, 1908), p. 13.

38. *Ibid.,* pp. 227–8.

39. *Ibid.,* p. 228.

40. *Ibid.,* p. 319.

41. *Ibid.,* p. 117.

42. *Ibid.,* pp. 130–1 & pp. 336–45.

43. *Ibid.,* pp. 387–8.

44. *Ibid.,* p. 391.

45. *Ibid.,* p. 391.

46. Herbert Spencer, *Principles of Psychology* (1st ed. 1855), *Principes de psychologie,* French trans. Baillière (1874–5), 2 Vols., see T. 1, pp. 295–6.

47. *Ibid.,* T. 2, p. 570.

48. *Ibid.,* p. 572.

49. The internal environment refers to "the organic fluid which circulates and bathes the various anatomical elements of the tissues; it is the lymph or plasma . . . which constitutes the overall interstitial tissue, the expression of all local nutritive substances, the source and confluence of basic cell exchanges"—Claude Bernard, *Leçons sur les phénomènes de la vie communs aux animaux et aux végétaux,* Baillière (Paris, 1878) new ed. Vrin (1966), pp. 113–4. See Mirko D. Grmek on this question, *Philosophie et méthodologie scientifiques chez Claude Bernard,* The Singer-Polignac Foundation Colloquium, Masson (Paris, 1967), pp. 117–50; see also Frederic Holmes, *Origins of the Concept of the Milieu Intérieur, Claude Bernard and Experimental Medicine,* F. Grade & M. Visscher, eds., Schenkam (Cambridge, Mass., 1967). The internal secretion concept was confirmed as a result of work on the liver's glycogenic functions. Brown-Séquard brought to light the secretion of the adrenals (1872) and studied the therapeutic uses of testicular extracts; he can thus be considered the founder of endocrinology. For a contemporary update on these points, see Christiane Sinding, *Le Clinicien et le Chercheur. Des grandes maladies de carence à la médecine moléculaire (1880–1980),* P.U.F. (Paris, 1991), pp. 89 ff. in particular.

50. Walter B. Cannon, *Bodily Changes in Pain, Hunger, Fear and Rage,* D. Appleton Company (New York & London, 1915), from the Preface, p. VII.

51. William James, "What is Emotion?" *Mind,* 9 (1884), pp. 188–205. Cannon criticised this theory in an article published in 1927, "The James-Lange Theory of Emotions," *American Journal of Physiology,* 39 (1927), pp. 106–24. See on this subject the excellent article by John R. Durant, "The Science of Sentiment: the Problems of the Cerebral Lo-

calization of Emotion," *Perspectives in Ethology*, Vol. 6, *Mechanisms*, ed. by P. P. G. Bateson and P. H. Klopfer, Plenum (New York & London, 1985), pp. 1–31.

52. W. B. Cannon, *op. cit.*, pp. 64–5.

53. *Ibid.*, Chap. XI, "The Utility of the Bodily Changes in Pain and Great Emotions," in particular pp. 185–205. For Mac Dougall, see *Introduction to Social Psychology* (London, 1908).

54. *Ibid.*, p. 212.

55. Flechsig, *op. cit.;* his method is described in Bechterew, *op. cit.*, p. 3, and in Roger & Binet, *op. cit.*, T. IX, pp. 266–71. Flechsig's laws are still considered valid though the areas circumscribed have been frequently redefined.

56. Léon Binet, *op. cit.*, T. XI, p. 347; the following question is highly significant, even if the answer is negative: "Does this therefore mean that the newborn child is reduced to being only a medullar or entirely vegetative entity?"

57. André-Thomas, *Études neurologiques sur le nouveau-né et le jeune nourrisson* (Neurological study of the newborn and young infant), Masson et Cie (Paris, 1952), p. 1.

58. Charles Darwin, *The Expression of Emotions in Man and Animals* (1st ed. 1872), *L'Expression des émotions, chez l'homme et les animaux* (1st Fr. trans. 1874, re-ed. from 2nd Eng. edition 1890), Verviers, Ed. Complexe (1981). We might mention the hypothesis that the decline of Darwinism, so often situated at the close of the 19th century, did not affect the psychology field. If we observe no interruption in the dissemination of Darwin's ideas between the end of the 19th century and the first half of the 20th century in this field, it may, among other things, be attributed to John Hughlings Jackson's ideas as he was a staunch follower of the evolutionist theory; see for examples the collection of articles entitled "Evolution and Dissolution of the nervous system," *Selected Writings*, T. II, ed. by James Taylor with the advice of Gordon Holmes & F. M. R. Walsh, Hodder & Stoughton (London, 1932), pp. 3–120.

59. André-Thomas, "Ontogénèse de la vie psycho-affective et de la douleur. Affect et Affectivité," *La Douleur et les douleurs*, supervised by P. Alajouanine, Masson (Paris, 1956), pp. 41–58, particularly p. 55. In this work, André-Thomas takes into account newly discovered knowledge and seems much more reserved about the absence of pain in the newborn child, insofar as he places considerable emphasis on affective elements in mental development.

60. Edgar Douglas Adrian, *The Basis of Sensation. The Action of Sense Organs*, Christophers (London, 1928), pp. 11–2.

61. Keith Lucas, *Nervous Impulse Conduction*, revised by Edgar Douglas Adrian, Longmans (London, 1917), p. 15. In the text, the traveller

metaphor is followed by: "The intensity of the impulse thus measured is relative to the impulse's transmission capacity, just as the traveller's strength can be calculated in relation to the distance he can cover before becoming exhausted."

62. *Ibid.*, p. 19. This concept that the intensity of the nervous impulse, within normal ranges of stimulation, was essentially an "all-or-nothing" relationship had already been expressed by Symes and Velcy (*Proc. Roy. Soc.*, B LXXXIII, 1910, p. 421) and even earlier by Gotch (*Journal of Psychology*, XXVIII, 1902, p. 395). It was taken up again by Adrian in his 1928 work, pp. 28 & ff.

63. Charles Richet, *Travaux du laboratoire de Marey* (1877), p. 97.

64. E. D. Adrian, *The Basis of Sensation, op. cit.,* p. 18.

65. *Ibid.*, p. 85. His allusion to the recent work of Erlanger and Gasser refers to articles published in 1921, 1924, and 1926, but most likely does not include the more important article which appeared in 1927, at the time Adrian was just finishing his work.

66. H. S. Gasser and J. Erlanger, "The Role played by the Sizes of the Constituent Fibres of a Nerve Trunk in Determining the Form of its Action Potential Wave," *The American Journal of Physiology,* LXXX (1927), pp. 522–47.

67. *Ibid.*, in particular p. 523.

68. Thomas Lewis, *Pain,* Lady Lorna Lewis (London, 1982), Fascimile edition, The Macmillan Press (London, 1981), p. 51.

69. Ranson, *Arch. Neurol. & Psychiat.,* 25 (1931), p. 1122.

70. Lewis, *op. cit.,* pp. 71–2.

71. *Ibid.*, p. 82.

72. Julian de Ajuriaguerra, *La Douleur centrale,* Couesland (Cahors, 1937), 150 p. This medical thesis can be considered to follow the tradition of Laennec's clinical approach and attempts to establish a theory to explain this type of pain (in terms of the spinal cord, the medulla oblongata, the pons, the thalamus, and the cortex) in the light of this type of observation.

73. René Thurel, *La Douleur en neurologie,* Masson (Paris, 1951), 191 p.

74. Thomas Alajouanine (under supervision), *La Douleur et les douleurs,* Masson (Paris, 1957).

75. Louis Lapicque, *L'Excitabilité en fonction du temps,* P.U.F. (Paris, 1926), p. 12.

76. See Jean-Claude Dupont's thesis on this point, *Les Origines de la théorie chimique de la neurotransmission. Contribution à l'histoire de la neurobiologie et du messager cellulaire,* Lille-III (1992).

77. Halsted had perfected an especially rigorous operative method, which was non-traumatic and had few post-operative complications; Harvey William Cushing (1869–1939), one of the first American physicians to specialise in neurosurgery, spent the major part of his career at the Harvard Medical School and headed the Peter Bent Brigham Hospital

between 1913 and 1932; he set up surgical research facilities both at
Harvard and at Johns Hopkins in Baltimore.

78. René Leriche, *Souvenirs de ma vie morte*, Seuil (Paris, 1956), pp. 175–86.

79. See Jean-Jacques Yvorel on this point, *Les Poisons de l'esprit. Drogues et drogués au XIX^e siècle*, Quai Voltaire Histoire (Paris, 1992), 322 p.

80. For a more detailed historical background, see Ernst Sachs, *The History and Development of Neurological Surgery*, Yale University School of Medicine, Paul B. Hoeber Inc. (1952), 158 p.

81. A. Dastre and Morat, *Recherches expérimentales sur le système vasomoteur*, Masson (Paris, 1884).

82. François Franck, "Grand Sympathique" article, Dechambre (1894), pp. 1–55.

83. William H. Gaskell, "On the Structure, Distribution and Function of the Nerves which Innervate the Visceral and Vascular System," *Journal of Physiology*, 7 (1886), 1–80, p. 2 in particular. The article discusses vasomotor, visceral, and secretory nerves.

84. The "splanchnic" root arises in Clarke's column and the cord's lateral horn (*ibid.*, pp. 58–9).

85. John Newport Langley (1852–1925), *The Autonomic Nervous System* (Cambridge, 1921), p. 16, French translation *Le Système nerveux autonome sympathique et parasympathique* by M. Tiffenau, Vigot (Paris, 1923).

86. Claude Bernard, *Leçons sur les propriétés physiologiques et les altérations pathologiques des liquides de l'organisme*, J. B. Baillère (Paris, 1859), 2 Vols., Vol. 1, p. 268–71.

87. James Mackenzie, *Symptoms and their Interpretations*, Shaw & Sons, Ltd. (London, 1909). The presentation of opposing viewpoints was undertaken by Lewis, *op. cit.*, pp. 136–51. See also François Lhermitte, "Les douleurs viscérales," in Alajouanine, *op. cit.*, pp. 141–70.

88. Henry Head, "On Disturbances of Sensation with Especial Reference to the Pain of Visceral Disease," *Brain* XVI (1893), pp. 1–133, note from p. 127.

89. Jaboulay, *Chirurgie du grand sympathique et du corps thyroïde*, original articles and commentaries assembled by Dr. E. Martin, Storck et Cie (Lyons, 1900), 358 p.; see preface p. II.

90. François Franck, "Signification physiologique de la résection du sympathique dans la maladie de Basedow, l'épilepsie, l'idiotie et le glaucome," *Bulletin de l'Académie de médecine*, T. XLI (1899), p. 587.

91. *Ibid.*, p. 590.

92. René Leriche and René Fontaine, "La chirurgie du sympathique," *Revue neurologique* (1929), pp. 1046–85.

93. René Leriche, *La Chirurgie à l'ordre de la vie*, Zeluck (1945), p. 37.

94. *Ibid.*, p. 58.

95. René Leriche, *La Chirurgie de la douleur*, Masson (Paris, 1937), p. 31.

96. Georges Duhamel, *Vie des martyrs*, Mercure de France (Paris, 1917).

See the work of Arlette Lafay on Duhamel, particularly "Le language de la douleur dans les livres de guerre de Georges Duhamel" (The language of pain in Georges Duhamel's war accounts), in *La Douleur, approches pluridisciplinaires, op. cit.*, pp. 111–34.

97. *La Chirurgie de la douleur, op. cit.*, p. 114.
98. *Ibid.*, p. 371.
99. *Ibid.*, p. 369.
100. *Ibid.*, p. 371.
101. *Ibid.*, pp. 27–8.

# ⇥ Selected Bibliography ⇤

Baszanger I., "Émergence d'un groupe professionnel et travail de légitimation. Le cas des médecins de la douleur," *Revue française de sociologie* (April–June 1990), XXXI–2.

Besson J. M., *La Douleur,* O. Jacob (Paris, 1992), 268 p.

Bourcau F., *Pratique du traitement de la douleur,* Doin (Paris, 1988), 402 p.

Brazier M. A. B., *A History of Neurophysiology in the 17th and 18th Century,* Raven Press (New York, 1984), 230 p.

Brazier M. A. B., *A History of Neurophysiology in the 19th Century,* Raven Press (New York, 1988), 265 p.

Brenot P., *Les Mots de la douleur,* L'Esprit du Temps (PUF diff., 1992), 104 p.

Clarke E. & Jacyna L. S., *Nineteenth-Century Origins of Neuroscientific Concepts,* University of California Press (Berkeley, 1987), 593 p.

Claverie B., Le Bars D., Zavialoff N. & Dantzer R. (under the direction of), *Douleurs, du neurone à l'homme souffrant,* Eshel (1991), 2 Vols.

Gauvain-Piquard A. & Meignier M., *La Douleur de l'enfant,* Calmann-Lévy (Paris, 1993), 268 p.

Haymaker W. & Schiller F. (eds.), *The Founders of Neurology,* Charles C. Thomas Publisher (Springfield, Illinois, 1970), 616 p.

Jelk S., "De la douleur à l'anesthésie: 1800–1850," *Équinoxe* 8 (1992), pp. 45–59.

Keele K. D., *Anatomies of Pain,* Blackwell Scientific Publications (Oxford, 1957), 206 p.

Keele K. D., "Some Historical Concepts of Pain," *in* C. A. Keele & Robert Smith (eds.), *The Assessment of Pain in Man and Animals,* Livingstone (London, 1962), pp. 12–27.

Lafay A. (under the direction of), *La Douleur,* L'Harmattan (Paris, 1992), 175 p.

Le Garrec É., *Mosaïque de la douleur,* Le Seuil (Paris, 1991), 318 p.

Mann R. D. (ed.), *The History of the Management of Pain: from Early Principles to Present Practice,* The Parthenon Publishing Group (Park Ridge, N.J., 1988), 205 p.

Melzack R. & Wall P. D., *The Challenge of Pain,* Penguin Books Ltd. (1982); French translation: *Le Défi de la douleur,* Chenelière et Stanké/Maloine (Montreal/Paris, 1982), 413 p.

Morris D. B., *The Culture of Pain,* University of California Press (Berkeley, 1991), 12, 342 p.

Moulin D. de, "A Historical-Phenomenological Study of Bodily Pain in Western Man," *Bulletin of the History of Medicine* 48 (1974), pp. 540–70.

Pichard-Leandri E. & Gauvain-Piquard A. (eds.), *La Douleur chez l'enfant,* Medsi McGraw-Hill (1989).

Riese W., *A History of Neurology,* M.D. Pub. (New York, 1959), 223 p.

Robinson V., *Victory over Pain. A History of Anesthesia,* Henry Schuman (New York, 1946), 338 p.

Rose F. C. & Bynum W. F. (eds.), *Historical Aspects of the Neurosciences,* A Festschrift for Macdonald Critchley, Raven Press (New York, 1981), 537 p.

Schiller F., "The History of Algology, Algotherapy and the Role of Inhibition," *History and Philosophy of the Life Sciences* 12 (1990), pp. 27–50.

Spillane J. D., *The Doctrine of the Nerves. Chapters in the History of Neurology,* Oxford University Press (Oxford, 1981), 467 p.

Szasz T., *Pain and Pleasure. A Study of Bodily Feelings,* Syracuse University Press (Syracuse, N.Y., 1988), 303 p.

## Dictionaries and Encyclopaedias

*Dictionary of Scientific Biography,* Ch. C. Gillipsie (ed.), Charles Scribner's Sons (New York, 1979–80), 15 Vols. + Index.

*Dictionnaire de médecine, de chirurgie, de pharmacie, des sciences accessoires et de l'art vétérinaire de P. H. Nysten,* re-edited by Émile Littré and Charles Robin, Baillière (Paris, 1855, 10th ed.), 2 Vols.

*Dictionnaire des sciences médicales* (by an association of physicians and surgeons: M. M. Adelon, Alard, Alibert *et alii*), Panckoucke (Paris, 1812–22), 60 Vols.

*Dictionnaire des sciences médicales. Biographie médicale,* Panckoucke (Paris, 1820–5), 7 Vols.

*Dictionnaire encyclopédique des sciences médicales* (under the direction of Amédée Dechambre), Asselin et Masson (Paris, 1864–89), 100 Vols.

*Encyclopédie ou Dictionnaire raisonné des sciences, des arts et des métiers,* by an association of scholars, collated and published by M. Diderot; the mathematical part by M. d'Alembert, Paris, Briasson, David, [. . .], then, Neufchâtel, Samuel Faulche (1751–65), 17 Vols.

*Encyclopédie méthodique, ou par ordre des matières, par une société de gens de lettres, de savants et d'artistes [. . .], Médecine,* Panckoucke, then Agasse (Paris, 1782–1830), 13 Vols. (the *Encyclopédie méthodique* includes 202 Vols.).

*Nouveau Dictionnaire de médecine et de chirurgie pratiques* (under the direction of Sigismond Jaccoud), Baillière (Paris, 1864–86), 40 Vols.

# ⇒ *Glossary* ⇐

**Afferent:** refers to neural fibres carrying nervous impulses from a sensory organ toward the spinal cord or the brain (central nervous system); syn.: centripetal (a broader definition than "sensory").

**Aesthesiometer:** a device for measuring sensibility.

**Animism:** a medical doctrine, particularly associated with G. E. Stahl, which considered that the rational soul intervened directly in organic processes; in this system, the body is the final instrument of the soul.

**Anodyne:** an anodyne remedy is one that relieves pain (lit. *a*, without, and *odunè*, the Greek for pain).

**Axon** (or axone): the cylindrical process of the nerve cell which ensures the centrifugal conduction of the nervous impulse.

**Causalgia:** pain which produces an intense burning sensation as a result of traumatic injury to a peripheral nerve.

**Central nervous system:** includes the brain and the spinal cord.

**Central pain:** pain of the central nervous system.

**Crisis:** etymologically speaking: to separate, to decide or "judge"; refers to the process whereby contaminated humours were separated from healthy humours and evacuated; in this sense, used until the 19th century, all crises were salutary.

**Decussation:** for certain nerve fibres, the crossing over the median line of the cerebrospinal axis and interlacing with similar fibres from the other side. This crossover occurs in the pons Varolii, in the cerebral peduncles, and in the cord's anterior region: it was the latter which Brown-Séquard studied.

**Efferent:** nervous fibre which conducts impulses from the central nervous system to the muscles or glands (or from a higher to a lower level); syn.: centrifugal (a broader definition than "motor").

**Etherise a patient:** to have him inhale ether.

**Funiculus** (in the cord): the white matter in the spinal cord is differentiated into the dorsal, lateral, and ventral funiculi.

**Galenism** (Galenic): Galen's philosophy, or that arising from his works which dominated Late Antiquity, the Middle Ages, and a part of the modern era.

**Grey matter:** part of the central nervous system made up of neurons, glia cells, and their processes.

**Hippocratic:** a number of common characteristics underlie the Hippocratic medicine handed down to us in the different treatises which make up the corpus (and which are not all by Hippocrates): practices based on observation, with very accurate descriptions of the symptoms and the prognosis of disorders; attention paid to the patient's individual traits and living conditions, confidence in nature's healing powers which the physician was to imitate, etc.

**Horn** (in the cord): the name conferred on a portion of the grey matter in the cord due to its shape.

**Hyperaesthesia:** abnormally intense cutaneous sensation in response to a normal stimulus.

**Ischemia:** reduced or interrupted arterial blood circulation in a localised area (contrary: hyperemia).

**Locomotor ataxia:** a voluntary movement disability affecting coordination, not caused by a motor disorder but by a deep-sensibility disorder. *See* "Tabes."

**Mechanism** (mechanist or, formerly, mechanic): dominant medical philosophy of the 17th century which tried to explain all physiological and pathological processes through mechanical laws.

**Medulla oblongata:** the lower portion of the brain stem.

**Monism** (as opposed to dualism): a philosophy which refuses to conceive of two different substances in human beings, *i.e.* body and soul, and which considers individuals as a single whole.

**Moxa:** a technique of Asiatic origin which consists in applying a cylinder or cone of fibre to the skin and setting it alight; it is used as a counterirritant in order to relieve pain.

**Peripheral nervous system:** includes all the organism's nerves (motor, sensory, sympathetic, and parasympathetic).

**Referred pain:** pain which is felt in a distinct and distant area from its region of origin (for example: pain of visceral origin felt in the skin).

**Sympathy:** term used until the 19th century meaning to "suffer with"; referred to the related sensibility and pain between distant organs or parts (for instance, when a disorder concerns the stomach but the patient has a headache).

**Tabes** *(tabes dorsalis):* 1. any form of wasting disease or consumption. 2. the degenerative nervous lesions which correspond to the third stage of syphilis, and which are due to the deterioration of the posterior nerve roots and the cord's posterior funiculi; characterised by shooting pains, locomotor ataxia, and gastric disorders.

**Vesicatories** (therapeutic): substances applied locally onto the skin (topically), to provoke irritation and blisters.

**Vitalism:** a medical philosophy which holds that the living entity and its laws cannot be reduced to physico-chemical processes, and though it does not exclude mechanism, it nonetheless reduces it to a different order, that of the whole being.

**White matter:** part of the central nervous system made up of myelinated nerve fibres but which has no nerve-cell bodies.

# ⤞ *Index* ⤝